Advanced Methods for Time Series Forecasting

Advanced Methods for Time Series Forecasting

Guest Editors

Camelia Delcea
Nora Monica Chirita

MDPI Basel • Beijing • Wuhan • Barcelona • Belgrade • Novi Sad • Cluj • Manchester

Guest Editors
Camelia Delcea
Department of Economic Informatics and Cybernetics
Bucharest University of Economic Studies
Bucharest
Romania

Nora Monica Chirita
Department of Economic Informatics and Cybernetics
Bucharest University of Economic Studies
Bucharest
Romania

Editorial Office
MDPI AG
Grosspeteranlage 5
4052 Basel, Switzerland

This is a reprint of the Special Issue, published open access by the journal *Applied Sciences* (ISSN 2076-3417), freely accessible at: https://www.mdpi.com/journal/applsci/special_issues/HZ8OX06EUE.

For citation purposes, cite each article independently as indicated on the article page online and as indicated below:

Lastname, A.A.; Lastname, B.B. Article Title. *Journal Name* **Year**, *Volume Number*, Page Range.

ISBN 978-3-7258-6940-4 (Hbk)
ISBN 978-3-7258-6941-1 (PDF)
https://doi.org/10.3390/books978-3-7258-6941-1

Contents

About the Editors

Camelia Delcea

Camelia Delcea received a Ph.D. degree in economic cybernetics and statistics from Bucharest University of Economic Studies, Bucharest, Romania. She is currently working in the Economic Informatics and Cybernetics Department, Bucharest University of Economic Studies. Her research interests include agent-based modeling, operations research (optimizing the airplane boarding methods and improving the evacuation process), grey systems theory, artificial intelligence systems, companies' financial and non-financial analysis, risk management, nonlinear and dynamic systems, consumer behavior, online social networks, and sentiment analysis. Dr. Delcea is an Active Member of the Grey Uncertainty Analysis Association and she is one of the organizers of the Cybernetics Scientific Circle within her department. She is the Editorial Advisory Board Member of the *Grey Systems* journal (Emerald) and academic editor of several journals such as *PLoS ONE, IEEE Access, Contemporary Mathematics, Neutrosophic Sets and Systems, Health & Social Care in the Community,* the *International Journal of Grey Systems, Advances in Civil Engineering, Frontiers in Public Health,* and *Operations Research Forum.* Since 2009, she has received more than 50 international and national Awards, such as Best Paper Award, "Georgescu Roegen" for excellent scientific research, "Opera Omnia" for excellent scientific research, Excellent Paper Award, and Top Reviewers Award. In 2025 she received the Romanian Academy Award "Virgil Madgearu" for the book *Advancements of Grey Systems Theory in Economics and Social Sciences*, published by Springer. She has been invited to deliver a keynote speech on grey systems themes at the IEEE GSIS Conference in 2013, 2016, and 2017; and at the GSUA Conference in 2018, 2020, 2022 and 2025.

Nora Monica Chirita

Nora Monica Chirita received a Ph.D. degree in economic cybernetics from Bucharest University of Economic Studies, Bucharest, Romania. She is currently a Professor with the Department of Economic Informatics and Cybernetics, Bucharest University of Economic Studies. Her research interests include economic cybernetics, econometric and macroeconomic modeling, systems diagnosis, risk management, circular economy, sustainable development, financial contagion, and the application of artificial intelligence and fuzzy logic in economic and financial systems. Dr. Chiriță is one of the organizers of the Cybernetics Scientific Circle within her department and has authored and co-authored numerous articles in high-impact international journals and conference proceedings, covering topics such as energy transition modeling, circular economy assessment, financial stability, bibliometric analysis, agent-based modeling, and complex adaptive systems. She has also contributed to several academic books and monographs in the field of economic cybernetics and modeling. Her scientific activity is complemented by sustained involvement in peer-review processes for international journals, as well as participation in national and international research projects. Through her interdisciplinary research, Dr. Chiriță actively contributes to advancing quantitative, data-driven approaches for understanding and managing complex economic and social systems.

Preface

This Reprint brings together a collection of research papers that explore the advancements in the field of time series forecasting. The papers succeed in providing evidence on the rapid evolution of this field as a response to increasing data availability, computational power, and methodological innovation. Time series forecasting plays a fundamental role in understanding complex and dynamic environments, while being able to provide guide informed decision making in various research fields, including economics, energy systems, transportation, environmental studies, finance, and digital management.

The scope of this Reprint is to provide a common ground for the researches in this field, while providing specific real-life applications in which time series forecasting and the advancements made in this field have proven their contribution, such as advanced machine learning and deep learning architectures, hybrid modeling frameworks, attention-based and transformer models, state–space approaches, signal decomposition techniques, and uncertainty-aware forecasting strategies. The selected papers illustrate how these modern approaches address key challenges encountered in real-life situations, such as nonlinearity, non-stationarity, long-term dependencies, noise, and heterogeneous data sources.

By assembling diverse perspectives and application contexts, this volume seeks to foster knowledge exchange between researchers and practitioners and to stimulate further innovation in the development of forecasting methods that can be both more robust and scalable.

This Reprint is addressed to researchers, academics, doctoral students, and professionals interested in time series analysis, artificial intelligence, data science, and applied forecasting. It is intended to serve both as a reference for shaping the current state-of-the-art in this field, as well as a source of inspiration for the interested parties, that can assist them in their future research directions.

Camelia Delcea and Nora Monica Chirita

Guest Editors

applied sciences

MDPI

Editorial

Recent Advances in Time Series Forecasting Methods

Camelia Delcea

Department of Economic Informatics and Cybernetics, Bucharest University of Economic Studies, 010552 Bucharest, Romania; camelia.delcea@csie.ase.ro; Tel.: +40-769-652-813

1. Introduction

Time series forecasting has become a key decision-support tool, with broad applicability within a series of domains within the economic field, among other fields. Given the observed growth in large-scale, high-frequency, heterogeneous temporal data, a real need for accuracy, robustness, and adaptability in the tools used has been observed, which is not properly covered by traditional statistical methods. As a result, in recent approaches to time series forecasting methods, researchers have tried to place more emphasis on overcoming issues such as nonlinearity, long-term dependence, and uncertainty. Furthermore, the shift towards Machine Learning (ML)- and Artificial Intelligence (AI)-based forecasting was motivated by the inherent limitations to the traditional statistical approaches in the field.

This Special Issue, "Advanced Methods for Time Series Forecasting", aims to capture the recent evolution in the field by bringing together recent methodological advances and innovative applications that illustrate the evolving state of the field. The contributions published in the Special Issue collectively demonstrate how modern forecasting research increasingly relies on hybrid modeling strategies, attention-based architectures, structured state-space formulations, and time–frequency analyses to address real-world challenges.

2. Advances Reflected in the Special Issue

A central theme in this Special Issue is the growing role of deep learning architecture enhanced with attention mechanisms. In the published works, transformer-based and attention-driven models prove to be effective in capturing complex temporal dependencies across multiple horizons in various situations, ranging from traffic flow prediction, where hierarchical and multi-view attention frameworks have been proposed [1,2], to meteorological forecasting tasks related to tropical cyclone track prediction, where frequency-aware attention mechanisms combined with dual-branch architectures have shown notable robustness [3].

The use of hybrid frameworks that integrate signal decomposition and learning-based forecasting models is another methodological direction. By decomposing complex time series into multi-scale or frequency-specific components prior to prediction, it has been observed that the proposed approaches succeed in addressing issues related to non-stationarity and noise in the case of financial time series [4]. Thus, it can be observed that combining domain-aware preprocessing with advanced neural architecture is a viable approach for improving the accuracy and stability of forecasts.

Long-term forecasting and memory retention are addressed in this Special Issue through recent advances based on state-space reformulations and selective memory mechanisms. These mechanisms offer the possibility of mitigating error accumulation and instability over extended forecasting horizons [5], which is particularly relevant for applications requiring reliable long-term predictions across diverse domains, such as, but not limited to, energy systems, finance, healthcare, and transportation.

The uncertainty quantification represents another critical contribution of this Special Issue. How prediction intervals and uncertainty estimates can be obtained alongside point forecasts is demonstrated through the use of probabilistic deep learning approaches—such as Monte Carlo dropout neural networks [6]. This approach has the advantage of offering more support for risk-sensitive decisions.

In addition to these contributions, this Special Issue also encompasses studies on short and irregular time series forecasting [7], hybrid statistical and learning-based approaches for mid-term energy load prediction in smart buildings [8], application-driven analyses in management and economic contexts [9], and demand forecasting in retail environments using optimized attention-based deep learning models [10].

Beyond the individual modeling contributions, this Special Issue also includes a bibliometric analysis of AI-driven time series forecasting research. The analysis maps research growth and trends, while highlighting key journals and collaboration networks [11].

3. Addressing Knowledge Gaps

Collectively, the papers published in this Special Issue address a series of gaps in the time series forecasting literature. First, contrary to other approaches in the field, which rely on isolated model comparisons, the papers in this Special Issue propose integrated and hybrid solutions that combine complementary modeling approaches. Second, the papers demonstrate how advanced forecasting methods can be shaped to better answer to domain-specific constraints, such as, but not limited to, issues related to short time series, managing irregular sampling, or the fusion of heterogeneous data sources. Third, the papers included in this Special Issue focus on uncertainty-aware forecasting, highlighting its importance in the context of high-impact application areas.

The diversity of approaches noted in this Special Issue further highlight that time series forecasting remains a debatable and evolving research problem due to the complexity of real-world systems, which continue to challenge existing methodologies.

4. Future Research Directions and Outlook Toward a Follow-Up Special Issue

Building upon the advances showcased in this Special Issue, a series of promising avenues for future research emerge, such as the development of general-purpose and pre-trained forecasting models capable of transferring knowledge across domains, reducing data requirements, and improving scalability; creating explainable and trustworthy forecasting systems by embedding explainability directly into forecasting architectures rather than treating it as a post hoc process; and integrating forecasting and decision-making for better support in risk management.

Finally, the growth of streaming and real-time data requires the development of adaptive and online learning approaches, which should be capable of handling concept drift. In this situation, hybrid models can be the most suitable way to address these challenges.

Taken together, these research directions motivate the launch of a new Special Issue on next-generation time series forecasting, in which the focus should be on foundation models, explainability, uncertainty-aware decision support, and adaptive learning in complex environments.

https://doi.org/10.3390/app16031417

Funding: This work was funded by the EU's NextGenerationEU instrument through the National Recovery and Resilience Plan of Romania—Pillar III-C9-I8, managed by the Ministry of Research, Innovation and Digitization, within the project entitled "JobKG—A Knowledge Graph of the Romanian Job Market based on Natural Language Processing", contract no. 760274/26 March 2024, code CF 178/31 July 2023.

Acknowledgments: The Guest Editors would like to express their sincere gratitude to all the authors and peer reviewers for their valuable contributions and constructive feedback, which were essential to the successful completion of the Special Issue "Advanced Methods for Time Series Forecasting".

Conflicts of Interest: The author declares no conflicts of interest.

References

1. Wu, H.; Teng, G.; Wu, H.; Qiu, Z.; Zhao, M. MMHFormer: Multi-Source and Multi-View Hierarchical Transformer for Traffic Flow Prediction. *Appl. Sci.* **2025**, *15*, 12804. [CrossRef]
2. Li, W.; Sun, Z.; Wan, Y. Spatio-Temporal Multi-Graph Convolution Traffic Flow Prediction Model Based on Multi-Source Information Fusion and Attention Enhancement. *Appl. Sci.* **2025**, *15*, 11295. [CrossRef]
3. Meng, F.; Xiong, X.; Zhao, L. A Time and Frequency Domain Based Dual-Attention Neural Network for Tropical Cyclone Track Prediction. *Appl. Sci.* **2025**, *16*, 436. [CrossRef]
4. Su, J.; Lau, R.Y.K.; Du, Y.; Yu, J.; Zhang, H. A Novel Hybrid Framework for Stock Price Prediction Integrating Adaptive Signal Decomposition and Multi-Scale Feature Extraction. *Appl. Sci.* **2025**, *15*, 12450. [CrossRef]
5. Tan, X.; Wang, L.; Wang, M.; Zhang, Y. KOSLM: A Kalman-Optimal Hybrid State-Space Memory Network for Long-Term Time Series Forecasting. *Appl. Sci.* **2025**, *15*, 12684. [CrossRef]
6. Kummaraka, U.; Srisuradetchai, P. Monte Carlo Dropout Neural Networks for Forecasting Sinusoidal Time Series: Performance Evaluation and Uncertainty Quantification. *Appl. Sci.* **2025**, *15*, 4363. [CrossRef]
7. Bakalis, E. Iterative Forecasting of Short Time Series. *Appl. Sci.* **2025**, *15*, 11580. [CrossRef]
8. Hussain, A.; Franchini, G.; Akram, M.; Ehtsham, M.; Hashim, M.; Fenili, L.; Messi, S.; Giangrande, P. Hybrid ML/DL Approach to Optimize Mid-Term Electrical Load Forecasting for Smart Buildings. *Appl. Sci.* **2025**, *15*, 10066. [CrossRef]
9. Cao, S.; Zhou, C. TFHA: A Time–Frequency Harmonic Attention Framework for Analyzing Digital Management Strategy Impact Mechanisms. *Appl. Sci.* **2025**, *15*, 9989. [CrossRef]
10. Fandi, I.; Khalifa, W. Enhancing Demand Forecasting Using the Formicary Zebra Optimization with Distributed Attention Guided Deep Learning Model. *Appl. Sci.* **2026**, *16*, 1039. [CrossRef]
11. Domenteanu, A.; Diaconu, P.; Delcea, C. Bibliometric Insights into Time Series Forecasting and AI Research: Growth, Impact, and Future Directions. *Appl. Sci.* **2025**, *15*, 6221. [CrossRef]

 https://doi.org/10.3390/app16031417

applied sciences

MDPI

Article

Bibliometric Insights into Time Series Forecasting and AI Research: Growth, Impact, and Future Directions

Adrian Domenteanu [1], Paul Diaconu [2] and Camelia Delcea [1,*]

[1] Department of Economic Informatics and Cybernetics, Bucharest University of Economic Studies, 010552 Bucharest, Romania
[2] Department of Accounting and Audit, Bucharest University of Economic Studies, 010552 Bucharest, Romania
* Correspondence: camelia.delcea@csie.ase.ro; Tel.: +40-769-652-813

Abstract: Considering that nowadays the economy plays a crucial role, time series forecasting has become an essential tool across various economic areas and industries. The process of predicting future trends based on historical values in a reliable and accurate manner has generated numerous benefits, such as simplified decision-making processes or strategic planning and reduced risk management. Furthermore, with the advancement made through the use of Artificial Intelligence (AI) methods, time series forecasting has quickly become more precise, adaptive, and scalable, being able to better overcome real-world challenges. In this context, the present paper analyzes the implications of artificial intelligence in time series forecasting by evaluating the scientific articles from the field indexed in Clarivate Analytics' Web of Science Core Collection database. Through a bibliometric approach, the research identifies key journals, affiliations, authors, and countries, as well as the collaboration networks among authors and countries. It also analyzes the most frequently used keywords and authors' keywords. The annual growth rate of 23.11% indicates sustained interest among researchers. Prominent journals such as *IEEE Access*, *Energies*, *Mathematics*, *Applied Sciences—Basel*, and *Applied Energy* have been the home for the most published papers in this field. Further, thanks to the Biblioshiny library in R, a variety of visualizations have been created, including thematic maps, three-field plots, and word clouds. A comprehensive review of the most cited papers has been performed to highlight the role of AI in time series forecasting. Research results and methods confirmed the versatility of the topics, which have been applied in various fields, such as, but not limited to, finance, energy, climate, and healthcare, and are further discussed. Cutting-edge methodologies and approaches that lead to the transformation of the field of time series analysis in the context of AI are uncovered and discussed through the use of thematic maps.

Keywords: time series forecasting; artificial intelligence; bibliometric analysis

1. Introduction

Historically, Artificial Intelligence (AI) has been defined in multiple ways, but due to the diverse topics of interest to researchers, it was difficult to formulate a definition accepted by everyone. Recently, AI has been described as a hot topic that offers the possibility of integrating social aspects into a machine that is able to solve tasks and communicate with society [1]. At the same time, AI provides autonomy and the possibility for users to automate activities and tasks.

The time series stands as one of the key topics of the academic community, representing a simple applicable method in domains such as medicine, decision-making, finance, climate modeling, and the biological sciences [2]. Initially, the domain focused on traditional models such as autoregressive models (AR), moving average (MA), seasonal autoregressive integrated moving average (SARIMA), exponential smoothing, and many other algorithms, while the current focus is to merge it with AI or deep learning methods.

Due to global environmental changes, it has become crucial to monitor and predict future trends in order to protect existing resources and minimize the negative impact on key sectors such as agriculture and hydrology [3]. Thanks to the development of remote sensing, the analysis of past and actual environmental characteristics has become much faster, facilitating easier monitoring of vegetation health and water quality and the identification of land cover changes [4].

Also, in the context of technological development, multiple time series models have evolved; the best examples include the self-exciting threshold autoregressive (SETAR) model, generalized autoregressive conditional heteroscedasticity (GARCH), random forest algorithm (RF), artificial neural networks (ANN), and multivariate adaptive regression splines (MARS) [5]. The application of time series models combined with AI opens new possibilities for implementation, the best example being in the hydrological field, where the algorithms have successfully predicted monthly river flows.

Furthermore, several metrics have been implemented, such as root mean square error (RMSE), coefficient of determination (R^2), Willmott Index (WI), and normalized root mean square error (NRMSE), which are also used to calculate the accuracy of the methods [6].

The scope of interest among researchers in recent decades has ranged from developing and improving algorithms that predict water resources to developing new models such as MARS or gene expression programming (GEP), which has shown strong results in real case scenarios. Some of the algorithms included time series models such as AR, MA, autoregressive conditional heteroscedasticity (ARCH), and non-linear time series methods [7]. According to the existing literature, AI methods have been more frequently implemented than time series models for streamflow estimation, thanks to the development of hybrid models that combine AI and time series [7]. The domain has recently been expanded to include algorithms such as support vector machine (SVM), feed-forward neural network, multilayer perception (MLP), and multiple linear regression (MLR).

Time series prediction has a significant impact on daily activities, thanks to the evolution of deep learning and AI. The development of convolutional neural networks and recurrent neural networks has led to an increased rate of implementation for deep learning and AI [8].

Rough sets represent an alternative for forecasting methods, with applications in decision-making, expert systems, and data mining, focusing on explaining the rule-based models and applying similar steps such as a time series analysis. The dataset is divided into multiple sections, with a lower approximation that contains the elements that correspond to a class, and an upper approximation where the elements could belong to the class [9].

In 2024, Bas et al. [10] published an analysis of the implementation of artificial neural networks in time series forecasting problems, evaluating the possibility of the dendritic neuron model to identify outliers and their impact. The model has been tested on closing prices from German, Spanish, and Italian stock exchanges for a specific timespan, resulting in an algorithm that is able to accurately predict the values even when outliers exist in the dataset.

Edalatpanah et al. [11] described the existing risks on the market, together with the economic pressures and how technological developments, time series forecasting, and artificial intelligence could improve scientific productivity and business environments. Multiple algorithms have been developed, such as particle swarm optimization, genetic algorithm, and quantum optimization, resulting in a hybrid model between neutrosophic time series and AI models as an optimal solution.

Furthermore, Dynamic Bayesian Networks (DBNs), included in the family of probabilistic graphical models (PGMs), have been used to model the evolution of parameters described as a time series, with a series of applications in the agricultural field, such as digital agriculture [12], and in transportation [13–15].

The purpose of the present bibliometric analysis was to evaluate the academic development of time series forecasting and AI domains and to understand the evolution of the algorithms and the level of inclusivity of the models. Thus, a bibliometric analysis has been conducted for a comprehensive overview of the main authors, affiliations, journals, international collaboration, thematic evolution, main contributors, and most used keywords, also taking into account the existing bibliometric analyses by the academic community [16–19]. According to Block and Fisch [17], a bibliometric analysis is different from a review analysis that focuses on the main findings in the investigated field; the bibliometric overview explores the structural elements of the topic, extracting the main factors of time series forecasting and AI. In order to extract representative information from the research, this paper will focus on answering the following scientific questions:

- SQ1: Are the authors contributing to this field likely to collaborate? If yes, what does the collaboration map look like?
- SQ2: What are the most relevant themes discussed in the time series and artificial intelligence topics?
- SQ3: Which countries are the most representative, taking into account the number of publications and citations?
- SQ4: What are the outcomes of the 10 most cited articles?
- SQ5: What are the sources with the highest impact based on the number of articles published?
- SQ6: What are the main keywords included in the thematic graphical representation?

As was already mentioned above, a detailed investigation of the 10 most cited papers will be carried out for a better understanding of the time series and AI domains. Furthermore, the time distribution of the publications included in the dataset is crucial in topic analysis evolution.

The decision to conduct a bibliometric analysis for evaluating the time series and AI domains is in accordance with the existing literature, focusing on extracting, in a proper manner, the main topics, key contributors, collaboration map, main research institutions, and the evolution of the research themes [16,20]. The bibliometric approach delves deeply into the analysis of the dataset using indicators such as the H-index or G-index, which will be further explained, together with graphical representations of Bradford's Law or Lotka's Law. Due to the nature of the scientific question, quantitative analysis is mandatory, evaluating the 10 most representative papers from the time series forecasting and AI fields.

The paper is divided into multiple sections as follows: the Section 1 explores the introductory information of the research, defining the scope of the article, while the Section 2 focuses on the main methods and techniques implemented in the bibliometric analysis, explaining also the dataset extraction process. The Section 3 describes the results of the analysis, highlighting the most representative countries, affiliations, authors, journals, keywords, and themes. The Section 4 presents the discussions and limitations of the analysis, while the Section 5 concludes the research.

2. Materials and Methods

The initial step was to investigate the existing literature in order to identify the most suitable database for the bibliometric analysis. ISI Web of Science, known also as

Clarivate Analytics' Web of Science Core Collection or WoS [21], was chosen as the database used for the papers' extraction [22–26]. Bakir et al. [27] explained why WoS is the best database to perform a bibliometric analysis, having a variety of journals and domains, most of them recognized and known in the academic community. Even if the level of inclusivity is smaller compared with other databases, the WoS database is still one of the best sources for researchers. Also, the choice of this database has been further highlighted by Koca [28]. The author mentions the number of publications included, their subject distribution, and their geographical distribution among the reasons for choosing this database. The majority of ISI Web of Science documents are published in English and have a wider journal coverage compared to Dimensions, IEEE, or Scopus databases, with a total of 74.8 million scholarly datasets and over 1.5 billion cited references, describing a diversity of academic topics [29]. Singh et al. [29] explained that WoS coverage improved significantly starting in 1990, indexing over 13,610 journals and 13 million publications as of June 2020. Even if the Scopus database has more papers (18 million as of June 2020), the impact of Scopus-indexed journals is smaller compared to those indexed by WoS. Furthermore, it should be stated that the WoS database offers, within the extracted dataset, a special feature related to the papers included in the dataset, called Keywords Plus [30]. This special class of keywords is extracted based on the references in the papers included in the dataset, helping in better shaping the information included in the studies.

A detailed analysis of the existing research has been carried out, in order to understand how to extract documents from Clarivate Analytics' Web of Science Core Collection. Liu [31,32] explained the main steps that should be taken in the extraction process. A crucial point that should be mentioned is that the WoS database is subscription-based, meaning that the results can be differentiated based on the subscription type. At the same time, some of the subscriptions offer limited access to specific indexes, while other subscriptions provide access to a variety of indexes and papers, and, as Liu [32] highlighted, it is crucial for authors to investigate and explain the indexes that are available with their subscription. The subscription influences the total number of papers that are extracted from WoS, and in our case, the indexes that were available are detailed below:

- Social Sciences Citation Index (SSCI)—1975–present;
- Book Citation Index—Science (BKCI-S)—2010–present;
- Emerging Sources Citations Index (ESCI)—2005–present;
- Science Citation Index Expanded (SCIE)—1900–present;
- Index Chemicus (IC)—2010–present;
- Book Citation Index—Social Sciences and Humanities (BKCI-SSH)—2010–present;
- Arts and Humanities Citation Index (A&HCI)—1975–present;
- Conference Proceedings Citation Index—Science (CPCI-S)—1990–present;
- Conference Proceedings Citation Index—Social Sciences and Humanities (CPCI-SSH)—1990–present;
- Current Chemical Reactions (CCR-Expanded)—2010–present;

In Figure 1 are detailed the main steps that should be performed for a bibliometric analysis, according to Preferred Reporting Items for Systematic Reviews and Meta-Analyses, also known as the PRISMA flowchart.

Sarkis-Onofre et al. [33] explained that PRISMA comprises a set of rules and recommendations used mainly for transparent and complete reporting of systematic reviews, offering the possibility to develop appropriate analyses of papers and include all aspects accurately.

The first step consists of dataset Identification, where the documents are related to time series forecasting and artificial intelligence. The keywords such as "time_serie*_forecast" and "artificial_intelligence" were applied to the existing WoS, searching in abstracts (AB), titles (TI), and authors' keywords (AK) in the most representative papers. In total, there were 207,977 papers related to artificial intelligence and 7686 papers related to time series forecasting. In order to investigate the common documents, a filter was applied, resulting in a total of 366 documents that contained artificial intelligence and time series forecasting. It should be mentioned that the query was performed using the singular forms of the filtering keywords, in order to be able to include also the plural forms of the terms by adding an asterisk ("*") at the end of the singular form. At the same time, the focus was to include also the combination of terms, rather than including only the individual forms, grouping the words using the underline ("_"). Thanks to the underline, the results contained sequences of the search terms and not just the individual forms. For instance, the keywords "artificial_intelligence" written in this form extracted from the database the papers that had "artificial intelligence" as a sequence either in the abstract, authors' keywords, or titles, and not only the individual term of "artificial" or "intelligence".

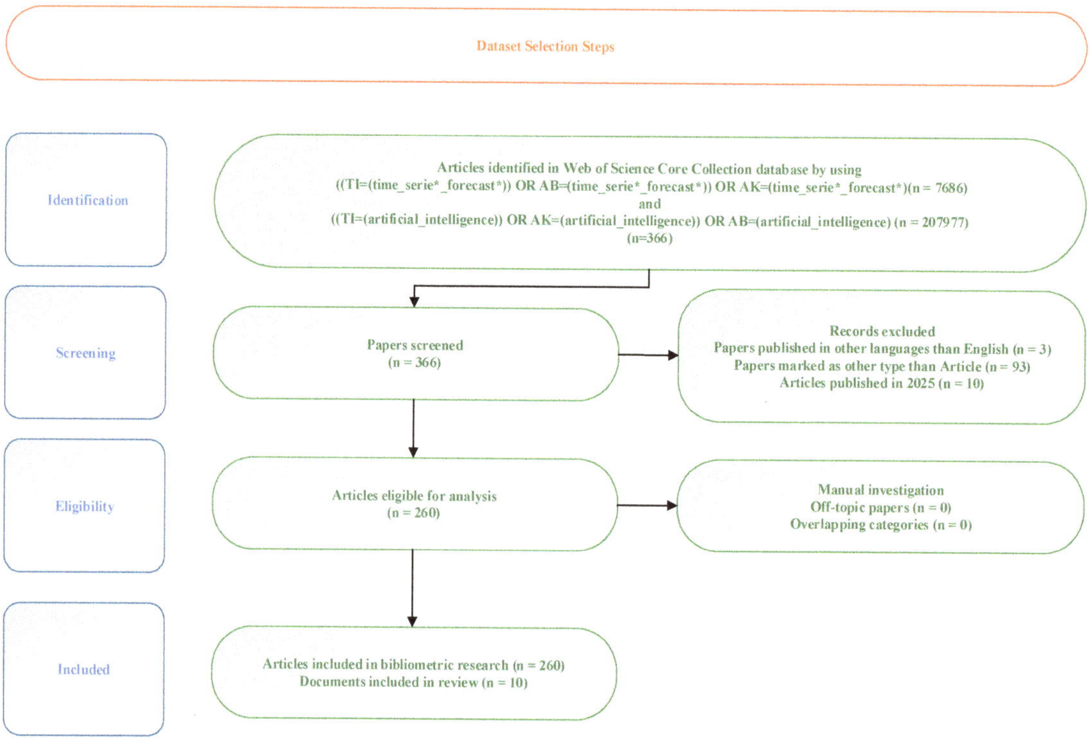

Figure 1. Paper Selection Process for Bibliometric Research Using PRISMA Flow Diagram.

The second step refers to the screening part, where the extracted papers were filtered in order to reduce the gap between documents and to define a relevant dataset. The documents published in languages other than English were excluded, reducing the size of the dataset by three articles. At the same time, the types of documents were restricted, including only papers marked as "Article", resulting in a total of 260 articles. At the same time, a manual analysis of paper with overlapping topics was performed during the eligibility step, as suggested by Oo and Rakthin [34], resulting in a homogenous dataset. According to Pham and Le [35], who performed a bibliometric analysis and review of papers published in the Scopus database using the PRISMA approach, the main steps they applied were similar to ours, including only specific types of documents, restricting the articles published in other languages than English, and restricting the timespan of the publications. The authors began with a dataset that contained 502 articles, and after the filters, the final dataset contained only 51 papers. Peixoto et al. [36] defined a set of inclusion and exclusion criteria. The exclusion criteria specified non-English papers and papers that were theoretical. The inclusion criteria specified the types of documents that should be taken into account in the research. From 705 documents at the beginning, the final dataset contained only 30 articles after screening.

In the end, our final dataset contained a total of 260 articles that would be evaluated from a bibliometric point of view, while a selection of the top-10 most cited papers would be used for the papers' review.

3. Overview of Results

The third chapter describes the main information about the dataset from different points of view, focusing initially on the timespan, number of documents, sources, and references; the chapter is divided into multiple sections, such as journals, authors, an evaluation of the literature, and mixed analysis. Using the R programming language and Biblioshiny library, a bibliometric analysis was performed. The current version of the R programming language is 4.4.1, while the Biblioshiny library version is 4.2.3.

Table 1 encompasses the most relevant information about the dataset that was extracted from WoS. The analyzed period was between 1997 and 2024, with a total of 158 sources and 260 documents with a mean of 3.91 years from publication and a mean of 20.51 citations per paper. In total, there were 4.08 co-authors per article and 11,079 references. Friou and Graa [37] explored the AI impact on e-commerce using a bibliometric approach, evaluating a total of 669 papers and 329 sources that had been published between 2014 and 2023, with an average of 3.24 co-authors per document and an average document age of 2.76 years. Bawack et al. [38] extracted from WoS papers published between 1991 and 2020, obtaining 4335 documents and 2599 sources in total with a mean of 8.645 citations per paper, 3.03 co-authors per article, and a total of 84,474 references. The description of the datasets differs from one paper to another due to the focus of the analysis, which was reflected in the filters that were applied to the databases.

Table 1. Main information about extracted dataset.

Indicator	Value
Timespan	1997–2024
Sources	158
Documents	260
Average years from publication	3.91
Average citations per documents	20.51
Co-authors per document	4.08
References	11,079

3.1. Journal Analysis

This sub-section focuses on the investigation of the main journals based on the number of published articles or citations. At the same time, a description of yearly scientific production and mean citations will be discussed.

Figure 2 details the yearly scientific production on the time series forecast and AI domains, between 1997 and 2024. During the timespan, a positive trend could be observed, especially starting with the 2019–2020 period. The increase in the number of papers starting with this period could have two main reasons. The first one could be related to the mainstream adoption of the transformer models across various disciplines as well as the creation of new transformer variants for time series forecasting, with impacts on the improvement of long-term forecasting accuracy. On the other hand, some of the increase could be due to the general trend observed in research by Ofer et al. [39], that underlined that the number of scientific papers had increased in recent decades. The increase in the number of publications could have multiple causes, such as, but not limited to, better automated data indexing observed through the use of various databases [39], the use of the keyword annotation scheme by research subjects [39], the advancements in computer science [39], the growth in research and development expenditures [40], and the overall number of researchers in the field [40].

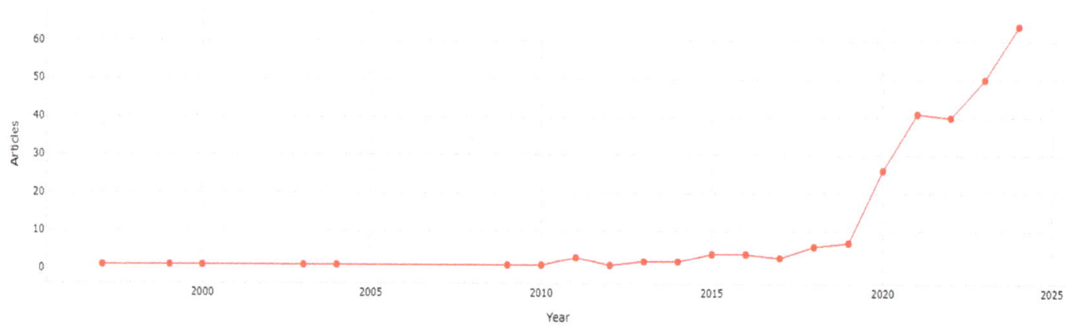

Figure 2. Annual Scientific Production.

The Annual Growth Rate was 23.11%, which points to the increased interest of researchers. It should be stated that the formula used for determining this indicator was the one offered by Biblioshiny [26,41]:

$$\text{Annual Growth Rate} = \left(\frac{\text{Final Year Production}}{\text{Initial Year Production}}\right)^{\frac{1}{\text{Time Span}}} - 1$$

in which the indicators are as follows:

- Initial Year Production refers to the initial year for the latest period of the timespan in which all years have at least one publication
- Final Year Production refers to the final year for the latest period of the timespan in which all years have at least one publication
- Timespan refers to the period between the initial and the final year of the latest period of the timespan in which all years have at least one publication

In the case depicted in Figure 2, the Annual Growth Rate was determined as follows:

$$\begin{aligned}\text{Annual Growth Rate} &= \left(\frac{\text{2024 Year Production}}{\text{2009 Year Production}}\right)^{\frac{1}{\text{Time Span between 2009 and 2024}}} - 1 \\ &= \left(\frac{64}{1}\right)^{\frac{1}{15}} - 1 = 1.2311 - 1 = 0.2311 = 23.11\%\end{aligned}$$

The first paper was published in 1997, which represents 0.38% of total articles released. In 1998, no papers were published, followed by two consecutive years with papers published. Until 2018, the number of publications was around three or four papers per year, which represented approximately 1.5% of total publications. Starting in 2018, the number increased to six (2.31%), followed by seven in 2019 (2.69%). In 2020, a significant spike was registered, with 26 articles published (10%), followed by 41 in 2021 (15.77%), 40 in 2022 (15.38%), 50 in 2023 (19.23%), and a peak in 2024, with 64 documents (24.62%) published. Monroy and Diaz [42] evaluated the time series of papers using a bibliometric approach. The authors focused on papers published between 1960 and 2020, and the trend was similar, starting to increase in 2000, with a peak in 2020. Palomero et al. [43] presented the evolution of fuzzy-based time series forecasting, applying a bibliometric approach for articles that had been published between 2017 and 2021. There was a positive trend between 2017 and 2020, similar to our findings, but in 2021, the value decreased (in a manner similar to what happened in 2022 in our dataset). According to the investigated research, there were similarities between our outcome and the existing literature reviews. Bawack et al. [38] described the AI impact on e-commerce, investigating the existing literature using a bibliometric analysis. The first paper was published in 1991, with a positive trend noted during the timespan. In 2019, a peak of 522 articles was achieved, followed by a small decrease in 2020 to 439 articles. Frioui and Graa [37] presented the impact of AI on e-commerce and discovered that there was a positive trend in the publication of articles starting in 2016, peaking in 2021 with 149 papers, followed by a small decrease in 2022, to 146 articles, and a more significant decrease in 2023, to 73.

Figure 3 illustrates the average citations per year on the domains of time series and AI. The first year with citations was 1997, with an average of 3.72 citations. A significant spike was observed in 2009, where the average citations per year reached 36.12, when the most cited article, with a total of 614 citations, was published by Wang [44]. During the remaining period, the mean citations were between 1 and 7 per year.

Figure 3. Average Citations per Year.

Figure 4 includes the most representative sources based on the number of publications. In first place was *IEEE Access* with 11 papers published, followed by *Energies* with nine documents, *Mathematics* with seven documents, and *Applied Sciences-Basel* with six articles. In fifth place, there were multiple journals: *Applied Energy, Energy, Knowledge-Based Systems, Sensors,* and *Sustainability* with five papers each, while *Artificial Intelligence Review, Computational Economics, Soft Computing, Journal of Hydrology, Engineering Applications of Artificial Intelligence,* and *Expert Systems with Applications* were the sources with four articles each. For a better understanding of the results, the existing articles that were published on similar topics were investigated. Palomero et al. [43] evaluated fuzzy-based time series forecasting using a bibliometric approach, discovering that *IEEE Access, Energies, Knowledge-Based Systems, Applied Sciences-Basel,* and *Applied Energy* were among the journals with the highest impact on the investigated field. Chen et al. [15] explored the AI impact on traffic flow prediction, where the time series method is one of the most implemented. The most significant journals identified were also among those in our findings: *Sustainability, Applied Sciences-Basel, Sensors,* and *IEEE Access,* but at the same time, there were sources that were not found in our results, such as *Neurocomputing* or *Neural Computing & Applications.*

Figure 5 shows the 10 most crucial local journals based on the number of citations. In first place are two journals, *Expert Systems with Applications* and *Journal of Hydrology,* each with 289 citations, followed by *ARXIV* with 225 citations, *Energy* with 219 citations, and

Applied Energy with 216 citations. *IEEE Access* had 175 citations and *International Journal of Forecasting* had a total of 169 citations, followed by *Neurocomputing* with 160 citations, *Energies* with 153 citations, and *Applied Soft Computing* with 150 citations. Shabani et al. [46] explored portfolio optimization using AI and discovered that *Expert Systems with Applications* was the journal with the highest number of publications, followed by *Journal of Portfolio Management* and *IEEE Access*. The rest of the journals had less impact and were not common in our outcome, demonstrating the variety of the domain and how the impact on the results changed if, during the extraction process, different filters were applied.

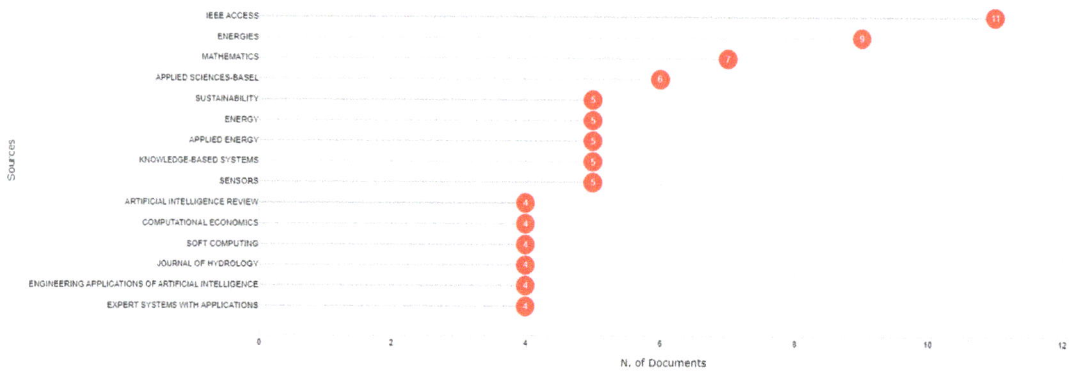

Figure 4. Most Relevant Sources.

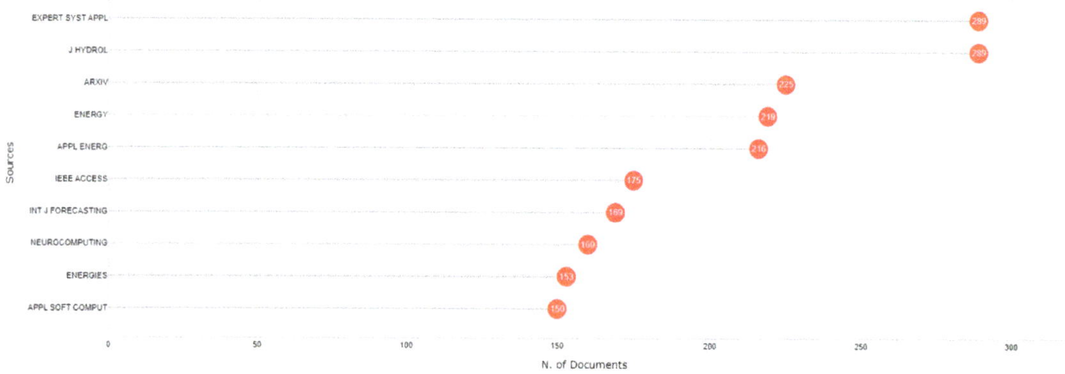

Figure 5. Most Relevant Local Sources.

Figure 6 presents the 10 most representative local sources using the H-index, also known as the Hirsch Index, which quantifies the total number of papers where the source has at least the same number of citations as the article [47].

Additionally, the G-index represents an improvement over the H-index, taking into account only the unique highest number where the top g documents obtained, in total, at least g^2 citations [48].

In our case, the most relevant source was *Energies*, with an H-index value of 5 and a G-index value of 9, followed by *Applied Sciences-Basel*, with an H-index value of 4 and a G-index value of 6. In third place was *Artificial Intelligence Review*, with an H-index of 4 and a G-index of 4. *Energy*, *IEEE Access*, *Journal of Hydrology*, *Mathematics*, *Sensors*, and *Soft Computing* each had an H-index value of 4, while *Applied Soft Computing*, *Engineering Applications of*

Artificial Intelligence, *Computational Economics*, *Neural Computing & Applications*, *Knowledge-Based Systems*, *Renewable Energy*, and *Expert Systems with Applications* had an H-index value of 3.

Figure 6. Sources with Local Impact.

Figure 7 incorporates the most cited sources, separating the journals with fewer citations.

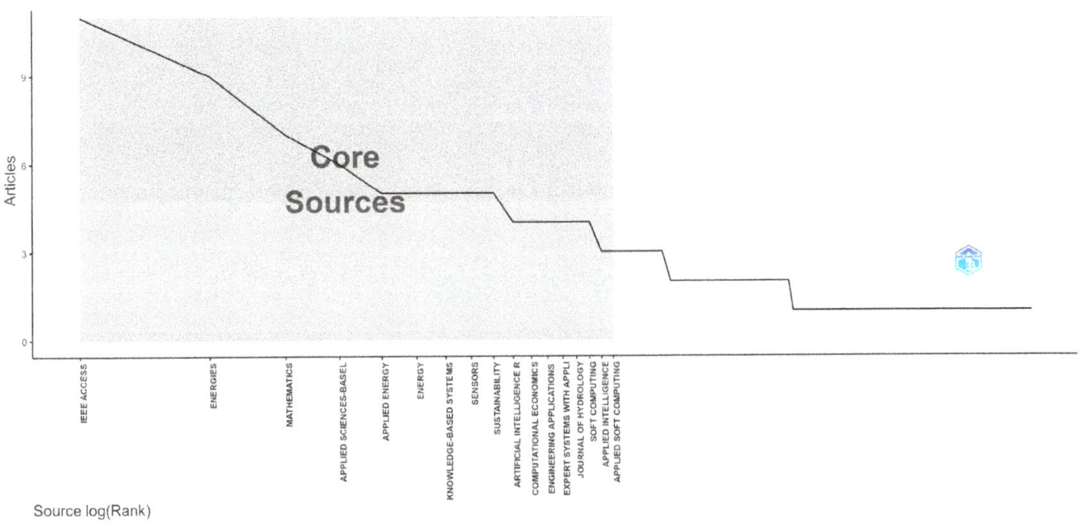

Figure 7. Core Sources by Bradford's Law.

The Bradford's Law method groups the journals into three different sections that should have the same number of papers, but each category has a different number of citations. According to Yang et al. [49], the Bradford's Law clusters journals using the formula 1:n:n^2. The most representative journals from Zone 1 were *IEEE Access* (11 articles), *Energies* (9 articles), *Mathematics* (7 articles), *Applied Sciences-Basel* (6 articles), and *Applied Energy* (5 articles). The most relevant sources in Zone 2 were *Journal of Hydrologic Engineering* (3 articles), *Journal of Intelligent & Fuzzy Systems* (3 articles), *Neural Computing & Applications* (3 articles), *Renewable Energy* (3 articles), and *Theoretical And Applied Climatology* (3 articles). In the last part, Zone 3, the most significant journals were *Economies* (1 article), *Energy and*

AI (1 article), *Energy and Buildings* (1 article), *Energy Conversion and Management* (1 article), and *Energy Exploration & Exploitation* (1 article).

Shabani et al. [46] implemented the Bradford's Law method and discovered that *Expert Systems with Applications*, *IEEE Access*, *Knowledge-Based Systems*, *Mathematics*, and *Computational Economics* were the main journals, which confirmed our results. At the same time, there were differences, with Bradford's Law identifying journals that were not included in our analysis, such as *Journal of Portfolio Management*, *Financial Markets and Portfolio Management*, and *Quantitative Finance*.

3.2. Authors and Affiliations Analysis

The second part of the third chapter explores the contributions of the most relevant researchers, presenting the total number of contributions, citations, and the main universities that published on the topic of time series forecasting and AI.

Figure 8 presents the authors with the highest impact, based on the number of published articles. In first place were two authors, Feng ZK. and Niu WJ., each with seven articles, followed by Wang JZ. and Wang Y., each with five documents, while Egrioglu E., Li HM., and Pinheiro CAM. published four documents each. The last authors, with three documents released, were Bueno-Crespo A., Cao Y., Cecilia JM., Yang HF., Xu YS., Morales-Garcia J., Feng BF., Jiang P., Khashei M., Lee J., and Martinez-Espana R. The article fractionalized value defines the contributions of authors based on the number of published articles [50], according to the Bibliometrix webpage. The author with the highest article fractionalized value was Egrioglu E. with 1.78, followed by Feng ZK. and Niu WJ. with 1.58 each, Wang JZ. with 1.33, Wang Y. with 1.25, Cao Y. with 1.03, and Li HM. and Pinheiro CAM. with 1.00 each, while Bueno-Crespo A. and Cecilia JM had the lowest values in the top 10, with only 0.58 each. Yang HF. had an article fractionalized value of 0.75, while Xu YS. had an article fractionalized value of 0.54, Morales-Garcia J. had a value of 0.58, Feng BF. had an article fractionalized value of 0.54, Jiang P. had an article fractionalized value of 1.00, Khashei M. had an article fractionalized value of 1.33, Lee J. had a value of 0.64, and Martinez-Espana R had a value of 0.58.

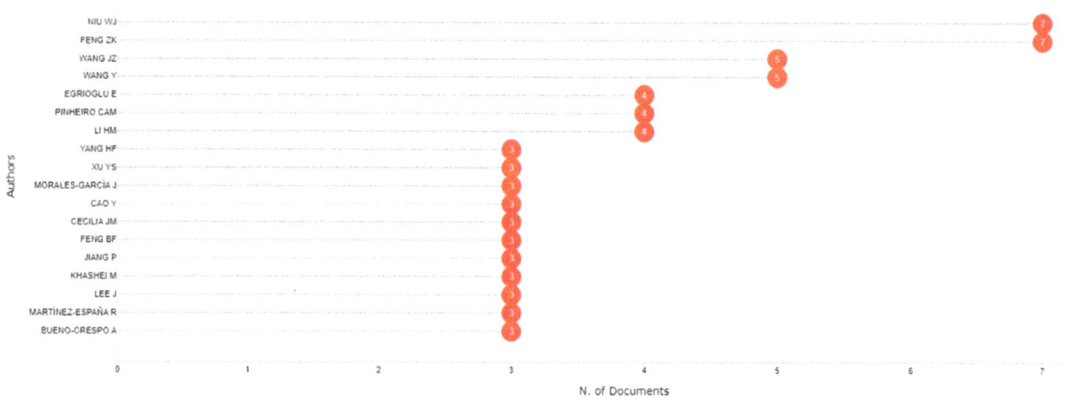

Figure 8. Most Relevant Authors.

According to the research conducted by Soliman et al. [51], the most relevant authors in the field of AI were Li X (148 papers), Wang Y. (144 papers), Zhang Y. (143 papers), Wang Z (139 papers), Liu Y. (133 papers) and Wang J. (131 papers). Comparing the authors' results with our outcome, only two authors were common, with the differences coming from the database and timespan selection, as Soliman et al. [51] focused on CiteScore Scopus journal papers that had been published between 2020 and 2021.

Figure 9 focuses on the major authors by taking into account the local citations. In first place are Feng ZK. and Niu WJ. with 11 citations each, while Cheng CT. is third with 8 citations, and Chau KW., Qiu L., and Wang WC., each have 7 citations. The last authors are Balog RS., Chou JS., Egrioglu E., Hajmirza S., Shardaga H., Tran DS., and Wang JZ., with 3 citations each.

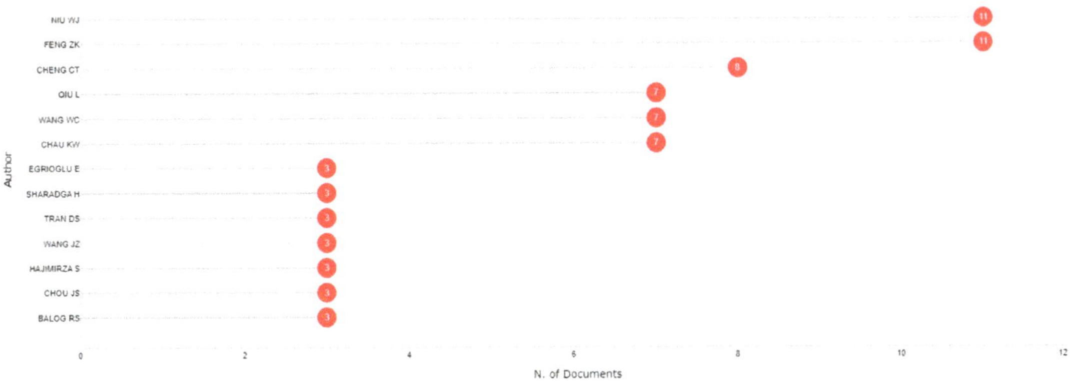

Figure 9. Authors with Most Local Citations.

Figure 10 reveals the productivity of the authors using Lotka's Law, which explains the number of publications of researchers, estimating that a reduced percentage of authors contributed to the majority of the articles [52].

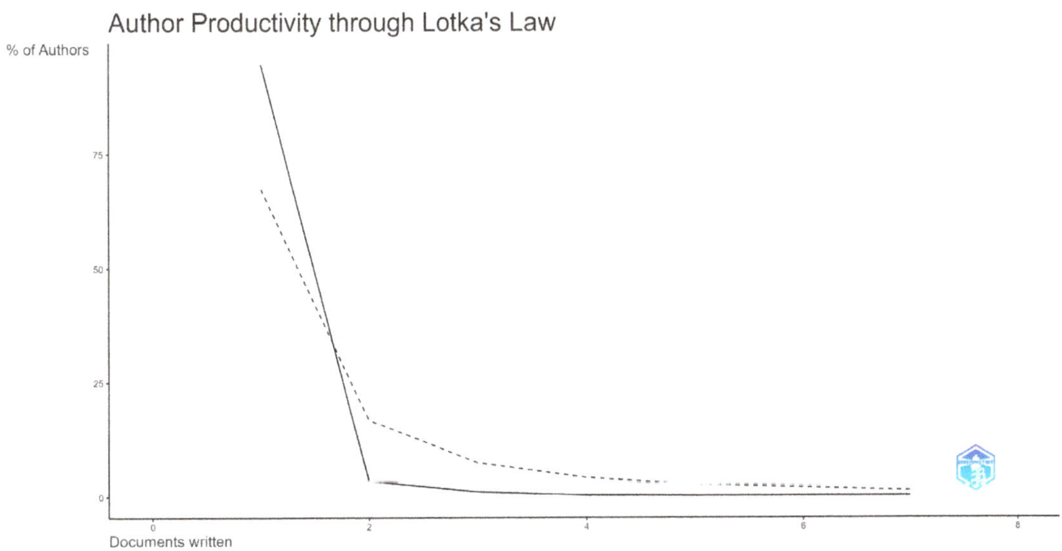

Figure 10. Author Productivity Using Lotka's Law.

The law takes into consideration a proportional rule of $1/n^2$, where n stands for the number of papers an author published, highlighting the contribution and productivity of high-performing researchers [52]. Most of the authors (94.7%) published only one article, and only 3.5% published two papers, 1.1% published three, 0.3% published four, and only 0.2% published five and seven articles. Lotka's distribution points out the complexity of the domains, since almost all authors published only once, but at the same time, the fields evolved significantly in recent years, and the imbalance can be reduced in the following period. According to Shabani et al. [46], the outcome with Lotka's law showed that 87.6% of the researchers contributed to a single paper, while 9% contributed to two documents, 1.9% to three articles, and only 0.7% contributed to four publications. Only 0.4% of the authors worked on five papers, while 0.1% worked on six and seven articles. The results were comparable with our outcome, showing the majority of researchers contributing to just one paper, while the percentage dropped sharply starting with two papers.

Figure 11 includes the 10 most representative universities that published in the fields of time series forecasting and AI by taking into account the number of publications. In first place was Egyptian Knowledge Bank (EKB) from Egypt with 13 documents, followed by Dongbei University of Finance and Economics and Hohai University, both from China with 10 documents each, while Chinese Academy of Sciences (China), Universidade de Aveiro (Portugal) and Unversity System of Georgia (United States of America) had eight publications each. King Abdulaziz University (Saudi Arabia) and University of Texas System (United States of America) published six papers each, while China University of Geosciences (China), Huazhong University of Science and Technology (China), Shanghai Jiao Tong University (China), University of Murcia (Spain), University of Science and Technology Beijing (China), University of Texas at San Antonio (USA) and Universitat Politecnica de Valencia (Spain) published five articles each.

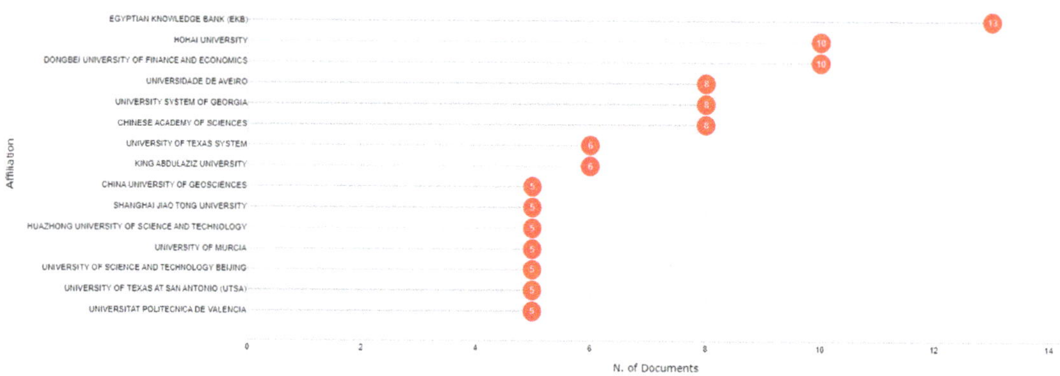

Figure 11. Most Representative Universities.

In total, there were seven universities from China, three from the United States of America (USA), two from Spain, one from Saudi Arabia, one from Portugal, and one from Egypt.

3.3. *Most Cited Documents*

This section presents the 10 most representative papers in the time series forecasting and Artificial Intelligence domains based on the total number of citations. A short description of each article will be described below in order to observe and understand the main objective of the analysis, extracting the major topics and methods that have been implemented.

Table 2 includes relevant information regarding the 10 most cited papers, such as the first author of the article, year when it was published, journal name, the number of researchers, region, total citations (TC), total citations per year (TCY), and normalized total citations (NTC). The NTC indicator presents importance mostly in the context of the other papers included in the dataset, as it presents how many times an article succeeds in receiving more citations than the average number of citations received by the papers published in the same year as the analyzed paper and included in the dataset [53,54].

AI and time series forecasting developed significantly in recent years, mainly thanks to the technological evolution. The most cited paper was published by Wang [44] and has been cited 614 times, having four authors, a total citations per year (TCY) of 36.12, and a normalized total citation (NTC) of 1. The authors explored the hydrological sector, focusing on estimating hydropower reservoir management and optimal scheduling, using past data, implementing multiple models such as the autoregressive moving average (ARMA), adaptive neural-based fuzzy inference system (ANFIS), artificial neural networks (ANN), genetic programming (GP), and support vector machine (SVM). The algorithms were trained on monthly river flow discharges and, in order to classify the models, multiple statistical metrics were calculated, such as the Nash–Sutcliffe efficiency coefficient (E), coefficient of correlation (R), mean absolute percentage error (MAPE), and root mean squared error (RMSE). The outcome showed that SVM, GP, and ANFIS were the algorithms with the highest accuracy based on various criteria that had been tested for training and validation processes.

Oil production was forecasted with the long short-term memory (LSTM) neural network approach for production by fractured horizontal wells, using two case studies from Xinjiang oilfield, China. Song et al. [55] also included Particle Swarm Optimization (PSO) in order to optimize the LSTM algorithm. The results were compared with traditional neural network methods, conventional decline curves, and time-series solutions. The outcome demonstrated that LSTM outperformed the traditional algorithms. In total, eight authors contributed, obtaining a total of 241 citations, with total citations per year of 40.17 and a normalized total citation of 5.70.

Solar power generated by photovoltaic panels plays a significant role in the electricity market, especially when fluctuations in the systems occur due to environmental factors. Sharadga et al. [56] predicted solar power generation using AI, neural networks, LSTM, and a multilayer perceptron (MLP) network. The objective of the research was to accurately predict solar power production, facilitating the operation, scheduling, and management of related power systems. At the same time, the effect of power forecasting was explored. The data included in the analysis contained over 3640 h of operation-related data points extracted from grid-connected photovoltaic stations in China, which produced a total of 20 megawatts (MW). There were three authors that investigated the topic, receiving a total of 225 citations, with a value of 37.50 for TCY and 5.32 for NTC.

The tourism sector plays a crucial role in the global economy, representing a priority for Cho, who evaluated multiple time series algorithms like exponential smoothing, Elman's Model of Artificial Neural Networks, and ARIMA. The objective of the research was to estimate tourism traffic based on historical data and to understand tourists' behavior. According to the results, the neural network algorithm was the most suitable solution for the prediction of visitor arrivals, especially if the data had no trend or pattern. Only one author contributed to the paper, receiving 193 citations, with a TCY value of 8.39 and an NTC value of 1.

A long short-term memory recurrent neural network (LSTM-RNN) and AI were implemented by Sahoo et al. for a gauging station located in the Mahanadi River basin, India, containing daily information. A comparison was performed between LSTM-RNN, RNN, and naïve methods. Multiple metrics, such as the Nash–Sutcliffe coefficient, correlation coefficient, and root mean squared error, were applied in order to test the accuracy of the algorithms. The outcome confirmed that LSTM-RNN was the model with the best metrics, representing a reliable AI solution for low-flow forecasting. The article had 171 citations, and four authors contributed to the research. The TCY value was 24.43, while the NTC value was 4.73.

The importance of accurately predicting photovoltaic power generation represents a crucial step for integrating, operating, and scheduling smart grid systems, as Qu et al. pointed out in their research. The authors focused on predicting the day-ahead hourly photovoltaic power supply, using historical data extracted from the DKASC website and taking into account existing constraints like volatility and intermittence of solar energy. A short-term temporal neural network model (ALSM), LSTM, and other algorithms were tested. The outcome confirmed that ALSM predicted the photovoltaic power with the highest accuracy. Three authors contributed to the article, resulting in a total of 154 citations, with a TCY value of 30.80 and an NTC value of 7.13.

Using time series and AI, Chou and Tran [57] predicted the energy consumption of buildings by taking into account the real-time data collected from an experimental construction project. The objective of the research was to improve sustainable development and energy efficiency, reducing the costs and environmental impact. A hybrid model that combines optimization and forecasting methods was defined, affording better accuracy than ensemble models and representing a significant support for planning energy efficiency and management. Two authors investigated the topic, and the article was cited 146 times, with a TCY value of 18.25 and an NTC value of 2.40.

A time series analysis, together with AI, was explored by Nourani [58] in order to predict daily rainfall runoff related to the hydrological domain. An emotional artificial neural network (EANN) was defined, trained, and evaluated, demonstrating the capability of the algorithm, compared with ANN, in understanding and differentiating rainless days and rainy days using artificial emotional system elements, such as hormonal criteria. Nourani was the only author that worked on the paper, obtaining 137 citations, with a TCY value of 15.22 and an NTC value of 1.96.

Published in 1997, the paper in nineth position based on the number of citations, authored by Ledoux [59], demonstrates an interest in the domain before the 21st century. Using simulated data, the author explored urban traffic flow based on neural networks. The initial approach focused on modeling a local neural network, then including the traffic flow. The objective of the analysis was to demonstrate the possibility of modeling traffic flow using AI. The outcome showed that the model predicted the queue lengths one minute ahead, representing a good starting point for further analysis. A total of 108 citations have been obtained by the paper published by Ledoux, with a TCY value of 3.72 and an NTC value of 1.

Self-organizing maps (SOMs) and least square support vector machine (LSSVM) have been combined, resulting in the SOM-LSSVM model, which was been tested by Ismail et al. [60] on two datasets, the Wolf yearly sunspot and monthly unemployed young women datasets. The outcome confirmed the accuracy and efficiency of the model, compared with LSSVM, with a much smaller mean absolute error (MAE) and root mean squared error (RMSE). At the same time, the algorithm was able to perform time series forecasting, representing a promising solution for time series topics. In total, three authors explored the topic, obtaining a total of 100 citations, with a TCY value of 6.67 and an NTC value of 2.29.

Table 3 contains information about the 10 most cited publications on the topic of time series forecasting and AI, presenting the first author name, year of publication, journal, title of the paper, the data used during the research, and the scope of the investigation.

Table 2. Top 10 most cited documents.

No.	Paper (First Author, Year, Journal, Reference)	Number of Authors	Total Citations (TC)	Total Citations per Year (TCY)	Normalized TC (NTC)
1	Wang WC., 2009, *Journal of Hydrology* [44]	4	614	36.12	1.00
2	Song XY., 2020, *Journal of Petroleum Science and Engineering* [55]	8	241	40.17	5.70
3	Sharadga H., 2020, *Renewable Energy* [56]	3	225	37.50	5.32
4	Cho V., 2003, *Tourism Management* [61]	1	193	8.39	1.00
5	Sahoo BB., 2019, *Acta Geophysica* [62]	4	171	24.43	4.73
6	Qu JQ., 2021, *Energy* [63]	3	154	30.80	7.13
7	Chou JS, 2018, *Energy* [57]	2	146	18.25	2.40
8	Nourani V., 2017, *Journal of Hydrology* [58]	1	137	15.22	1.96
9	Ledoux C., 1997, *Transportation Research Part C: Emerging Technologies* [59]	1	108	3.72	1.00
10	Ismail S., 2011, *Expert Systems with Applications* [60]	3	100	6.67	2.29

Table 3. Brief summary of the content of the top 10 most globally cited documents.

No.	Paper (Primary Author, Year, Journal, Reference)	Title	Data	Purpose
1	Wang WC., 2009, *Journal of Hydrology* [44]	A comparison of performance of several artificial intelligence methods for forecasting monthly discharge time series	Hydrological data from January 1953 to December 2004 from Manwan Hydropower on Lancanjiang River	To develop a hydrological forecasting model based on historical values, in order to manage and schedule the hydropower reservoir in an optimal way,
2	Song XY., 2020, *Journal of Petroleum Science and Engineering* [55]	Time-series well performance based on long short-term memory (LSTM) neural network model	Two use cases from Xianjiang oilfield, China, with production rate over time	To predict the oil production by implementing a long short-term memory neural network algorithm, including the existing production limitations
3	Sharadga H., 2020, *Renewable Energy* [56]	Time series forecasting of solar power generation for large-scale photovoltaic plans	Time series data with historical production of solar power	To estimate the solar power for grid-connected photovoltaic systems
4	Cho V., 2003, *Tourism Management* [61]	A comparison of three different approaches to tourist arrival forecasting	Historical data from the USA, UK, Japan, Singapore, Korea regarding travel demand	To evaluate three time-series algorithms, ARIMA, Elman's Model of Artificial Neural Networks, and exponential smoothing, to forecast travel demand such as the number of arrivals in various countries
5	Sahoo BB., 2019, *Acta Geophysica* [62]	Long short-term memory (LSTM) recurrent neural network for low-flow hydrological time series forecasting	Daily data from Basantapur, India gauging station which is situated in Mahanadi River basin	To explore the long short-term memory recurrent neural network (LSTM-RNN) algorithm and low-flow time series forecasting of Basantapur gauging station from India

Table 3. *Cont.*

No.	Paper (Primary Author, Year, Journal, Reference)	Title	Data	Purpose
6	Qu JQ., 2021, *Energy* [63]	Day-ahead hourly photovoltaic power forecasting using attention-based CNN-LSTM neural network embedded with multiple relevant and target variables prediction pattern	Historical data on photovoltaic systems extracted from DKASC website	To implement short-term and long-term temporal neural network algorithms that predict the photovoltaic power values based on historical data
7	Chou JS, 2018, *Energy* [57]	Forecasting energy consumption time series using machine learning techniques based on usage patterns of residential households	Real-time information extracted from a smart grid of an experimental building	To predict the energy consumption of buildings using machine learning algorithms and based on past data
8	Nourani V., 2017, *Journal of Hydrology* [58]	An emotional ANN (EANN) approach to modeling rainfall-runoff process	Data from two watersheds have been selected	To define an emotional artificial neural network approach for daily rainfall runoff
9	Ledoux C., 1997, *Transportation Research Part C: Emerging Technologies* [59]	An urban traffic flow model integrating neural networks	Simulated data	To explore the neural network technique in traffic management systems, in order to be integrated into a real-time adaptive urban traffic system if the results are in accordance with expectations
10	Ismail S., 2011, *Expert Systems with Application* [60]	A hybrid model of self-organizing maps (SOMs) and least square support vector machine (LSSVM) for time-series forecasting	Wolf yearly sunspot and monthly unemployed young women datasets have been taken into consideration	To develop self-organizing maps and least square support vector machine (SOM-LSSVM) algorithm and to test the accuracy of the model

3.4. Country Analysis

This section includes information about the countries with the highest number of publications and citations, noting also the collaboration between states.

Figure 12 explores the main 10 countries that contributed the most to the development of time series forecasting and AI topics. The countries are classified based on the number of published papers, single-country publications (SCPs), and multiple-country publications (MCPs). In first place was China with 74 papers, which represented 28.5% of the total in articles published, with 57 SCPs (77%) and 17 MCPs (23%). In second place was Spain with only 17 articles, representing 6.5% of the total in articles published. In total, Spanish authors published 12 SCPs (70.6%) and five MCPs (29.4%). In third place was the United States of America (USA), with 16 publications (6.2% of total publications), comprising 13 SCPs (81.2%) and three MCPs (18.8%). In fourth place was Brazil, which released a total of 13 documents, representing 5% of the article total, with 11 SCPs (74.6%) and two MCPs (15.4%), followed by Italy with 10 papers (3.8% of total papers), including five SCPs (50%) and five MCPs (50%). The remaining countries each had less than 10 documents published but were still worthy of mention: India with nine papers (3.5% of total articles), including eight SCPs (88.9%) and only one MCPs (11.1%); Iran with nine documents (3.5% of total articles), comprising six SCPs (66.7%) and three MCPs (33.3%); Korea with nine papers (3.5% of total articles), consisting of eight SCPs (88.9%) and one MCP (11.1%); the United Kingdom (UK) with eight documents (3.1% of total published documents), comprising four SCPs (50%) and four MCPs (50%); and Saudi Arabia with only seven publications (2.7% of total papers), consisting of four SCPs (57.1%) and three MCPs (42.9%).

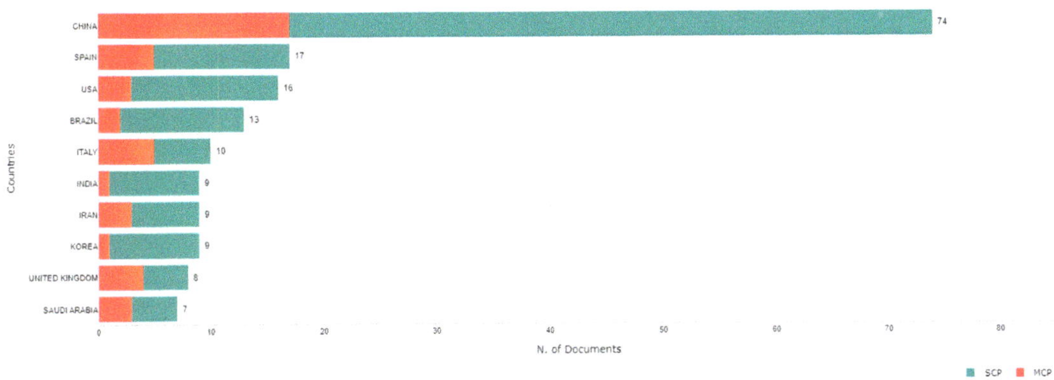

Figure 12. Top 10 Most Representative Corresponding Author's Countries.

According to Frioui and Graa [37], the country with the most articles published was China, with a total of 98 papers, 79 of them SCPs and 19 MCPs, followed by India with 55 papers, including 49 SCPs and six MCPs, while the USA was third with 38 articles, consisting of 31 SCPs and seven MCPs. Also in the top 10 were the UK, Korea, Australia, Germany, Spain, France, and Greece. Compared with our results, half of the countries included in top 10 were common, with China being the most representative country in both cases, and the USA third. The differences could appear due to the different filters that were applied to dataset extraction.

Shabani et al. [46] investigated the scientific evolution and contribution of AI in the field of portfolio optimization, evaluating the most representative countries. Regarding the number of publications, China was in first place with a total of 233 papers, including 187 SCPs and 46 MCPs, followed by the USA with 78 articles, consisting of 57 SCPs and 21 MCPs, and India with 39 articles, including 36 SCPs and only three MCPs. The rest of the countries included in the top 10 were Korea, the UK, Iran, Canada, Brazil, Japan, and Italy. According to our results, there were numerous similarities with the paper published by Shabani et al. [46], with China being the most relevant country based on the number of publications and a total of eight countries in common.

Figure 13 focuses on the 10 most cited countries. In first place was China, with a total of 2415 citations, having a mean article citation count of 32.60, followed by the USA, with 421 citations and an average article citation count of 26.30. In third place was India, with 259 citations and an average article citation count of 28.80, while Iran had 205 citations and a mean article citation count of 22.80. The remaining countries had less than 200 citations: Brazil (195 citations, with a mean of 15.00 article citations), Malaysia (182 citations, with a mean of 60.70 article citations), Spain (144 citations, with a mean of 8.50 article citations), Germany (110 citations, with a mean of 18.30 article citations), France (109 citations, with a mean of 36.30 article citations) and Italy (108 citations, with a mean of 10.80 article citations).

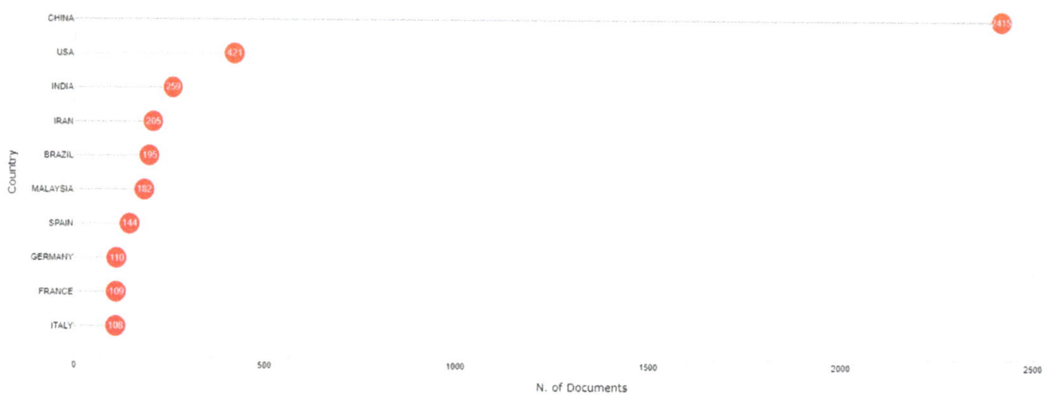

Figure 13. Top 10 Most Cited Countries.

Soliman et al. [51] evaluated the main trends and themes in AI that had been discussed by the academic community during the COVID-19 era, finding that China was the most cited country with 2874 citations, followed by the USA with 895 citations, the UK with 430 citations, Australia with 426 citations, Canada with 218 citations, France with 182 citations, Germany with 181 citations, South Korea with 169 citations, and Singapore with 164 citations. In total, the findings for four countries were similar to our results, with China being, in both cases, the most representative country, followed by the USA by a significant margin, while Germany and France stood out as relevant countries in the time series and AI domains.

Shabani et al. [46] found that China had the highest number of citations, with a total of 2169, followed by the USA with 1171, Brazil with 589, India with 500, and Singapore with 432. The remaining countries included in the top 10 were Korea, Greece, the UK, Spain, and Italy, highlighting that the findings for the majority of the countries were similar with our results, but at the same time, there were significant differences due to the different topics, timespan and focus of the research.

Figure 14 presents a map of scientific collaboration between countries. First, it should be mentioned that the blue color used for the countries in Figure 14 depicts the contribution of each country in the selected domain. Furthermore, the countries shaded in grey made no contribution to the selected field, according to the extracted dataset. Consequently, the density of the lines marked in red in Figure 14 symbolizes the intensity of the collaboration between two countries, measured by the number of papers written through collaboration between the two countries at the end of each line. Therefore, it can be observed that the most fruitful collaboration was between China and the USA, which together published a total of six papers, followed by China–Saudi Arabia with four documents, China–UK with four articles, and Saudi Arabia–Egypt, Saudi Arabia–Tunisia, and the USA–UK with three articles each; and Brazil–Italy, China–Australia, China–Canada, and China–Japan with two documents each.

To better understand the impact of countries in international collaboration, Table 4 contains the number of publications for the most representative 20 states in terms of number of publications. In first place was China with 42 articles, followed by the USA with 26 documents, the UK with 21 papers, and Spain and Saudi Arabia with 20 each. Among the remainder of the top-20 most productive countries listed in Table 4, Malaysia and Ireland were at the bottom, with only six papers each.

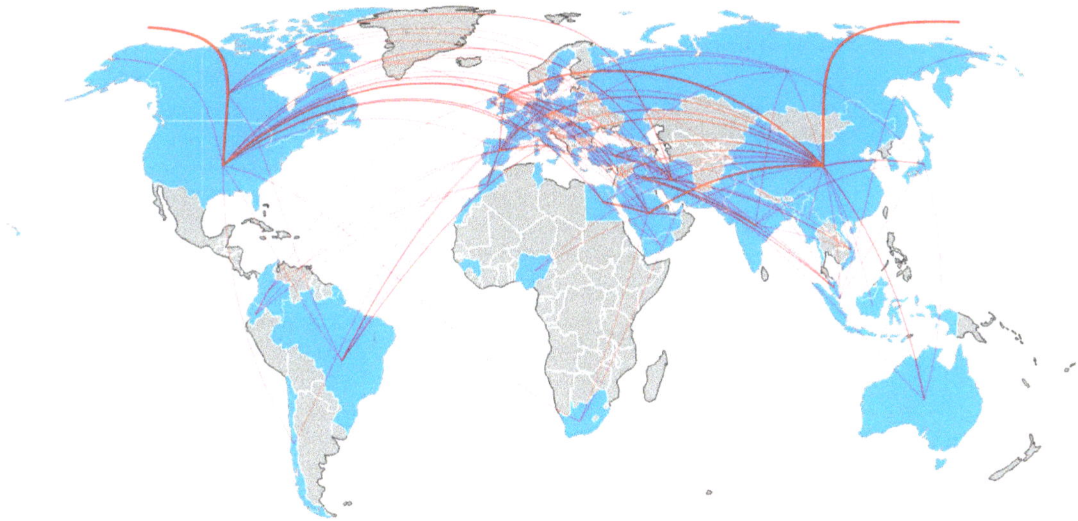

Figure 14. Country Collaboration Map.

Table 4. Countries with Most Papers published.

Country	Number of Papers
China	42
USA	26
UK	21
Spain	20
Saudi Arabia	20
Iran	15
India	12
Turkey	12
France	11
Canada	10
Egypt	10
Serbia	10
Tunisia	9
Lebanon	9
Brazil	8
Iraq	8
Italy	7
United Arab Emirates	7
Malaysia	6
Ireland	6

3.5. Mixed Analysis

This section focuses on the thematic evolution and analysis, noting the major topics that have been discussed by the academic community, at the same time highlighting the most frequently used Keywords Plus and author's keywords. A detailed analysis of collaboration networks for the 30 most representative authors has been performed.

Figure 15 presents the 150 most used Keywords Plus grouped in multiple circles, based on the thematic context. In order to obtain the graph below, the minimum cluster frequency was set to 27. Two metrics are available in the thematic map: centrality, which expresses the impact of the keywords in the thematic context, investigating the external associations; and the density, which shows the degree of development, exploring the internal association of keywords in various thematic contexts [64]. The most representative cluster, colored in blue, on the bottom right part of the graph with a high centrality and a reduced density, contains information about ML algorithms and time series methods such as "prediction" (52 occurrences), "models" (41 occurrences), "models" (25 occurrences), "neural-networks" (40 occurrences), "algorithm" (20 occurrences), "optimization" (20 occurrences), "system" (19 occurrences), "time-series" (18 occurrences), and "regression" (16 occurrences). In the middle of the graph is a cluster colored in orange, with a medium density and centrality, referring to the ML networks: "machine" (10 occurrences) and "networks" (8 occurrences). The last cluster, in the top left part of the graph and highlighted in pink, contains two of the most used time series algorithms: "lstm" (8 occurrences) and "arima" (7 occurrences).

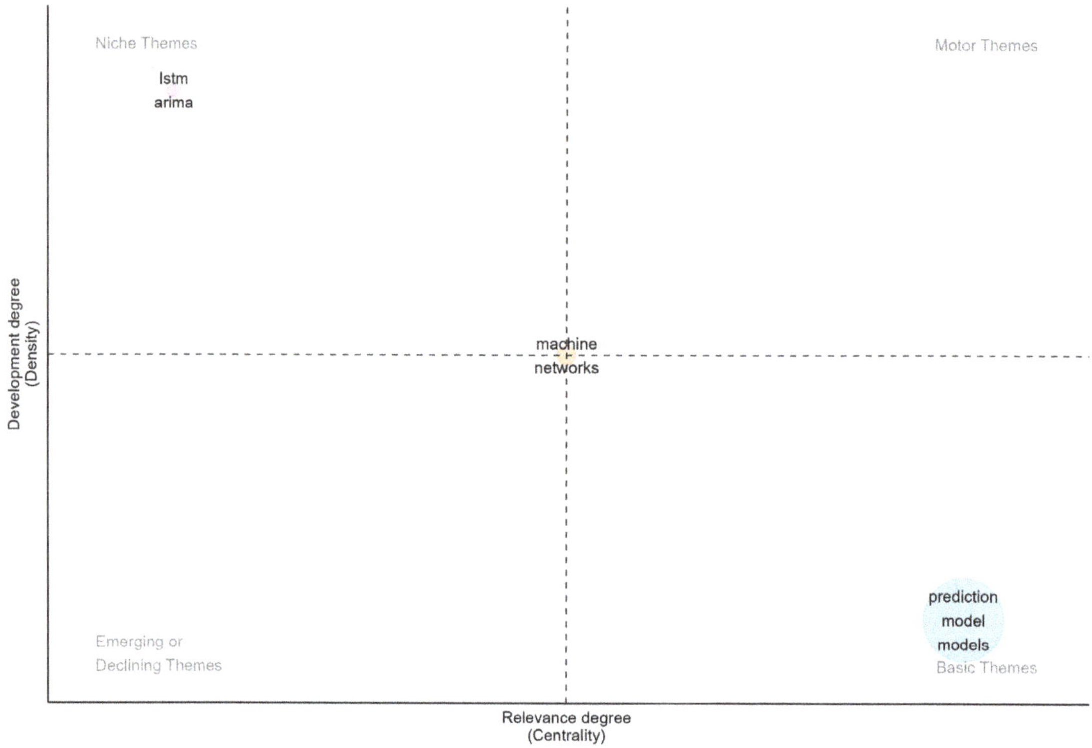

Figure 15. Thematic Map of Keywords Plus.

Figure 16 includes the 150 most representative keywords used by the main authors, grouped in four different themes. The minimum cluster frequency setting was set to 25. In the bottom right part of the graph is the most representative circle, highlighted in orange, referring to machine learning integration with time series and AI, featuring a high centrality and small density.

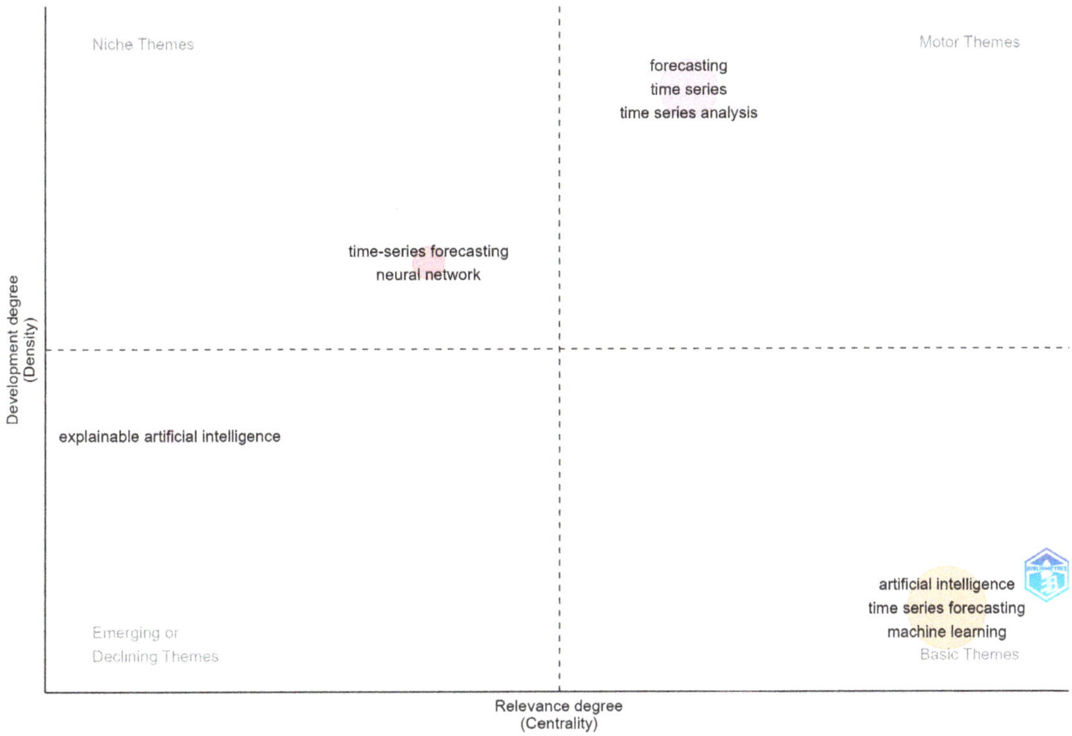

Figure 16. Thematic Map of Authors' Keywords.

The most used keywords in the orange cluster are "artificial intelligence" (93 occurrences), "time series forecasting" (67 occurrences), "machine learning" (41 occurrences), "deep learning" (40 occurrences), "time series" (17 occurrences), "neural networks" (15 occurrences), "COVID-19" (13 occurrences), "explainable artificial intelligence" (11 occurrences), "artificial neural network" (9 occurrences), and "lstm" (8 occurrences). The second most relevant cluster is colored in pink and presents the time series analysis, together with forecasting, containing the following keywords: "forecasting" (39 occurrences), "time series" (17 occurrences), "time series analysis" (14 occurrences), "predictive models" (11 occurrences), "lstm" (8 occurrences), and "recurrent neural network" (8 occurrences). The last two clusters, colored in red and pink, have a relatively small centrality value, while the red circle has a higher density compared to the pink circle. The focus on the red circle is "time-series forecasting" (21 occurrences) and "neural network" (6 occurrences), while the pink circle contains "explainable artificial intelligence" (11 occurrences). Soliman et al. [51] performed a bibliometric analysis of the main themes of AI in the leading scientific journals during the COVID-19 pandemic, discovering that the main authors' keywords were clustered in three groups, with the first cluster focusing on "training", "deep learning", and "classification", the second cluster presenting "functions", "uncertainty", and "neural networks", and the third cluster containing information about "robots", "neurons", and "computer simulation". There are similarities between the existing literature and our results, COVID-19 being one of the main factors that contributed to the development of the fields, together with "neural networks" and "deep learning", according to Soliman et al. [51]. At the same time, there are significant differences, which is normal, thanks to the different databases that were used, different software tools, different topics, and different time periods.

Figure 17 presents the thematic evolution for Keywords Plus, divided into three different periods: 1997–2015, 2016–2020, and 2021–2024. Between 1997 and 2015, the main topics discussed in the contexts of time series and AI were related to "neural networks" and "model", while, starting in 2016, they evolved to "prediction", "particle swarm optimization", "classification", "market", "models", and "artificial neural network", demonstrating the development of the domains, making it more versatile. Between 2021 and 2024, the main focus of the researchers was on improving the prediction, since numerous models had been defined between 2016 and 2019. At the same time, the "artificial neural network" remained a key topic, together with "system" and "network".

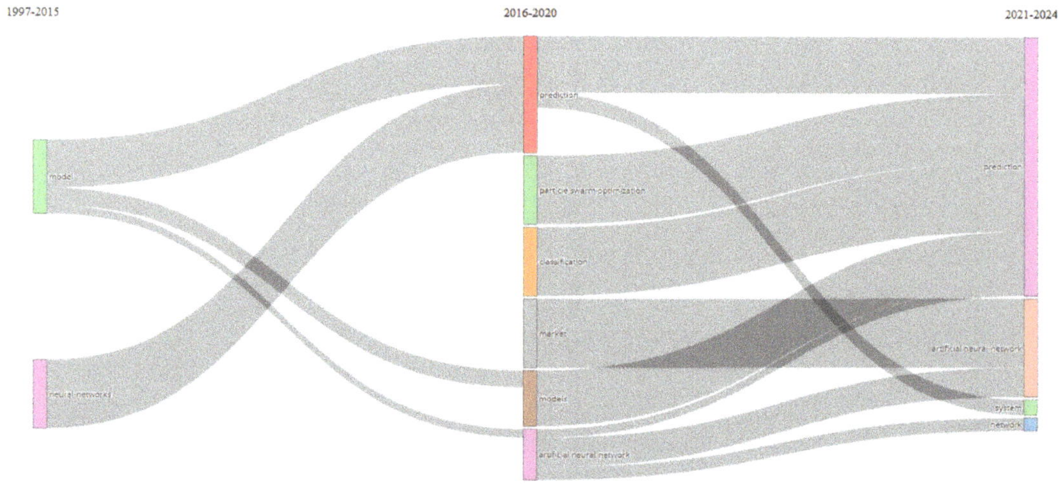

Figure 17. Thematic Evolution of Keywords Plus.

In their research, Romero-Riano et al. [65] defined the most used AI keywords between 2000 and 2020. Initially, the majority of terms were related to "optimization", "prediction", "models", and "neural networks", while researcher interest shifted during the timespan to a broader variety of topics, such as "complexity", "agents", "prediction", "architecture", and "behavior".

Figure 18 includes a factorial analysis of the most frequently used 100 Keywords Plus, grouped in three clusters.

The blue cluster contains the majority of terms, which define the applicability of time series and AI: "wavelet transform", "empirical mode decomposition", "optimization", "power", "system", "impact", "runoff", "river", "streamflow", and "reservoir system", while the red cluster focuses on "artificial neural networks", "genetic algorithms", and "forecasting enrollments". The last cluster, highlighted in green, encapsulates the strategies that should be taken into account: "optimal operation", "basin", "synchronization", "strategies", and "intelligence".

Articles that investigated the time series forecasting and AI domains were evaluated to compare the results with the existing literature, and Romero-Riano et al. [65] grouped the keywords into multiple clusters, with one focusing on "wavelet transform", "decomposition", "memory", and "validation, another cluster exploring "genetic algorithms", "regression", "svm", and "prediction", and the last cluster focusing on "reliability" and "complexity".

According to Sarikoc [66], the most frequently used keywords were "artificial neural network", "runoff", "neural networks", "forecasting", "hydrological modeling", "artificial intelligence", "stream flow" and "prediction". Comparing our results with the existing literature, there are similarities for the majority of keywords, but at the same time, there are differences due to the analyzed timespan or the domain the articles focusing on.

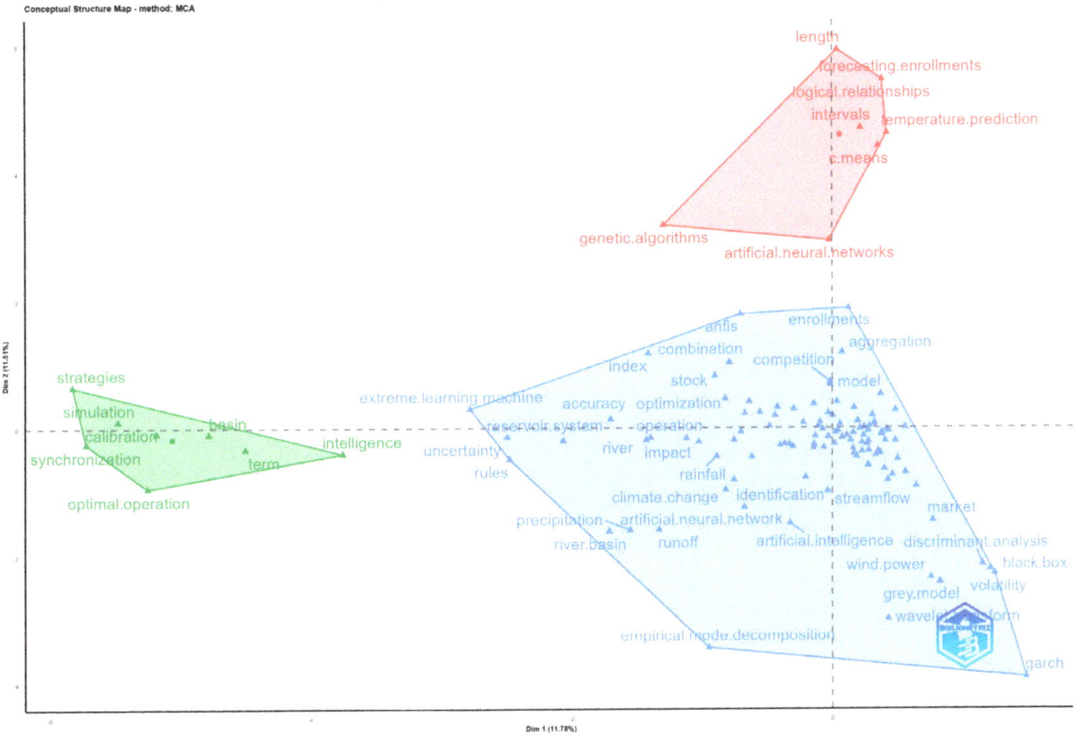

Figure 18. Factorial Analysis of Keywords Plus.

Figure 19 illustrates a factorial analysis for the most representative 100 bigrams used in abstracts, grouped into two clusters. The blue cluster contains terms referring to time series and AI algorithms: "AI techniques", "deep learning", "memory lstm", "energy consumption", "regression svr", "machine learning", "forecasting accuracy", "convolutional neural", and "time series", while the red cluster, which is smaller, contains information about the time series and AI error metrics, such as "percentage error", "error rmse", "squared error", "absolute error", and "error mape".

Palomero et al. [43] evaluated the time series forecasting and modeling domain using a bibliometric analysis. The most relevant keywords discovered were divided into multiple groups: "mean square error", "root mean square errors", and "mean absolute percentage error", while the second group focused on "deep neural networks", "time series forecasting", "prediction", "long short-term memory", and "artificial neural network".

Loan et al. [67] investigated the applied artificial intelligence of an international journal using a bibliometric approach, discovering that "model", "system", "neural networks", "artificial neural networks", and "genetic algorithms" were the most representative terms. Our outcome correlates with the existing literature results, which confirm that the approach was correct, but at the same time, there were significant differences, which makes sense, due to the database differences, time period, and fields that were investigated.

Figure 20 presents the 30 most relevant authors, grouped into five clusters. The most representative cluster is colored in blue, and it contains six authors: Niu W.J., Feng Z.K.,

Xu Y.S., ChenJ., Feng B.F., and Cheng C.T. The focus of the authors was on AI algorithms for hydrological time series analysis, using parallel and swarm intelligence methods, signal decomposition, twin support vector machine, and river flow time series analysis using artificial neural networks for humid and semi-humid regions [68–71].

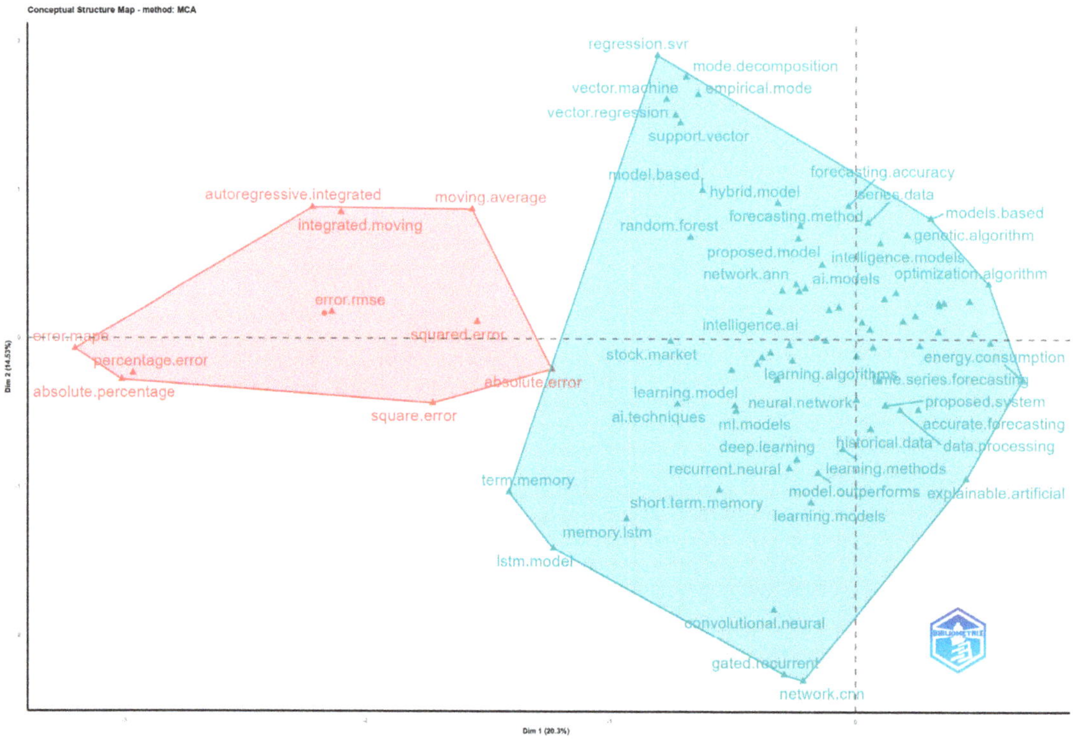

Figure 19. Factorial Analysis of Abstract Bigrams.

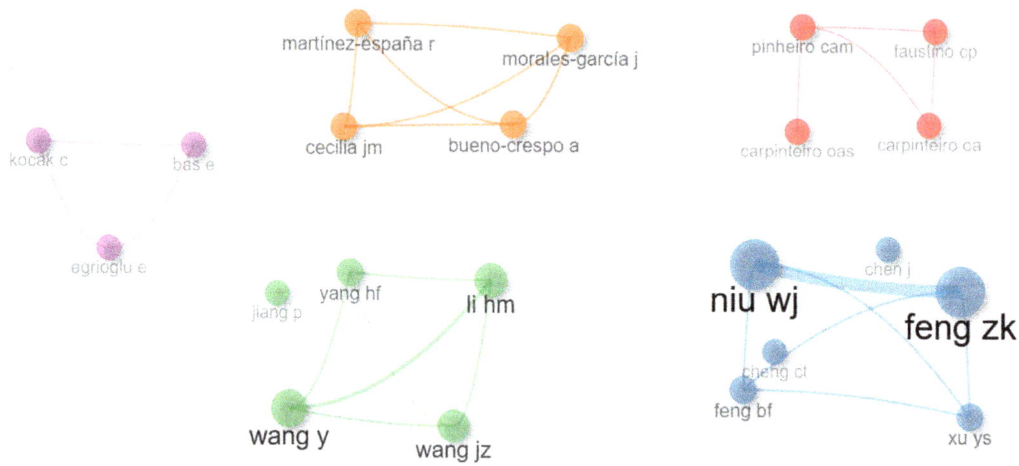

Figure 20. Collaboration Network of 30 authors.

The second cluster, colored in red, is represented by four authors: Carpinteiro O.A.S., Pinheiro C.A.M., Carpinteiro O.A., and Faustino C.P. The main topics discussed by the researchers incorporated financial prediction using support vector regression, improving the accuracy of fuzzy forecasting of fuzzy c-means, explaining the rough sets, and testing time series algorithms for improving the prediction rate [9,72–74].

The third group, delimited in green, is formed by five authors: Wang Y., Wang J.Z., Li H.M., Jiang P., and Yang H.F., who focused mainly on predicting traffic accident economic loss and air pollution, wind speed prediction using defuzzificaition, and multi-objective optimization methods, energy consumption forecasting, and air quality prediction with nonlinear robust outliers and fuzzy sets [75–78].

The fourth cluster, highlighted in purple, contains three authors: Egrioglu E., Bas E., and Kocak C., who explored Istanbul stock market prediction using fuzzy time series methods, ARMA, artificial neural networks, and fuzzy time-series evaluation [79–81].

The fifth cluster, colored in yellow, is formed by four authors: Bueno-Crespo A., Cecilia J.M., Martinez-Espana R., and Morales-Garcia J. The main subjects explored by these authors were machine learning models for operational smart greenhouses, evaluation of smart greenhouses, evolution of low-power devices, and the investigation of data generated by intelligent climate control of greenhouses [82–84].

Based on the mixed analysis, the major topics discussed in the analyzed papers were related to artificial neural network algorithms, deep learning models, long short-term memory algorithms, random forest, support vector machines, recurrent neural networks, network CNN, explainable AI, vector regression, and support vector regression. At the same time, the main steps of AI and ML implementation were presented, such as data processing, learning methods from historical data, forecasting, and accuracy estimation. The last step was also explained by the following terms: squared error, absolute percentage error, absolute error, root mean squared error, and mean absolute percentage error.

The main domains were the time series algorithms and AI, focusing on energy consumption, climate change, stock analysis, decision making, and wind power estimation.

4. Discussion and Limitations

The Section 4 discusses the results, together with the existing limitations of the research.

4.1. Bibliometric Research Summary

The bibliometric approach presented in this paper evaluated the development of time series forecasting and AI domains from the beginning of the period in which papers have been written and indexed in the WoS database, including the first paper published in 1997, which had an impact if we take into consideration the number of citations. Thanks to technological and scientific developments, it became easier to investigate the topics, and a positive trend was observed in the number of publications, especially in recent years. As observed during the bibliometric analysis process, a series of results, such as the most impactful sources, were also identified in similar studies from the field, underlining once more the stability of the selected research field over time.

To demonstrate the outcome, the research addressed several scientific questions in the first chapter of the paper, to which, during the analysis, answers have been provided. First, it was observed that the collaboration network among the 30 most representative authors could be presented in five different clusters, each one focusing on different topics related to time series forecasting and artificial intelligence. Niu W.J., Feng Z.K., Xu Y.S., Feng B.F., Chen J., and Cheng C.F. explored hydrological time series prediction. Carpinteiro O.A.S., Pinheiro C.A.M., Faustino C.P., and Carpinteiro O.A. evaluated the time series and artificial intelligence algorithms for financial predictions. Yang H.F., Wang Y., Jiang P.,

Li H.M., and Wang J.Z. focused on traffic accident loss and air pollution estimations, while Egrioglu E., Bas E., and Kocak C. presented Istanbul stock market prediction using time series algorithms. Morales-Garcia J., Bueno-Crespo A., Martinez-Espana R., and Cecilia J.M. described smart greenhouse methods for the evaluation of low-power devices.

Second, the main topics extracted during the scientific research present the methods and domains where time series and artificial intelligence have been successfully implemented. The algorithms that have been implemented often were "lstm", "neural networks", "support vector regression", "ARIMA", "particle swarm optimization", and "regression", with great success especially in the hydrological, financial, and environmental fields, where the prediction of future values stands out as a key factor in the evolution of these domains.

In terms of the most representative countries based on the number of publications and citations, the following countries were ranked at the top: China with 74 articles and 2415 citations, followed by Spain with 17 documents and 144 citations, the USA with 16 articles and 421 citations, Brazil with 13 documents and 195 citations, Italy with 10 documents and 108 citations, India with 9 papers and 259 citations, Iran with 9 articles and 205 citations, Korea with 9 documents and 17 citations, the UK with 8 articles and 75 citations, and Saudi Arabia with 7 papers and 52 citations.

The 10 most cited documents explore the implementation of new methods of time series and artificial intelligence using adaptive neural-based fuzzy systems or genetic programming for monthly river flow discharges, oil production estimation using LSTM, and prediction of the number of tourists with ARIM or Elman's Model of Artificial Neural Networks. At the same time, detailed analyses were provided for electricity, energy consumption, and urban traffic flow, implementing multiple methods in order to obtain high accuracy and a reduced error rate.

The sources with the highest impact based on the number of publications were IEEE Access with 11 documents, Energies with 9 articles, Mathematics with 7 papers, Applied Sciences-Basel with 6 papers, and Applied Energy, Energy, Knowledge-Based Systems, Sensors, and Sustainability with 5 documents each.

The main keywords used by the authors on time series forecasting and artificial intelligence topics refer to the main algorithms that have been implemented, such as "ARIMA", "LSTM", "regression", "neural-networks", and "artificial neural network", or present the main factors that contributed to the evolution of the fields, such as "COVID-19", "uncertainty", "optimization", "strategies", and "impact".

4.2. Discussion of Specific Topics

In this section, the focus is on analyzing the main topics where time series and AI have been successfully implemented, presenting also five of the most representative articles from each field.

4.2.1. Implications of Time Series Forecasting and Artificial Intelligence in the Energy and Electricity Sector

The energy and electricity topics have become more and more relevant not only for the academic community, but also for stakeholders and governments, since it is crucial now to provide sustainable and affordable sources of energy and power for the population. Using time series and AI, an estimation of production and consumption levels can be achieved.

Estimates of solar energy production have been explored by Al-Ali et al. [85], implementing a combination of algorithms, convolutional neural networks (CNN), LSTM, and a transformer, due to the evolution of AI and time series, which now facilitate this approach. The Fingrid open dataset was extracted for training and testing the model, and the outcome demonstrated high efficiency. The results were compared with existing methods, and the accuracy of the hybrid algorithm was higher, positioning it as a new technique that could be integrated into the solar energy domain.

Power system safety and stability represent crucial aspects for short-term electricity load time and, according to Niu et al. [86], electricity demand is marked by high randomness and volatility, with the consumption levels being almost impossible to predict with traditional algorithms. The objective of the authors was to define a machine learning model capable of creating an empirical decomposition that would be divided into subcomponents, and, using a support vector machine and cooperation search algorithm, an estimation of the short-term load data would be performed. A case study has been performed on provincial grid centers in China, showing great results compared to conventional algorithms.

The sustainable development of the environment and economy is a key topic that was discussed by Jiang et al. [77], and considers energy consumption a key element. The scope of the research was focused on fuzzy time series and fuzzy c-means clustering for fuzzification and to predict energy consumption using the previous algorithms. The outcome confirms the good stability and accuracy of the algorithms, which are able to predict the structure of energy consumption in an optimal manner.

A dynamic transformer for energy power prediction has been developed by Zhou et al. [87], as power estimation is considered to be one of the key elements nowadays for carbon neutrality. Thanks to AI algorithms that have been recently developed, predictions can now be achieved more easily. The authors proposed a dynamic self-differential transformer method (Diformer) that can be used for univariate time series analysis and predictions. The algorithm has been tested on six power time series datasets related to the energy and power domain, showing great results and metrics and outperforming existing models.

The prediction of energy consumption using time series data was discussed by Hamdoun et al. [88], focusing on an analytical evaluation of artificial intelligence, deep learning, and machine learning algorithms that are used for energy consumption estimation and at the same time on implementing the methods in real use cases, extracting datasets related to natural gas, electric power, or energy consumption. The authors underlined the main challenges and potential developments for the research, emphasizing the importance of deep learning and machine learning algorithms in the energy sector.

Based on the descriptions provided above for the five most representative papers on energy and electricity areas, the main applicability of time series and AI is focused in the estimation of energy consumption [77,88], energy power prediction [87], short-term loads [86], and solar energy production [85].

4.2.2. Implications of Time Series Forecasting and Artificial Intelligence for Tourism and Economic Activities

The impact of time series forecasting and artificial intelligence on economic activities will be discussed further, investigating five relevant papers.

An evaluation of portfolio management using time series and AI was accomplished by Bareith et al. [89], who also took inflation into account. The scope of the research was focused on using AI to predict the inflation rate and stock index evolution and to define a portfolio of indexes that would protect investments against inflation. Two algorithms were tested: Long Short-Term Memory (LSTM), Gated Recurrent Unit (GRU), and a hybrid model that contained both algorithms for inflation forecasting. The outcome showed that errors in inflation rate forecasting account for a quarter of stock market index estimation, recommending investments in the national economy for economic rebalancing.

A forecast for sustainable tourism in Saudi Arabia was performed by Louati et al. [90], who investigated the behavior of tourists during the COVID-19 pandemic. The dataset was obtained from the Saudi Tourism Authority and contained information from 2015 through 2021. Multiple AI algorithms, such as K-Neighbors Classifiers, Gaussian Naïve Bayes,

Support Vector Classifiers, Decision Trees, ARIMA, and Random Forests, were tested. The results portrayed a positive trend in the number of tourists, providing insights regarding future decisions and plans for sustainable development of the tourism sector. The mean expenditure of tourists was estimated as SAR 89,443 per trip, with a value of SAR 9198 per night, according to the statistical tests that were performed.

The LSTM algorithm was implemented by Zhang et al. [91] for time series forecasting in the tourism sector, focusing on daily consumer and tourist datasets from Jiuzhaigou, a popular tourist attraction in China, containing information for the period from 8 October 2013 to 7 August 2017. A prediction for the next 150 days was performed, obtaining good results according to statistical metrics that had been applied for training and testing, with the authors confirming that LSTM stands as one of the most suitable models for tourism prediction.

Supply chain management is one of the main elements of the economic sector in each country, and it is crucial in predicting future sales with high accuracy. Sohrabpour et al. [92] proposed a framework using Genetic Programming, an AI model that estimates export sales for the next 6 weeks. A causal estimation was included, facilitating extraction of the behavior between features, together with the previous trends and behavior. A variable sensitivity analysis was performed for the algorithm, and the results were compared with real sales data.

Economic recession was the focus of Tang et al. [93], who also included share prices in the logistics sector for the purpose of improving decision-making and estimating future share prices in the logistics industry. The LSTM algorithm was tested on historical data from five logistics companies in Hong Kong. The Mean Absolute Percentage Error (MAPE) and Root Mean Square Error (RMSE) were the metrics used, demonstrating that LSTM can efficiently predict future share prices. The LSTM model was trained with multiple hyperparameters, and the optimum algorithm obtained an RMSE value of 0.43%. Economic recession forecasts were performed using the stock predictions and LSTM model.

Based on the scientific approaches discussed above, time series forecasting and AI have significant implications for economic activities, and their use is being sought more and more by stakeholders for predicting the number of tourists [90,91], stock price trends [89,93], and sales trends [92].

4.2.3. Implications of Time Series Forecasting and Artificial Intelligence for Hydrologic Predictions

The hydrologic domain has recently adopted time series forecasting and AI, thanks to technological developments that facilitate their implementation in a variety of fields. Five of the most representative papers on the implementation of time series forecasting and AI in the hydrological field will be discussed.

Streamflow prediction was discussed by Feng et al. [94], who developed an algorithm that estimates the nonlinear streamflow for hydrological stations in China. A parallel cooperation search algorithm (PCSA) and extreme learning machine (ELM) were tested. The PCSA method was optimized with multiple input-hidden weights and biases, obtaining better results than the ELM model. Initially, a large population was stored, then divided into multiple subswarms that would be further investigated using the PCSA algorithm.

The domain of water resource management is very sensitive to disasters, and it is crucial to optimally predict the hydrologic metrics in order to minimize risks. Christian et al. [95] explored multiple time series algorithms, a support vector machine, and a seasonal autoregressive moving average (SARIMA) for estimation of Pemali River Basin discharges in Indonesia. Using an autocorrelation function (ACF), the analysis demonstrated a correlation between historical values up to 12 months prior, which facilitated the prediction of future values and reduced the risks.

In the hydrological domain, a crucial step is to optimally predict the streamflow. Yaseen et al. [96] investigated the literature and implemented the non-tuned machine learning algorithm ELM. The model's effectiveness and reliability were tested on predicting one-step-ahead streamflow of the Johor River in Malaysia, using three time-scale datasets (daily, average weekly, and average monthly). The outcome demonstrated that ELM has higher accuracy than an artificial neural network (ANN) algorithm, with a higher R^2 value (0.94 compared to 0.90 for ANN) and a lower RMSE (2.78 compared to 11.63) and MAE (0.10 compared to 0.43). The authors recommended the ELM algorithm for streamflow prediction.

Using a Least-Squares Support Vector Machine algorithm, decomposition reconstruction, and swarm intelligence, Niu et al. [97] developed an improved method for hydrologic time series prediction. Four indexes that contained monthly information on two hydrologic stations in China were analyzed, resulting in multiple scenarios. The hybrid algorithm that contained signal swarm intelligence and decomposition reconstruction provided optimum results, improving the accuracy with an RMSE of 58.9%, compared to support vector machine and neural network algorithms.

Water wave oscillation, a topic in the hydrological domain, was evaluated by Sang et al. [98], detailing the preprocessing, modeling, and problems regarding wavelet analysis, such as the choice of mother wavelet, selection of temporal scale, uncertainty evaluation of the predicted wavelet, and differences between discrete and continuous wavelet methods. The authors explained the key differences between continuous and discrete wavelet algorithms, when they can be implemented, and what the main advantages are with time series forecasting, also taking into consideration the uncertainty.

The hydrological domain benefits from AI models, which facilitate the implementation of solutions that predicts water wave oscillation, hydrologic prediction [97,98], streamflow [94,96], and discharge estimation [95].

Nevertheless, apart from the areas of applicability for time series forecasting and artificial intelligence, a focused discussion should be conducted on the evaluation metrics considered in these cases. Even though a wide range of statistical metrics are commonly employed, including Mean Squared Error (MSE), Root Mean Squared Error (RMSE), Mean Absolute Error (MAE), Mean Absolute Percentage Error (MAPE), and the coefficient of determination (R^2), distinct attention should be given to the inconsistencies that may arise in model performance assessment due to the different sensitivities and mathematical properties of each metric. As Mishra et al. [99] pointed out, even though some of these metrics are well-known and widely used in evaluating the predictive accuracy of the models, attention should be paid to various issues such as non-convergent assessments that might derive from the fact that different methods favor or penalize models when outliers are encountered. Furthermore, as Zhou et al. [100] showed, RMSE is more suitable for scenarios in which large errors are of primary concern, as the metric squares them; MAE is more appropriate for cases in which large errors should not be unduly penalized; MAPE emphasizes the importance of relative errors, while R^2 evaluates the model's ability to explain the variation in the dependent variable. As a result, in the scientific literature, recently, heavy emphasis has been placed on using composite measures that combine RMSE, MAE, MAPE, and R^2 and offer a more balanced approach to model evaluations [100].

4.3. Limitations

This research faced several limitations that could have had an impact on the obtained results, most of them related to the process of dataset extraction.

The main limitation is the inclusion of only one database, namely WoS, into the analysis. The argument for taking into consideration only one database was based on the prestige of WoS, which offers a variety of sources, according to papers in the academic community that also had been investigated before the decision was made. The indexes that are very popular, such as Social Sciences Citations Index (SSCI) or Science Citation Index Expanded (SCIE), are part of the WoS database, providing a robust analysis of the citations. At the same time, the Keywords Plus [30] variable is offered by WoS and identifies the terms derived from the references in the document, highlighting the scope of the analysis, a feature that is not offered by other databases. In order to further support the choice for a single database for the bibliometric analysis, a fact that should also be stated is that using multiple databases would have interfered with the analysis conducted in terms of the number of citations, as it would have been hard to decide which database to use for these analyses because various situations might arise, such as papers missing from one or another database or duplicated in multiple databases. To support this point of view, in Table 5, we included the number of citations recorded for the five most cited papers included in the dataset, according to each database.

Table 5. Example of citations for top-five most cited papers included in the dataset.

Paper	Number of Citations	
	ISI Web of Science (Database Used in This Study)	Scopus
Wang et al. [44]	614	714
Song et al. [55]	241	-
Sharadga et al. [56]	225	-
Cho [61]	193	238
Sahoo et al. [62]	171	-

Furthermore, it should be mentioned that, as access to the WoS database is based on a paid subscription, the resulting dataset might be different depending on the level of the subscription. As stated at the beginning of the paper, in this case, we had access to all 10 indexes offered by WoS.

The exclusion of any other language apart from English stands as a constraint for the research, limiting the temporal coverage of papers. The decision to include only English language papers was intentional, in order to have a relevant n-grams and themes analysis. At the same time, we were restricted to English due to the lack of proficiency in any other languages. In our case, only three papers among those initially extracted were published in a language other than English, showing that the decision to exclude papers in other languages would not affect the outcome in a decisive manner.

The software tool that was used for the bibliometric approach, the Biblioshiny library from the R programming language, is a popular tool used, together with VOSViewer for analysis of specific datasets extracted from databases [41]. For example, Biblioshiny is able to analyze Scopus, Dimensions, Cochrane Library, WoS, Lens, and PubMed, while VOSViewer is able to work with Lens, Scopus, WoS, PubMed and Dimensions [101].

The inclusion of only the documents that were marked as "Article" in the WoS database [102] represents another limitation, but the step was mandatory in order to obtain a representative comparison among papers from the number of citations and content perspectives.

Lastly, somewhat different results might be obtained if the steps provided in the paper are slightly different when selecting the dataset.

5. Conclusions

This research focuses on the main elements of a bibliometric analysis of the time series forecasting and artificial intelligence fields, presenting the evolution from 1997 to 2024. The major factor that aided the evolution of these domains was technological advancement, and the analysis that we performed highlights the main papers, authors, sources, collaborative networks, and thematic maps.

Therefore, in terms of collaboration networks among authors in the selected research areas, it was observed that the 30 most representative authors based on the number of publications were grouped into five major clusters addressing various research topics, such as hydrological time series prediction; the evaluation of time series and artificial intelligence algorithms on financial predictions; traffic accident loss and air pollution estimations; stock market predictions using time series algorithms; and smart greenhouse methods for the evaluation of low-power devices.

The analysis of the most used keywords, main topics, and most cited documents revealed interests in various research areas such as environmental and hydrological forecasting, energy consumption and electricity demand modeling, oil production estimation, tourism demand forecasting, and urban traffic flow analysis. These areas often rely on advanced time series forecasting techniques and artificial intelligence methods, including adaptive neuro-fuzzy systems, genetic programming, ARIMA models, LSTM networks, and Elman's recurrent neural networks.

Lastly, the sources with the highest impact were extracted, as well as the most representative countries and authors considering the number of publications.

Based on the papers included in the dataset, some research directions that were highlighted related to the use of time series forecasting and artificial intelligence in the energy and electricity sector, in tourism and economic activities, as well as in hydrologic predictions. Considering the works in the above mentioned fields, it was noticed that the main areas of applicability for time series forecasting and AI were in energy consumption estimation [77,88], energy power prediction [87], short-term load [86], solar energy production [85], prediction of water wave oscillations, hydrologic prediction [97,98], streamflow [94,96] and discharge estimation [95], and in forecasting the number of tourists [90,91], stock prices [89,93], and sales trends [92].

The research community associated with the time series forecasting and artificial intelligence research areas can benefit from the results presented in this paper by gaining a general view on the evolution of the field, a better understanding of the main areas of research that have captured the interest of researchers—highlighted by the most cited works from the field, as well as by easily identifying the journals that have been dedicated to works in the field. In terms of authors, the most prominent authors have been presented, which might help the interested parties in easily following the research conducted by these authors. Furthermore, by discussing some of the research directions in various fields, as well as the main themes, the interested parties can orient their research into the emerging or motor themes.

While the process of article information collection has improved significantly in recent years, there are still some inherent limitations that could have impacted the outcome of the analysis, such as the filtering terms used for dataset collection, together with the restrictions applied over the data, such as the inclusion of only English papers that were marked as "Article" in WoS and not published in 2025. At the same time, the decision to include only the WoS database in the analysis could have had an impact on the results of the research, but the decision was adopted based on an investigation of the existing academic literature. The graphical representation of the data could also represent a limitation, as only some of the results were provided, generally those referring to the top or the most prominent results.

Future research in time series forecasting and artificial intelligence could continue the exploration of scientific articles, extracting the outcomes from a wider variety of databases and comparing the results with other software tools. At the same time, a comparative analysis of algorithm performance could be achieved by testing the models on various datasets from different domains in order to classify them. The current integration of time series forecasting and artificial intelligence into the activities of stakeholders and individuals is a reality that should be noticed by the academic community.

Author Contributions: Conceptualization, A.D., P.D. and C.D.; Data curation, A.D., P.D. and C.D.; Formal analysis, A.D., P.D. and C.D.; Investigation, A.D., P.D. and C.D.; Methodology, A.D. and C.D.; Project administration, P.D. and C.D.; Resources, C.D.; Software, A.D.; Supervision, C.D.; Validation, A.D. and P.D.; Visualization, A.D., P.D. and C.D.; Writing—original draft, A.D.; Writing—review and editing, P.D. and C.D. All authors have read and agreed to the published version of the manuscript.

Funding: The work is supported by a grant of the Romanian Ministry of Research, Innovation and Digitalization, project CF 178/31.07.2023—'JobKG—A Knowledge Graph of the Romanian Job Market based on Natural Language Processing'.

Institutional Review Board Statement: Not applicable.

Informed Consent Statement: Not applicable.

Data Availability Statement: The original contributions presented in the study are included in the article, further inquiries can be directed to the corresponding author.

Conflicts of Interest: The authors declare no conflicts of interest.

Abbreviations

AB	abstract
ACF	autocorrelation function
AI	Artificial Intelligence
AK	authors' keywords
ANN	artificial neural network
AR	autoregressive models
ARIMA	autoregressive integrated moving average
A&HCI	Arts and Humanities Citation Index
BKCI-S	Book Citation Index-Science
CCR	Current Chemical Reactions
CNN	Convolutional Neural Networks
CPCI-S	Conference Proceedings Citation Index—Science
DBN	Dynamic Bayesian Networks
DIFORMER	dynamic self-differential transformer method
EANN	emotional artificial neural network
ELM	extreme learning machine
ESCI	Emerging Sources Citations Index
GARCH	generalized autoregressive conditional heteroscedasticity
GEP	gene expression programming
GRU	Gated Recurrent Unit
IC	Index Chemicus
LSSVM	least square support vector machine
LSTM	Long Short-Term Memory
LSTM-RNN	Long-Short Term Memory Recurrent Neural Network
MA	moving average

MAE	mean absolute error
MAPE	mean absolute percentage error
MARS	multivariate adaptive regression splines
MCP	Multiple-Country Publications
ML	machine learning
MLP	multilayer perception
MLR	multiple linear regression
NRMSE	normalized root mean square error
NTC	normalized total citations
PCSA	Parallel cooperation search algorithm
PGM	probabilistic graphical models
PRISMA	Preferred Reporting Items for Systematic Reviews and Meta-Analysis
RF	random forests
RMSE	root mean squared error
SARIMA	seasonal autoregressive integrated moving average
SCIE	Science Citation Index Expanded
SCP	Single-Country Publications
SETAR	self-exciting threshold autoregressive
SOM	Self-organizing Maps
SSCI	Social Sciences Citations Index
SVM	support vector machine
TC	total citations
TCY	total citations per year
TI	titles
UK	United Kingdom
USA	United States of America
WI	Willmott Index
WoS	Clarivate Analytics' Web of Science Core Collection database

References

1. Abbass, H. Editorial: What Is Artificial Intelligence? *IEEE Trans. Artif. Intell.* **2021**, *2*, 94–95. [CrossRef]
2. Lim, B.; Zohren, S. Time-Series Forecasting with Deep Learning: A Survey. *Philos. Trans. R. Soc. A Math. Phys. Eng. Sci.* **2021**, *379*, 20200209. [CrossRef] [PubMed]
3. Yang, J.; Gong, P.; Fu, R.; Zhang, M.; Chen, J.; Liang, S.; Xu, B.; Shi, J.; Dickinson, R. Erratum: The Role of Satellite Remote Sensing in Climate Change Studies. *Nat. Clim. Change* **2013**, *3*, 1001. [CrossRef]
4. Naciri, H.; Ben Achhab, N.; Ezzaher, F.E.; Raissouni, N. AIRS: A QGIS Plugin for Time Series Forecasting Using Deep Learning Models. *Environ. Model. Softw.* **2024**, *177*, 106045. [CrossRef]
5. Fathian, F.; Mehdizadeh, S.; Kozekalani Sales, A.; Safari, M.J.S. Hybrid Models to Improve the Monthly River Flow Prediction: Integrating Artificial Intelligence and Non-Linear Time Series Models. *J. Hydrol.* **2019**, *575*, 1200–1213. [CrossRef]
6. Aghelpour, P.; Varshavian, V. Evaluation of Stochastic and Artificial Intelligence Models in Modeling and Predicting of River Daily Flow Time Series. *Stoch. Environ. Res. Risk Assess.* **2020**, *34*, 33–50. [CrossRef]
7. Mehdizadeh, S.; Fathian, F.; Safari, M.J.S.; Adamowski, J.F. Comparative Assessment of Time Series and Artificial Intelligence Models to Estimate Monthly Streamflow: A Local and External Data Analysis Approach. *J. Hydrol.* **2019**, *579*, 124225. [CrossRef]
8. Wang, M.; Qin, F. A TCN-Linear Hybrid Model for Chaotic Time Series Forecasting. *Entropy* **2024**, *26*, 467. [CrossRef]
9. Faustino, C.P.; Pinheiro, C.A.M.; Carpinteiro, O.A.; Lima, I. Time Series Forecasting through Rule-Based Models Obtained via Rough Sets. *Artif. Intell. Rev.* **2011**, *36*, 299–310. [CrossRef]
10. Bas, E.; Egrioglu, E.; Cansu, T. Robust Training of Median Dendritic Artificial Neural Networks for Time Series Forecasting. *Expert Syst. Appl.* **2024**, *238*, 122080. [CrossRef]
11. Edalatpanah, S.A.; Hassani, F.S.; Smarandache, F.; Sorourkhah, A.; Pamucar, D.; Cui, B. A Hybrid Time Series Forecasting Method Based on Neutrosophic Logic with Applications in Financial Issues. *Eng. Appl. Artif. Intell.* **2024**, *129*, 107531. [CrossRef]
12. Kocian, A.; Massa, D.; Cannazzaro, S.; Incrocci, L.; Di Lonardo, S.; Milazzo, P.; Chessa, S. Dynamic Bayesian Network for Crop Growth Prediction in Greenhouses. *Comput. Electron. Agric.* **2020**, *169*, 105167. [CrossRef]
13. Li, C.; Mahadevan, S.; Ling, Y.; Choze, S.; Wang, L. Dynamic Bayesian Network for Aircraft Wing Health Monitoring Digital Twin. *AIAA J.* **2017**, *55*, 930–941. [CrossRef]

14. Fan, H.; Jia, H.; He, X.; Lyu, J. Navigating Uncertainty: A Dynamic Bayesian Network-Based Risk Assessment Framework for Maritime Trade Routes. *Reliab. Eng. Syst. Saf.* **2024**, *250*, 110311. [CrossRef]
15. Guo, J.; Ma, K. Risk Analysis for Hazardous Chemical Vehicle-Bridge Transportation System: A Dynamic Bayesian Network Model Incorporating Vehicle Dynamics. *Reliab. Eng. Syst. Saf.* **2024**, *242*, 109732. [CrossRef]
16. Delcea, C.; Oprea, S.-V.; Dima, A.M.; Domenteanu, A.; Bara, A.; Cotfas, L.-A. Energy Communities: Insights from Scientific Publications. *Oecon. Copernic.* **2024**, *15*, 1101–1155. [CrossRef]
17. Block, J.H.; Fisch, C. Eight Tips and Questions for Your Bibliographic Study in Business and Management Research. *Manag. Rev. Q.* **2020**, *70*, 307–312. [CrossRef]
18. Sandu, A.; Ioanăș, I.; Delcea, C.; Florescu, M.-S.; Cotfas, L.-A. Numbers Do Not Lie: A Bibliometric Examination of Machine Learning Techniques in Fake News Research. *Algorithms* **2024**, *17*, 70. [CrossRef]
19. Sandu, A.; Cotfas, L.-A.; Stănescu, A.; Delcea, C. A Bibliometric Analysis of Text Mining: Exploring the Use of Natural Language Processing in Social Media Research. *Appl. Sci.* **2024**, *14*, 3144. [CrossRef]
20. Sandu, A.; Cotfas, L.-A.; Delcea, C.; Crăciun, L.; Molănescu, A.G. Sentiment Analysis in the Age of COVID-19: A Bibliometric Perspective. *Information* **2023**, *14*, 659. [CrossRef]
21. WoS Web of Science. Available online: http://webofknowledge.com (accessed on 10 October 2024).
22. Ciuverca, A.-C.-D.; Oprea, S. The Influence of AI on Price Forecasting. The View of the Academic Community. *JBEM* **2025**, *26*, 231–254. [CrossRef]
23. Întorsureanu, I.; Oprea, S.-V.; Bâra, A.; Vespan, D. Generative AI in Education: Perspectives Through an Academic Lens. *Electronics* **2025**, *14*, 1053. [CrossRef]
24. Oprea, S.-V.; Bâra, A. Generative Literature Analysis on the Rise of Prosumers and Their Influence on the Sustainable Energy Transition. *Util. Policy* **2024**, *90*, 101799. [CrossRef]
25. Cișmașu, I.-D.; Cibu, B.R.; Cotfas, L.-A.; Delcea, C. The Persistence Puzzle: Bibliometric Insights into Dropout in MOOCs. *Sustainability* **2025**, *17*, 2952. [CrossRef]
26. Sandu, A.; Cotfas, L.-A.; Stănescu, A.; Delcea, C. Guiding Urban Decision-Making: A Study on Recommender Systems in Smart Cities. *Electronics* **2024**, *13*, 2151. [CrossRef]
27. Bakır, M.; Özdemir, E.; Akan, Ş.; Atalık, Ö. A Bibliometric Analysis of Airport Service Quality. *J. Air Transp. Manag.* **2022**, *104*, 102273. [CrossRef]
28. Koca, T.T. Bibliometric Analysis of "Sarcopenia" in Web of Science. *Egypt. Rheumatol. Rehabil.* **2023**, *50*, 26. [CrossRef]
29. Singh, V.K.; Singh, P.; Karmakar, M.; Leta, J.; Mayr, P. The Journal Coverage of Web of Science, Scopus and Dimensions: A Comparative Analysis. *Scientometrics* **2021**, *126*, 5113–5142. [CrossRef]
30. Clarivate. KeyWords Plus Generation, Creation, and Changes. Available online: https://support.clarivate.com/ScientificandAcademicResearch/s/article/KeyWords-Plus-generation-creation-and-changes?language=en_US (accessed on 2 March 2024).
31. Liu, F. Retrieval Strategy and Possible Explanations for the Abnormal Growth of Research Publications: Re-Evaluating a Bibliometric Analysis of Climate Change. *Scientometrics* **2023**, *128*, 853–859. [CrossRef]
32. Liu, W. The Data Source of This Study Is Web of Science Core Collection? Not Enough. *Scientometrics* **2019**, *121*, 1815–1824. [CrossRef]
33. Sarkis-Onofre, R.; Catalá-López, F.; Aromataris, E.; Lockwood, C. How to Properly Use the PRISMA Statement. *Syst. Rev.* **2021**, *10*, 117. [CrossRef] [PubMed]
34. Khin Khin Oo, N.C.; Rakthin, S. Integrative Review of Absorptive Capacity's Role in Fostering Organizational Resilience and Research Agenda. *Sustainability* **2022**, *14*, 12570. [CrossRef]
35. Pham, X.-L.; Le, T. Bibliometric Analysis and Systematic Review of Research on Expert Finding: A PRISMA-Guided Approach. *Int. Arab J. Inf. Technol.* **2024**, *21*, 661–674. [CrossRef]
36. Peixoto, B.; Pinto, R.; Melo, M.; Cabral, L.; Bessa, M. Immersive Virtual Reality for Foreign Language Education: A PRISMA Systematic Review. *IEEE Access* **2021**, *9*, 48952–48962. [CrossRef]
37. Frioui, S.; Graa, A. Bibliometric Analysis of Artificial Intelligence in the Scope of E-Commerce: Trends and Progress over the Last Decade. *Manag. Econ. Rev.* **2024**, *9*, 5–24. [CrossRef]
38. Bawack, R.E.; Wamba, S.F.; Carillo, K.D.A.; Akter, S. Artificial Intelligence in E-Commerce: A Bibliometric Study and Literature Review. *Electron. Mark.* **2022**, *32*, 297–338. [CrossRef] [PubMed]
39. Ofer, D.; Kaufman, H.; Linial, M. What's next? Forecasting Scientific Research Trends. *Heliyon* **2024**, *10*, e23781. [CrossRef]
40. Vincent-Lancrin, S. What Is Changing in Academic Research? Trends and Futures Scenarios. *Eur. J. Educ.* **2006**, *41*, 169–202. [CrossRef]
41. Aria, M.; Cuccurullo, C. Bibliometrix: An R-Tool for Comprehensive Science Mapping Analysis. *J. Informetr.* **2017**, *11*, 959–975. [CrossRef]
42. Monroy, S.E.; Diaz, H. Time Series-Based Bibliometric Analysis of the Dynamics of Scientific Production. *Scientometrics* **2018**, *115*, 1139–1159. [CrossRef]

43. Palomero, L.; García, V.; Sánchez, J.S. Fuzzy-Based Time Series Forecasting and Modelling: A Bibliometric Analysis. *Appl. Sci.* **2022**, *12*, 6894. [CrossRef]
44. Wang, W.-C.; Chau, K.-W.; Cheng, C.-T.; Qiu, L. A Comparison of Performance of Several Artificial Intelligence Methods for Forecasting Monthly Discharge Time Series. *J. Hydrol.* **2009**, *374*, 294–306. [CrossRef]
45. Chen, Y.; Wang, W.; Chen, X.M. Bibliometric Methods in Traffic Flow Prediction Based on Artificial Intelligence. *Expert Syst. Appl.* **2023**, *228*, 120421. [CrossRef]
46. Shabani, M.; Ghousi, R.; Mohammadi, E. Exploring the Evolution of Scientific Publication on Portfolio Optimization in the Light of Artificial Intelligence: A Bibliometric Study. *Accounting* **2025**, *11*, 71–90. [CrossRef]
47. Hirsch, J.E. An Index to Quantify an Individual's Scientific Research Output. *Proc. Natl. Acad. Sci. USA* **2005**, *102*, 16569–16572. [CrossRef]
48. Egghe, L. Theory and Practise of the G-Index. *Scientometrics* **2006**, *69*, 131–152. [CrossRef]
49. Yang, J.-M.; Tseng, S.-F.; Won, Y.-L. A Bibliometric Analysis on Data Mining Using Bradford's Law. In Proceedings of the 3rd International Conference on Intelligent Technologies and Engineering Systems (ICITES2014) 2016, Kaohsiung, Taiwan, 19–21 December 2014; pp. 613–620. [CrossRef]
50. Bibliometrix—Home. Available online: https://www.bibliometrix.org/home/ (accessed on 3 July 2024).
51. Soliman, M.; Fatnassi, T.; Elgammal, I.; Figueiredo, R. Exploring the Major Trends and Emerging Themes of Artificial Intelligence in the Scientific Leading Journals amidst the COVID-19 Era. *Big Data Cogn. Comput.* **2023**, *7*, 12. [CrossRef]
52. Newby, G.B.; Greenberg, J.; Jones, P. Open Source Software Development and Lotka's Law: Bibliometric Patterns in Programming. *J. Am. Soc. Inf. Sci. Technol.* **2003**, *54*, 169–178. [CrossRef]
53. Cibu, B.; Delcea, C.; Domenteanu, A.; Dumitrescu, G. Mapping the Evolution of Cybernetics: A Bibliometric Perspective. *Computers* **2023**, *12*, 237. [CrossRef]
54. Jigani, A.-I.; Delcea, C.; Florescu, M.-S.; Cotfas, L.-A. Tracking Happiness in Times of COVID-19: A Bibliometric Exploration. *Sustainability* **2024**, *16*, 4918. [CrossRef]
55. Song, X.; Liu, Y.; Xue, L.; Wang, J.; Zhang, J.; Wang, J.; Jiang, L.; Cheng, Z. Time-Series Well Performance Prediction Based on Long Short-Term Memory (LSTM) Neural Network Model. *J. Pet. Sci. Eng.* **2020**, *186*, 106682. [CrossRef]
56. Sharadga, H.; Hajimirza, S.; Balog, R.S. Time Series Forecasting of Solar Power Generation for Large-Scale Photovoltaic Plants. *Renew. Energy* **2020**, *150*, 797–807. [CrossRef]
57. Chou, J.-S.; Tran, D.-S. Forecasting Energy Consumption Time Series Using Machine Learning Techniques Based on Usage Patterns of Residential Householders. *Energy* **2018**, *165*, 709–726. [CrossRef]
58. Nourani, V. An Emotional ANN (EANN) Approach to Modeling Rainfall-Runoff Process. *J. Hydrol.* **2017**, *544*, 267–277. [CrossRef]
59. Ledoux, C. An Urban Traffic Flow Model Integrating Neural Networks. *Transp. Res. Part C Emerg. Technol.* **1997**, *5*, 287–300. [CrossRef]
60. Ismail, S.; Shabri, A.; Samsudin, R. A Hybrid Model of Self-Organizing Maps (SOM) and Least Square Support Vector Machine (LSSVM) for Time-Series Forecasting. *Expert Syst. Appl.* **2011**, *38*, 10574–10578. [CrossRef]
61. Cho, V. A Comparison of Three Different Approaches to Tourist Arrival Forecasting. *Tour. Manag.* **2003**, *24*, 323–330. [CrossRef]
62. Sahoo, B.B.; Jha, R.; Singh, A.; Kumar, D. Long Short-Term Memory (LSTM) Recurrent Neural Network for Low-Flow Hydrological Time Series Forecasting. *Acta Geophys.* **2019**, *67*, 1471–1481. [CrossRef]
63. Qu, J.; Qian, Z.; Pei, Y. Day-Ahead Hourly Photovoltaic Power Forecasting Using Attention-Based CNN-LSTM Neural Network Embedded with Multiple Relevant and Target Variables Prediction Pattern. *Energy* **2021**, *232*, 120996. [CrossRef]
64. Wilczewski, M.; Alon, I. Language and Communication in International Students' Adaptation: A Bibliometric and Content Analysis Review. *High Educ.* **2023**, *85*, 1235–1256. [CrossRef]
65. Romero-Riaño, E.; Rico-Bautista, D.; Martinez-Toro, M.; Medina-Cárdenas, Y.; Rico-Bautista, N. Artificial Intelligence Theory: A Bibliometric Analysis. *J. Phys. Conf. Ser.* **2021**, *2046*, 012078. [CrossRef]
66. Özdoğan Sarıkoç, G. Artificial Intelligence Applications in the Field of Streamflow: A Bibliometric Analysis of Recent Trends. *Hydrol. Sci. J.* **2024**, *69*, 1141–1157. [CrossRef]
67. Loan, F.A.; Bashir, B.; Nasreen, N. Applied Artificial Intelligence: A Bibliometric Study of an International Journal. *COLLNET J. Scientometr. Inf. Manag.* **2021**, *15*, 27–45. [CrossRef]
68. Feng, Z.; Niu, W. Hybrid Artificial Neural Network and Cooperation Search Algorithm for Nonlinear River Flow Time Series Forecasting in Humid and Semi-Humid Regions. *Knowl. Based Syst.* **2021**, *211*, 106580. [CrossRef]
69. Fu, X.; Feng, Z.; Cao, H.; Feng, B.; Tan, Z.; Xu, Y.; Niu, W. Enhanced Machine Learning Model via Twin Support Vector Regression for Streamflow Time Series Forecasting of Hydropower Reservoir. *Energy Rep.* **2023**, *10*, 2623–2639. [CrossRef]
70. Feng, Z.; Niu, W.; Wan, X.; Xu, B.; Zhu, F.; Chen, J. Hydrological Time Series Forecasting via Signal Decomposition and Twin Support Vector Machine Using Cooperation Search Algorithm for Parameter Identification. *J. Hydrol.* **2022**, *612*, 128213. [CrossRef]
71. Niu, W.; Feng, Z.; Feng, B.; Xu, Y.; Min, Y. Parallel Computing and Swarm Intelligence Based Artificial Intelligence Model for Multi-Step-Ahead Hydrological Time Series Prediction. *Sustain. Cities Soc.* **2021**, *66*, 102686. [CrossRef]

72. Carpinteiro, O.A.S.; Leite, J.P.R.R.; Pinheiro, C.A.M.; Lima, I. Forecasting Models for Prediction in Time Series. *Artif. Intell. Rev.* **2012**, *38*, 163–171. [CrossRef]
73. Faustino, C.P.; Novaes, C.P.; Pinheiro, C.A.M.; Carpinteiro, O.A. Improving the Performance of Fuzzy Rules-Based Forecasters through Application of FCM Algorithm. *Artif. Intell. Rev.* **2014**, *41*, 287–300. [CrossRef]
74. Vilela, L.F.S.; Leme, R.C.; Pinheiro, C.A.M.; Carpinteiro, O.A.S. Forecasting Financial Series Using Clustering Methods and Support Vector Regression. *Artif. Intell. Rev.* **2019**, *52*, 743–773. [CrossRef]
75. Li, H.; Wang, J.; Yang, H.; Wang, Y. Air Quality Deterministic and Probabilistic Forecasting System Based on Hesitant Fuzzy Sets and Nonlinear Robust Outlier Correction. *Knowl. Based Syst.* **2022**, *237*, 107789. [CrossRef]
76. Wang, J.; Li, H.; Wang, Y.; Lu, H. A Hesitant Fuzzy Wind Speed Forecasting System with Novel Defuzzification Method and Multi-Objective Optimization Algorithm. *Expert Syst. Appl.* **2021**, *168*, 114364. [CrossRef]
77. Jiang, P.; Yang, H.; Li, H.; Wang, Y. A Developed Hybrid Forecasting System for Energy Consumption Structure Forecasting Based on Fuzzy Time Series and Information Granularity. *Energy* **2021**, *219*, 119599. [CrossRef]
78. Wang, J.; Li, H.; Wang, Y.; Yang, H. A Novel Assessment and Forecasting System for Traffic Accident Economic Loss Caused by Air Pollution. *Environ. Sci. Pollut. Res.* **2021**, *28*, 49042–49062. [CrossRef] [PubMed]
79. Kocak, C.; Egrioglu, E.; Bas, E. A New Explainable Robust High-Order Intuitionistic Fuzzy Time-Series Method. *Soft Comput.* **2023**, *27*, 1783–1796. [CrossRef]
80. Kocak, C.; Dalar, A.Z.; Cagcag Yolcu, O.; Bas, E.; Egrioglu, E. A New Fuzzy Time Series Method Based on an ARMA-Type Recurrent Pi-Sigma Artificial Neural Network. *Soft Comput.* **2020**, *24*, 8243–8252. [CrossRef]
81. Yolcu, U.; Aladag, C.H.; Egrioglu, E.; Uslu, V.R. Time-Series Forecasting with a Novel Fuzzy Time-Series Approach: An Example for Istanbul Stock Market. *J. Stat. Comput. Simul.* **2013**, *83*, 599–612. [CrossRef]
82. Morales-García, J.; Bueno-Crespo, A.; Martínez-España, R.; Cecilia, J.M. Data-Driven Evaluation of Machine Learning Models for Climate Control in Operational Smart Greenhouses. *J. Ambient Intell. Smart Environ.* **2023**, *15*, 3–17. [CrossRef]
83. Morales-García, J.; Bueno-Crespo, A.; Martínez-España, R.; Posadas, J.-L.; Manzoni, P.; Cecilia, J.M. Evaluation of Low-Power Devices for Smart Greenhouse Development. *J. Supercomput.* **2023**, *79*, 10277–10299. [CrossRef]
84. Morales-García, J.; Bueno-Crespo, A.; Terroso-Sáenz, F.; Arcas-Túnez, F.; Martínez-España, R.; Cecilia, J.M. Evaluation of Synthetic Data Generation for Intelligent Climate Control in Greenhouses. *Appl. Intell.* **2023**, *53*, 24765–24781. [CrossRef]
85. Al-Ali, E.M.; Hajji, Y.; Said, Y.; Hleili, M.; Alanzi, A.M.; Laatar, A.H.; Atri, M. Solar Energy Production Forecasting Based on a Hybrid CNN-LSTM-Transformer Model. *Mathematics* **2023**, *11*, 676. [CrossRef]
86. Niu, W.; Feng, Z.; Li, S.; Wu, H.; Wang, J. Short-Term Electricity Load Time Series Prediction by Machine Learning Model via Feature Selection and Parameter Optimization Using Hybrid Cooperation Search Algorithm. *Environ. Res. Lett.* **2021**, *16*, 055032. [CrossRef]
87. Zhou, C.; Che, C.; Wang, P.; Zhang, Q. Diformer: A Dynamic Self-Differential Transformer for New Energy Power Autoregressive Prediction. *Knowl. Based Syst.* **2023**, *281*, 111061. [CrossRef]
88. Hamdoun, H.; Sagheer, A.; Youness, H. Energy Time Series Forecasting-Analytical and Empirical Assessment of Conventional and Machine Learning Models. *J. Intell. Fuzzy Syst.* **2021**, *40*, 12477–12502. [CrossRef]
89. Bareith, T.; Tatay, T.; Vancsura, L. Navigating Inflation Challenges: AI-Based Portfolio Management Insights. *Risks* **2024**, *12*, 46. [CrossRef]
90. Louati, A.; Louati, H.; Alharbi, M.; Kariri, E.; Khawaji, T.; Almubaddil, Y.; Aldwsary, S. Machine Learning and Artificial Intelligence for a Sustainable Tourism: A Case Study on Saudi Arabia. *Information* **2024**, *15*, 516. [CrossRef]
91. Zhang, B.; Li, N.; Shi, F.; Law, R. A Deep Learning Approach for Daily Tourist Flow Forecasting with Consumer Search Data. *Asia Pac. J. Tour. Res.* **2020**, *25*, 323–339. [CrossRef]
92. Sohrabpour, V.; Oghazi, P.; Toorajipour, R.; Nazarpour, A. Export Sales Forecasting Using Artificial Intelligence. *Technol. Forecast. Soc. Change* **2021**, *163*, 120480. [CrossRef]
93. Tang, Y.M.; Chau, K.-Y.; Li, W.; Wan, T.W. Forecasting Economic Recession through Share Price in the Logistics Industry with Artificial Intelligence (AI). *Computation* **2020**, *8*, 70. [CrossRef]
94. Feng, Z.; Shi, P.; Yang, T.; Niu, W.; Zhou, J.; Cheng, C. Parallel Cooperation Search Algorithm and Artificial Intelligence Method for Streamflow Time Series Forecasting. *J. Hydrol.* **2022**, *606*, 127434. [CrossRef]
95. Christian, K.; Roy, A.F.V.; Yudianto, D.; Zhang, D. Application of Optimized Support Vector Machine in Monthly Streamflow Forecasting: Using Autocorrelation Function for Input Variables Estimation. *Sustain. Water Resour. Manag.* **2021**, *7*, 29. [CrossRef]
96. Yaseen, Z.M.; Allawi, M.F.; Yousif, A.A.; Jaafar, O.; Hamzah, F.M.; El-Shafie, A. Non-Tuned Machine Learning Approach for Hydrological Time Series Forecasting. *Neural Comput. Appl.* **2018**, *30*, 1479–1491. [CrossRef]
97. Niu, W.; Feng, Z.; Xu, Y.; Feng, B.; Min, Y. Improving Prediction Accuracy of Hydrologic Time Series by Least-Squares Support Vector Machine Using Decomposition Reconstruction and Swarm Intelligence. *J. Hydrol. Eng.* **2021**, *26*, 04021030. [CrossRef]
98. Sang, Y.-F.; Singh, V.P.; Sun, F.; Chen, Y.; Liu, Y.; Yang, M. Wavelet-Based Hydrological Time Series Forecasting. *J. Hydrol. Eng.* **2016**, *21*, 06016001. [CrossRef]

99. Mishra, M.; Byomakesha Dash, P.; Nayak, J.; Naik, B.; Kumar Swain, S. Deep Learning and Wavelet Transform Integrated Approach for Short-Term Solar PV Power Prediction. *Measurement* **2020**, *166*, 108250. [CrossRef]
100. Zhou, Y.; He, X.; Montillet, J.-P.; Wang, S.; Hu, S.; Sun, X.; Huang, J.; Ma, X. An Improved ICEEMDAN-MPA-GRU Model for GNSS Height Time Series Prediction with Weighted Quality Evaluation Index. *GPS Solut.* **2025**, *29*, 113. [CrossRef]
101. Tay, A. Using VOSviewer as a Bibliometric Mapping or Analysis Tool in Business, Management & Accounting | Singapore Management University (SMU). Available online: https://library.smu.edu.sg/topics-insights/using-vosviewer-bibliometric-mapping-or-analysis-tool-business-management (accessed on 7 March 2025).
102. WoS Document Types. Available online: https://webofscience.help.clarivate.com/en-us/Content/document-types.html (accessed on 3 December 2023).

applied sciences

MDPI

Article

Monte Carlo Dropout Neural Networks for Forecasting Sinusoidal Time Series: Performance Evaluation and Uncertainty Quantification

Unyamanee Kummaraka [1] and Patchanok Srisuradetchai [2,*]

[1] Department of Statistics, Faculty of Science, Chiang Mai University, Chiang Mai 50200, Thailand; unyamanee.k@cmu.ac.th

[2] Department of Mathematics and Statistics, Faculty of Science and Technology, Thammasat University, Pathum Thani 12120, Thailand

* Correspondence: patchanok@mathstat.sci.tu.ac.th

Abstract: Accurately forecasting sinusoidal time series is essential in various scientific and engineering applications. However, traditional models such as the seasonal autoregressive integrated moving average (SARIMA) rely on assumptions of linearity and stationarity, which may not adequately capture the complex periodic behaviors of sinusoidal data, including varying amplitudes, phase shifts, and nonlinear trends. This study investigates Monte Carlo dropout neural networks (MCDO NNs) as an alternative approach for both forecasting and uncertainty quantification. The performance of MCDO NNs is evaluated across six sinusoidal time series models, each exhibiting different dynamic characteristics. Results indicate that MCDO NNs consistently outperform SARIMA in terms of root mean square error, mean absolute percentage error, and the coefficient of determination, while also producing more reliable prediction intervals. To assess real-world applicability, the airline passenger dataset is used, demonstrating MCDO's ability to effectively capture periodic structures. These findings suggest that MCDO NNs provide a robust alternative to SARIMA for sinusoidal time series forecasting, offering both improved accuracy and well-calibrated uncertainty estimates.

Keywords: sinusoidal time series; Monte Carlo dropout; neural networks; forecasting; prediction intervals

1. Introduction

Time series data play a crucial role in various fields, including economics, finance, environmental sciences, and engineering, as they provide valuable insights into temporal patterns and trends. Among these, sinusoidal time series are particularly significant due to their ability to model periodic behaviors observed in both natural and artificial systems [1,2]. For instance, in astronomy, periodic variations in solar time and astronomical measurements arise from Earth's rotation and orbit around the Sun, which can be effectively captured using sinusoidal models [3]. Similarly, in electrical engineering, sinusoidal waveforms form the foundation for understanding alternating current behaviors, as they accurately represent the periodic fluctuations in voltage and current in real-world circuits [4]. In management, sinusoidal functions assist in predicting and managing demand variations in inventory systems, as demonstrated by Tawadros [5] in forecasting dynamic demand shifts over time.

In biological sciences, sinusoidal models are instrumental in analyzing circadian rhythms, which regulate sleep cycles, hormone secretion, and metabolism [6]. Oceanography relies on sinusoidal patterns to predict tidal movements [7], while quantum mechanics and wave theory employ sinusoidal functions to describe particle behaviors and wave propagation, driving advancements in modern physics [8].

Recent advancements in machine learning (ML), particularly in neural networks, have shown significant promise in addressing the challenges of time series forecasting [9,10]. Neural networks excel at capturing nonlinear patterns and dependencies, making them well-suited for modeling sinusoidal time series. However, traditional neural networks primarily provide point estimates, often failing to quantify the uncertainty inherent in their predictions. To address this limitation, Gal and Ghahramani [11] introduced the use of dropout in neural networks as a method for approximate Bayesian inference, allowing for uncertainty estimation in predictions. Building upon this, Kendall and Gal [12] extended the application of Monte Carlo dropout (MCDO) to explicitly distinguish between aleatoric uncertainty, which arises from inherent data noise, and epistemic uncertainty, which stems from limited model knowledge. Additionally, Srivastava et al. [13] originally proposed dropout as a regularization technique, which later became a foundational tool for uncertainty quantification in deep learning models.

Further advancements in uncertainty quantification have explored various techniques to improve reliability in forecasting. Zhang et al. [14] applied Bayesian networks for short-term traffic flow prediction, demonstrating their potential in time series modeling. Pearce et al. [15] combined Bayesian neural networks (BNNs) with ensemble methods to improve uncertainty estimation in time series forecasting. Similarly, Blundell et al. [16] employed variational inference for neural network weights, supporting dropout as an effective tool for uncertainty quantification. In the domain of medical image analysis, Lemay et al. [17] showed that MCDO enhances model consistency and repeatability, particularly in classification and regression tasks. These contributions underscore the importance of integrating uncertainty estimation into forecasting models to improve reliability across diverse applications.

To address computational challenges, Atencia et al. [18] integrated echo state networks (ESNs) with MCDO, enabling uncertainty estimation with minimal computational overhead. Similarly, Sheng et al. [19] proposed the bootstrapping reservoir computing network ensemble (BRCNE), which blends Bayesian linear regression with ensemble learning to generate robust prediction intervals for noisy, nonlinear time series.

In a practical application, Srisuradetchai and Phaphan [10] demonstrated the effectiveness of MCDO neural networks (NNs) in interval forecasting for Thailand's durian export data. Their study found that MCDO-NN outperformed the seasonal autoregressive integrated moving average (SARIMA) model, achieving a lower root mean squared error (RMSE) and producing more reliable uncertainty estimates. Additionally, Khosravi et al. [20] emphasized the significance of prediction intervals in uncertain forecasting, leveraging Bayesian and delta methods to assess interval reliability through coverage probability and width. Recent work has also explored ensemble-based regression approaches, such as random kernel k-nearest neighbors, to improve prediction robustness and model reliability in nonparametric forecasting contexts [21].

Although MCDO has achieved success in various areas such as computer vision, medical imaging, and general time series forecasting, its application to sinusoidal time series has not been thoroughly investigated. Two main reasons likely explain this gap in literature. First, the adoption of MCDO in deep learning is relatively recent, with initial research focusing on broader applications before narrowing to specific forecasting problems. Second, while sinusoidal time series are common in scientific and engineering disciplines, they have only recently been recognized as a distinct category for MCDO-based forecasting, prompting the need for targeted investigations.

This study aims to bridge this gap by applying MCDO-NNs to sinusoidal time series and rigorously evaluating their performance in generating accurate forecast intervals under diverse conditions. The study compares MCDO-NNs with SARIMA, a widely used benchmark model for handling autocorrelated errors in time series analysis. The evaluation focuses on forecast accuracy, interval reliability, and forecast interval width, offering a comprehensive assessment of MCDO's utility for sinusoidal time series forecasting.

2. Monte Carlo Dropout Neural Networks

2.1. Monte Carlo Dropout

MCDO extends dropout, originally designed as a regularization technique, into a method for approximate Bayesian inference, enabling neural networks to quantify prediction uncertainty. MCDO captures two key types of uncertainty:

- Aleatoric uncertainty, caused by inherent data noise such as measurement errors, is irreducible and can be modeled through predictive distributions obtained via multiple forward passes [11].
- Epistemic uncertainty, on the other hand, arises from limited training data and represents uncertainty in the model's parameters.

MCDO addresses epistemic uncertainty by applying dropout at inference time, effectively creating a committee of sub-networks, where each forward pass samples from a distribution over model weights. The variance of these predictions serves as a confidence estimate, making MCDO a crucial tool in financial forecasting, medical diagnosis, and other high-risk applications where interpretable uncertainty is essential for decision-making [22]. Recent studies have advanced uncertainty-aware deep learning approaches for time series forecasting across a range of domains, including industrial systems, smart cities, and hybrid statistical frameworks [23–25].

2.2. Neural Network Model

A neural network model consists of mathematical transformations that map input features to output predictions. Consider a feedforward neural network with L layers, where each layer l performs a transformation on its input to produce an output that serves as the input to the next layer.

Let $x^{(0)}$ denote the input feature vector. The operation of each layer is expressed as:

$$x^{(l)} = f^{(l)}\left(W^{(l)}x^{(l-1)} + b^{(l)}\right), \quad (1)$$

where $x^{(l)}$ is the output of layer l, $W^{(l)}$ and $b^{(l)}$ represent the weight matrix and bias vector, respectively, and $f^{(l)}$ is the activation function, commonly chosen from ReLU, sigmoid, or tanh functions.

2.3. Dropout Mechanism

A dropout layer is applied after the activation function of each hidden layer during both training and inference. In training, neurons are randomly "dropped" with a dropout rate p, which refers to the probability of a neuron being deactivated. This simulates a sparse network and helps prevent overfitting. Mathematically, this is represented as:

$$x^{(l)}_{\text{dropout}} = d^{(l)} \odot x^{(l)}, \quad (2)$$

where $d^{(l)}$ is a binary mask vector where each element is drawn independently from a Bernoulli distribution with parameter $1 - p$, and $\odot$ denotes element-wise multiplication. The original and dropout networks are depicted in Figure 1, where neurons are randomly removed during each forward pass based on the dropout rate. This stochastic process can be interpreted as sampling from a variational distribution, forming the basis for Monte Carlo inference [26].

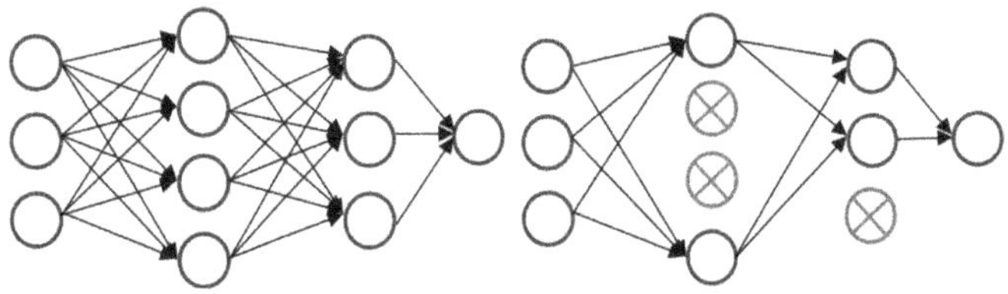

Figure 1. Standard neural network (**left**) and Monte Carlo dropout (MCDO) neural network (**right**). Arrows indicate signal flow. Grey crossed-out nodes in the MCDO network represent neurons dropped during inference for uncertainty estimation.

2.4. MCDO as Approximation

A Bayesian neural network (BNN) extends traditional neural networks by placing a probability distribution over weights, enabling a comprehensive quantification of uncertainty in predictions. Using Bayes' theorem, the posterior distribution over the weights W, given the dataset $D = \{(x_i, y_i)\}_{i=1}^{N}$ is computed as:

$$P(W|D) = P(W)P(D|W)/P(D). \tag{3}$$

However, directly computing $P(W|D)$ is computationally intractable for deep neural networks due to the high dimensionality of W and the complexity of $P(D|W)$ [27].

To approximate this, MCDO applies dropout during both training and inference, randomly deactivating a subset of activations. This effectively samples from a distribution of thinned networks, which can be viewed as drawing from an approximate posterior distribution $q(W)$. The predictive distribution for a new input x^* is then obtained by marginalizing over this approximate posterior [28]:

$$p(y^*|x^*, D) \propto \int p(y^*|x^*, W)\ q(W)\ dW. \tag{4}$$

Since evaluating this integral is computationally expensive, Monte Carlo integration is used to approximate it. By performing M stochastic forward passes with dropout, MCDO generates a set of predictions $\{y_1^*, y_2^*, \ldots, y_M^*\}$. The empirical mean and variance of these predictions serve as approximations for the predictive mean and uncertainty, respectively [29]:

$$\overline{y}^* = \frac{1}{M}\sum_{m=1}^{M} y_m^*,\ \sigma_{y^*}^2 = \frac{1}{M}\sum_{m=1}^{M}(y_m^* - \overline{y}^*)^2. \tag{5}$$

The 95% confidence intervals for predictions are then computed as:

$$\overline{y}^* \pm 1.96\sqrt{\sigma_{y^*}^2} \tag{6}$$

2.5. Network Input and Implementation Details

In our time series forecasting framework, the input to the neural network at time step t consists of the twelve most recent lagged observations, defined as

$$\mathbf{x}^{(0)} = [y_{t-1}, y_{t-2}, \ldots, y_{t-12}], \tag{7}$$

and the network is trained to predict the next value, y_t. This one-step-ahead prediction structure is used during training. To forecast multiple future steps, such as the final 12 months of the series, the model is applied recursively: after predicting one value, the output is appended to the input window, and the oldest lag is removed, allowing the model to generate the next input for the following time step.

The input features are standardized using z-score normalization, while the target values remain in their original scale. At inference time, MCDO is applied by performing $M = 20,000$ stochastic forward passes with dropout layers active, yielding a distribution of predictions:

$$\left\{y_t^{*(1)}, y_t^{*(2)}, \dots, y_t^{*(M)}\right\}, \tag{8}$$

from which the predictive mean and standard deviation are estimated.

3. Sinusoidal Models in Simulation Studies

This section explores six sinusoidal time series models, highlighting their structures, applications, and integration of trend components [30,31].

3.1. Constant Amplitude with Trend

The constant amplitude model represents a fundamental form of sinusoidal time series, characterized by regular oscillations of a fixed amplitude around a central level:

$$y_t = A\sin(\omega t + f) + C + \beta t + \varepsilon_t, \tag{9}$$

where A is the amplitude, ω is the angular frequency, f is the phase shift, C is the central level, βt introduces a linear trend, and ε_t is a random error component. The inclusion of βt allows the model to capture long-term directional movements, making it applicable to phenomena exhibiting both cyclic behavior and overall growth or decline trends.

3.2. Varying Amplitude with Trend

To model changing intensity in periodic behavior, the varying amplitude model integrates a time-dependent amplitude function $M(t)$:

$$y_t = (A + \delta t)\sin(\omega t + f) + C + \beta t + \varepsilon_t. \tag{10}$$

This model is particularly useful for capturing seasonal effects, where periodic fluctuations strengthen or weaken over time.

3.3. Mixed Frequency with Trend

The mixed frequency model accounts for scenarios where multiple cyclical components overlap within a single time series:

$$y_t = A_1\sin(\omega_1 t + f_1) + A_2\sin(\omega_2 t + f_2) + C + \beta t + \varepsilon_t. \tag{11}$$

By incorporating sine waves of different frequencies (ω_1, ω_2) and amplitudes (A_1, A_2), this model captures complex cyclical behaviors, such as those observed in economic data with overlapping business and seasonal cycles.

3.4. Phase Shift with Trend

The phase shift model simulates temporal changes in periodic patterns:

$$y_t = A\sin(\omega t + f(t)) + C + \beta t + \varepsilon_t. \tag{12}$$

Here, the phase shift $f(t)$ evolves over time, enabling the model to represent phenomena such as biological rhythm shifts or adjustments in seasonal patterns.

3.5. Nonlinear Modulation with Trend

The nonlinear modulation model addresses cases where the amplitude of oscillations is influenced by additional cyclical factors, such as modulated signal strength in telecommunications or seasonal businesses with varying peak demand. The model is defined as:

$$y_t = [A\sin(\omega t + f) + C] \times [1 + B\sin(\theta t + \psi)] + \beta t + \varepsilon_t. \quad (13)$$

3.6. Noise Variability with Trend

This model extends phase shift dynamics by incorporating stochastic fluctuations in signal strength, often observed in telecommunications and environmental systems:

$$y_t = A\sin(\omega t + f(t)) + C + \beta t + \sigma(t)\varepsilon_t, \quad (14)$$

where $\sigma(t)$ modulates the variance of the noise component over time.

Figure 2 illustrates the six sinusoidal models over a five-year period, each comprising 300 data points (one observation every six days). This resolution captures finer seasonal variations, allowing for a detailed analysis of trend and cyclic patterns. In practical applications, datasets are often aggregated to monthly intervals, aligning them with real-world use cases while preserving key patterns and trends.

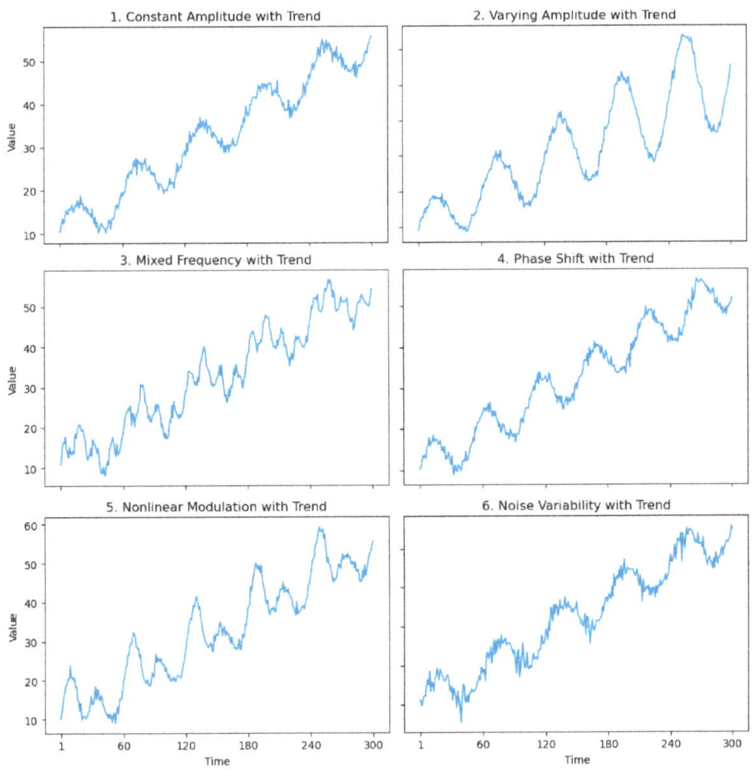

Figure 2. Simulated sinusoidal time series models over a five-year period, each showing synthetic data (blue lines) with distinct periodic patterns and a consistent upward trend.

4. Studied Settings

4.1. Simulation Models

In our simulation study, we examine six distinct sinusoidal time series models, allowing for a detailed analysis of their behaviors under a consistent upward trend ($\beta = 0.75$). The model settings are as follows:

- Model 1: Constant amplitude with trend, using a fixed amplitude of 5 units, representing a stable cyclic pattern.
- Model 2: Varying amplitude with trend, starting with an amplitude of 3 units, increasing linearly by 0.05 units per month, simulating intensifying seasonal effects.
- Model 3: Mixed frequency with trend, combining a primary cycle amplitude of 5 units with a secondary cycle amplitude of 3 units, capturing overlapping periodic patterns.
- Model 4: Phase shift with trend, initially set to zero for simplicity, but adaptable to simulate shifts in periodic events.
- Model 5: Nonlinear modulation with trend, incorporating a modulation depth of 0.5 to reflect additional cyclical influences.
- Model 6: Noise variability with trend, using a fixed amplitude of 5 units with stochastic noise modulation, modeling fluctuations in predictability over time.

Across all models, the angular frequency is set to $2\pi/12$ to reflect a yearly cycle in monthly measurements. The mean level remains at 10 units, ensuring a standardized basis for comparison.

4.2. Hyperparameter Grid

The simulation begins with a basic neural network architecture consisting of two layers with 60 and 30 neurons, establishing a baseline for model behavior. The architecture is then expanded incrementally to assess the effect of increasing complexity. A three-layer network consists of 120 neurons, while a four-layer network contains 180 neurons.

Activation functions such as ReLU, Sigmoid, and Tanh are tested to evaluate their effectiveness in capturing nonlinear relationships. To mitigate overfitting, dropout regularization is applied during training. Initial models use dropout rates between 0.2 and 0.3, with deeper networks employing rates as high as 0.7.

Training is conducted over 25, 50, or 100 epochs. Batch sizes of 16, 32, 64, and 128 are tested to examine their impact on computational efficiency and learning stability. By systematically adjusting these parameters, the study provides insights into how network depth, activation functions, and training configurations influence forecasting performance.

4.3. Model to Be Compared with MCDO NN

To assess forecast uncertainty, MCDO-NN predictions are compared with those of the SARIMA model, defined as [32,33]:

$$\left(1-\sum_{i=1}^{p} f_i L^i\right)\left(1-\sum_{i=1}^{P} \Phi_i L^{is}\right)(1-L)^d(1-L^s)^D y_t = \left(1+\sum_{i=1}^{q} \theta_i L^i\right)\left(1-\sum_{i=1}^{Q} \Theta_i L^{is}\right)\varepsilon_t, \quad (15)$$

where p is the number of autoregressive terms (AR part), d is the number of nonseasonal differences needed for stationarity (integrated part), q is the number of lagged forecast errors in the prediction equation (moving average part), P is the number of seasonal autoregressive terms, D is the number of seasonal differences, Q is the number of seasonal moving average terms, and s is the number of time steps for a single seasonal period. A SARIMA model is typically denoted as SARIMA$(p, d, q)(P, D, Q)_s$.

4.4. Comparison Criteria

In this study, the evaluation of forecasting models considers both point prediction accuracy and uncertainty quantification. For MCDO, each forecast is obtained by performing 20,000 stochastic forward passes with dropout enabled at inference. This yields a distribution of predictions at each time step, from which the predictive mean and standard deviation are calculated. The resulting statistics are then used to compute forecast intervals and assess model reliability. To evaluate the forecasting models, four key criteria are considered:

- Root Mean Square Error (RMSE): RMSE measures the average magnitude of the errors between predicted and actual values [34]. It is defined as

$$RMSE = \sqrt{\sum_{t=1}^{M} (y_t - \hat{y}_t)^2 / M}, \tag{16}$$

where $\hat{y}_t$ represents the forecasted value at time t, y_t is the actual observed value, and M is the total number of forecasts. A lower RMSE indicates higher prediction accuracy.

- Mean Absolute Percentage Error (MAPE): MAPE measures the average absolute percentage difference between the predicted and actual values, expressed as a percentage. It is defined as

$$MAPE = \sum_{t=1}^{M} |(y_t - \hat{y}_t)/y_t| / M \times 100. \tag{17}$$

This metric provides a scale-independent measure of forecast accuracy [35]. Lower MAPE values reflect better model performance.

- Coverage of Interval Forecasts (CIF): CIF refers to the proportion of times the actual values fall within the model's predicted confidence intervals. For instance, a 95% interval should ideally contain the true values 95% of the time. This metric assesses how well the model quantifies uncertainty [36].
- Width of Forecast Intervals (WIF): WIF quantifies the average width of the prediction intervals. Narrower intervals suggest more precise forecasts, while wider intervals may indicate greater uncertainty [37]. When interpreted alongside CIF, WIF helps evaluate the sharpness and reliability of the interval predictions.

All performance metrics are computed on the test set, which consists of the final segment of the time series excluded from training. This setup ensures that forecast quality reflects generalization to unseen data. Due to the temporal dependency in time series, standard cross-validation is not applicable. Therefore, a fixed train-test split is adopted to maintain the chronological order of observations [38,39].

In the simulation study, the time series is sampled every six days, producing 300 data points across a five-year span. The last 12 months, corresponding to approximately 60 observations, are held out as a test set to evaluate model performance.

5. Results

5.1. Optimal Parameters for MCDO NNs

A grid search was conducted to determine the optimal configurations for MCDO neural networks, evaluating the number of hidden layers, activation functions, dropout rates, epochs, and batch sizes.

For Model 1 (constant amplitude with trend), the best-performing neural network consists of five hidden layers with 240, 180, 120, 60, and 30 nodes, using the ReLU activation function and dropout rates of 0.6, 0.5, 0.4, 0.3, and 0.2. It was trained for 50 epochs with a batch size of 16. This configuration achieved an RMSE of 1.5049 and an R^2 of 0.6331, as shown in Figure 3.

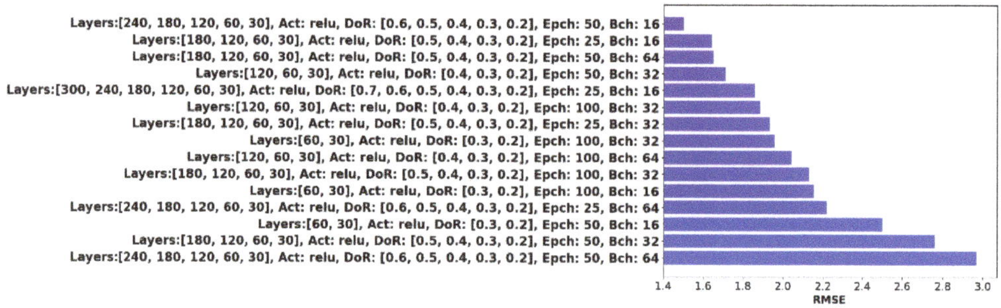

Figure 3. Lowest 15 RMSEs for MCDO configurations for Model 1.

For Model 2 (varying amplitude with trend), the top-performing architecture follows the same structure as Model 1. This configuration resulted in an RMSE of 2.0204 and an R^2 of 0.9528, demonstrating its ability to capture the increasing amplitude variations effectively. The results for Model 2 are depicted in Figure 4.

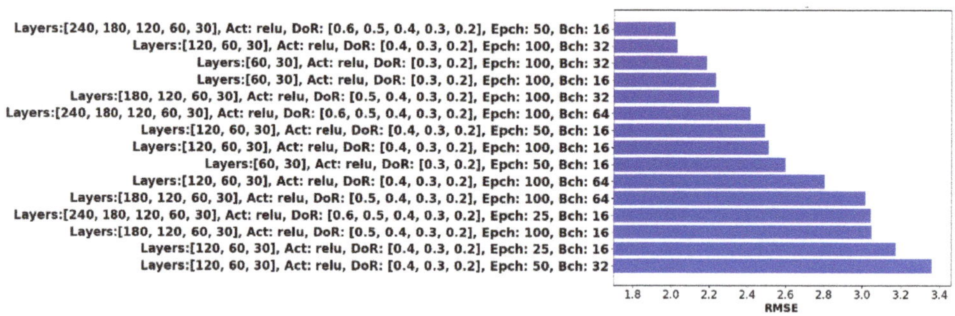

Figure 4. Lowest 15 RMSEs for MCDO configurations for Model 2.

For Model 3 (mixed frequency with trend), the optimal configuration features three hidden layers with 120, 60, and 30 nodes, trained for 100 epochs with dropout rates of 0.4, 0.3, and 0.2. The batch size is set to 16, and this setup achieves an RMSE of 1.8352 and an R^2 of 0.5340, indicating a reasonable performance in capturing multiple overlapping periodic components. The results for this configuration are presented in Figure 5.

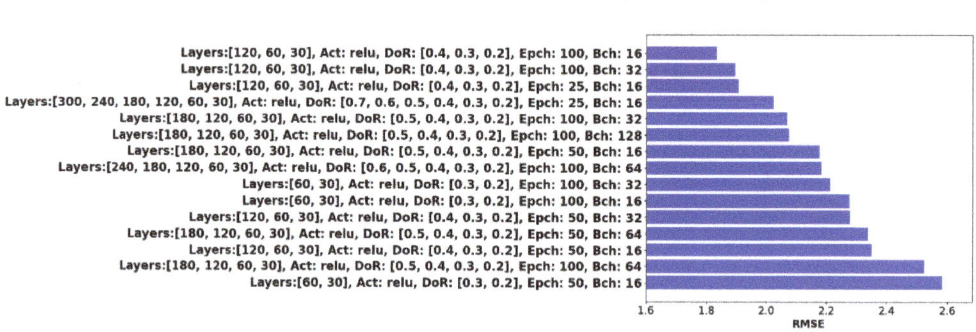

Figure 5. Lowest 15 RMSEs for MCDO configurations for Model 3.

For Model 4 (phase shift with trend), the network shares the same architecture as Model 3 but undergoes training for 50 epochs instead of 100. This model achieves an RMSE of 1.7906 and an R^2 of 0.8562, indicating improved accuracy in forecasting time series with evolving phase shifts. The results are shown in Figure 6.

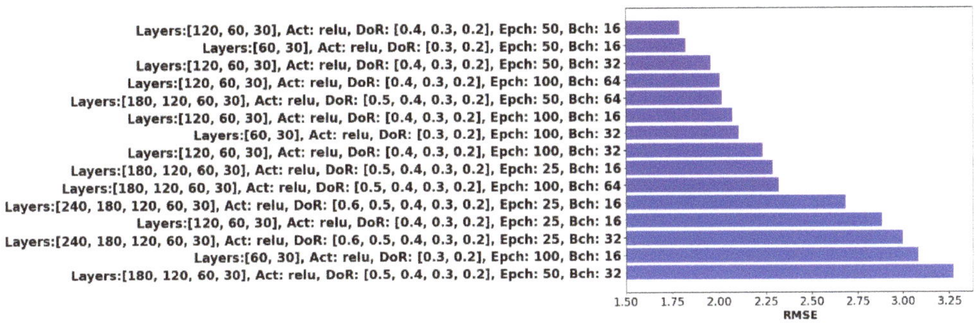

Figure 6. Lowest 15 RMSEs for MCDO configurations for Model 4.

For Model 5 (nonlinear modulation with trend), the best-performing neural network consists of four hidden layers, utilizing the ReLU activation function and trained for 100 epochs with a batch size of 64. This configuration achieves an RMSE of 1.4387 and an R^2 of 0.8512, suggesting strong predictive performance in scenarios with nonlinear amplitude modulations. The results for this setup are illustrated in Figure 7.

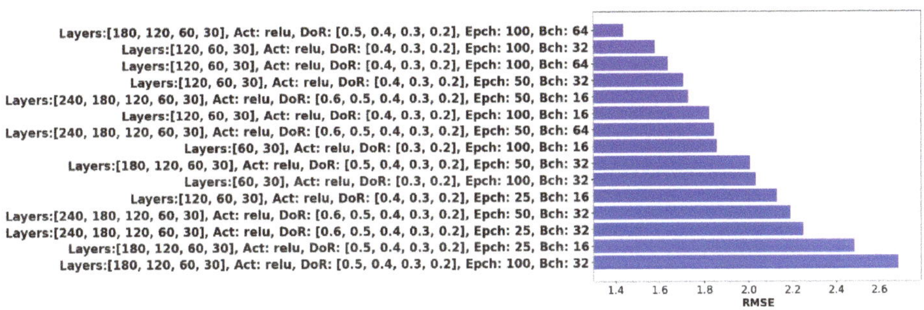

Figure 7. Lowest 15 RMSEs for MCDO configurations for Model 5.

For Model 6 (noise variability with trend), the optimal neural network consists of two hidden layers with 60 and 30 nodes, trained for 100 epochs with a batch size of 16, using the ReLU activation function. This model attains an RMSE of 1.8831 and an R^2 of 0.5969, reflecting moderate predictive accuracy given the added uncertainty from noise fluctuations. The results for this model are displayed in Figure 8.

The range of optimal configurations shows how well MCDO neural networks can adapt to different data characteristics. The depth and size of the networks, along with choices like dropout rates, batch sizes, and training epochs, are fine-tuned to handle varying levels of trend complexity, amplitude changes, frequency shifts, phase adjustments, nonlinearity, and noise. This flexibility ensures the networks are capable of capturing the unique patterns in each dataset. The grid search process is key here, as it identifies the best setup for each model.

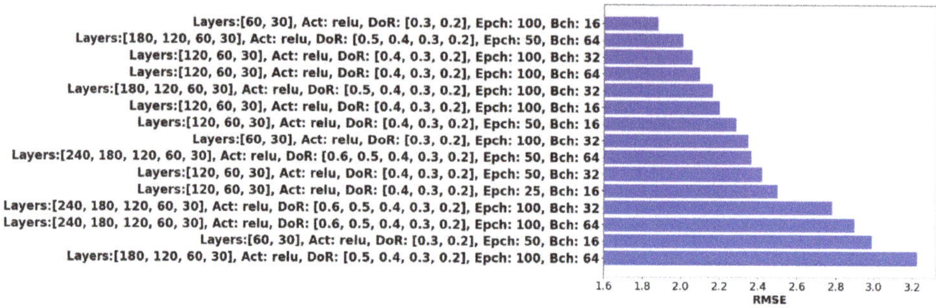

Figure 8. Lowest 15 RMSEs for MCDO configurations for Model 6.

5.2. Performance of MCDO NNs

Table 1 presents a comparative analysis of the forecasting performance of MCDO and SARIMA models across six sinusoidal trend scenarios. The results suggest that MCDO generally achieves better performance than SARIMA in terms of forecast accuracy (RMSE and MAPE) and interval width reduction.

Table 1. Performance comparison of MCDO and SARIMA across models (test data).

Model	RMSE		MAPE		WIF
	MCDO	SARIMA	MCDO	SARIMA	(MCDO vs. SARIMA)
1. Constant Amplitude	2.50	4.57	4.30%	7.44%	22.94 vs. 33.58
2. Varying Amplitude	3.40	8.84	6.14%	15.97%	21.62 vs. 28.56
3. Mixed Frequency	1.78	1.79	2.93%	2.70%	8.95 vs. 10.07
4. Phase Shift	2.00	12.56	3.12%	21.78%	18.56 vs. 36.88
5. Nonlinear Modulation	1.79	3.68	2.90%	6.13%	28.85 vs. 48.86
6. Noise Variability	2.51	2.74	4.05%	4.56%	13.90 vs. 31.23

The RMSE of MCDO models ranges from 1.58 to 2.82, whereas SARIMA's RMSE varies between 1.79 and 8.84, indicating that MCDO tends to produce lower errors across most models. Similarly, MCDO achieves lower MAPE values in the majority of cases. For instance, in Model 2 (varying amplitude with trend), MCDO's MAPE is 6.14%, substantially lower than SARIMA's 15.97%.

Additionally, MCDO often produces narrower forecast intervals, helping reduce uncertainty while maintaining comparable or higher coverage probabilities. For example, in Model 2, MCDO achieves an interval width of 21.62 compared to 28.56 for SARIMA, suggesting more concentrated forecast bounds. Figures 9–14 visualize these results for each model and are discussed individually.

For Model 1 (constant amplitude with trend), the MCDO model outperformed SARIMA$(1,1,2)(0,0,0)_{12}$, which was selected based on its lowest AIC value of 776.248. The MCDO model achieved a lower RMSE (2.50 vs. 4.57) and MAPE (4.30% vs. 7.44%). Additionally, the MCDO model produced significantly narrower WIF (22.94 vs. 33.58) while maintaining a 100% CIF for both models, as shown in Figure 9.

For Model 2 (varying amplitude with trend), the MCDO model demonstrated superior accuracy, achieving a lower RMSE (3.40 vs. 8.84) and MAPE (6.14% vs. 15.97%) compared to SARIMA. Additionally, MCDO produced a narrower prediction interval width (21.62 vs. 28.56) while maintaining a slightly higher CIF of 100% compared to SARIMA's 96.67%, as shown in Figure 10.

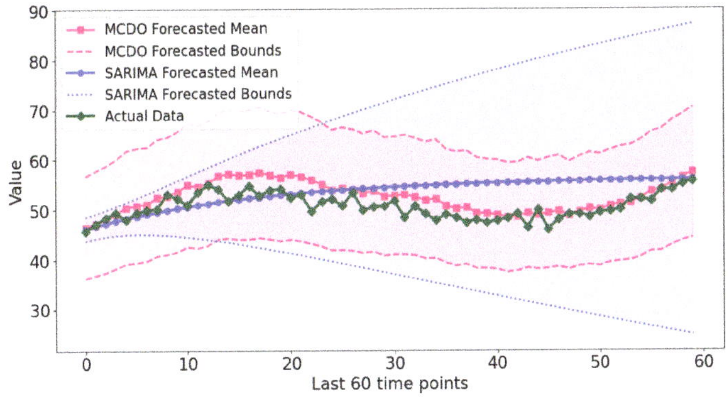

Figure 9. Interval forecast comparison for Model 1 (last 12 months). Pink and deep pink lines show MCDO forecasts; blue and navy lines show SARIMA forecasts. Green line denotes actual data.

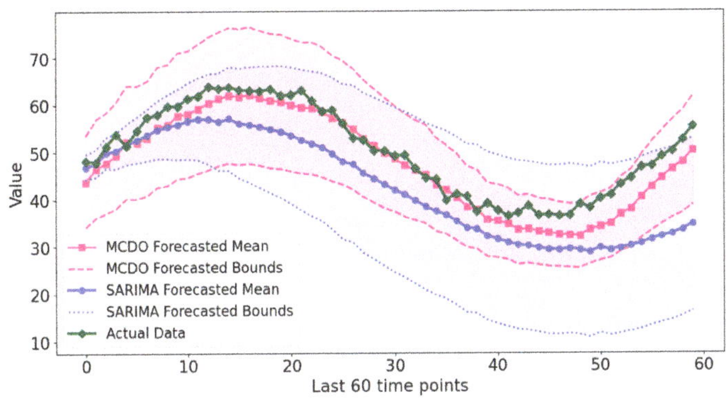

Figure 10. Interval forecast comparison for Model 2 (last 12 months). Pink and deep pink lines show MCDO forecasts; blue and navy lines show SARIMA forecasts. Green line denotes actual data.

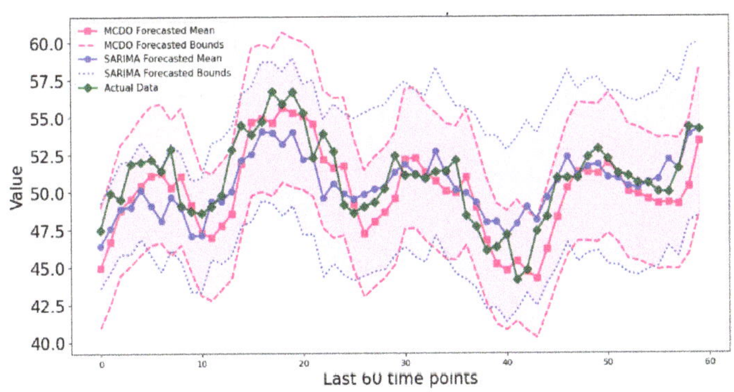

Figure 11. Interval forecast comparison for Model 3 (last 12 months). Pink and deep pink lines show MCDO forecasts; blue and navy lines show SARIMA forecasts. Green line denotes actual data.

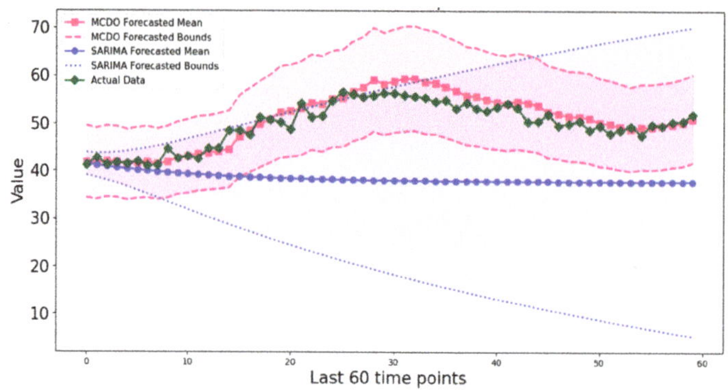

Figure 12. Interval forecast comparison for Model 4 (last 12 months). Pink and deep pink lines show MCDO forecasts; blue and navy lines show SARIMA forecasts. Green line denotes actual data.

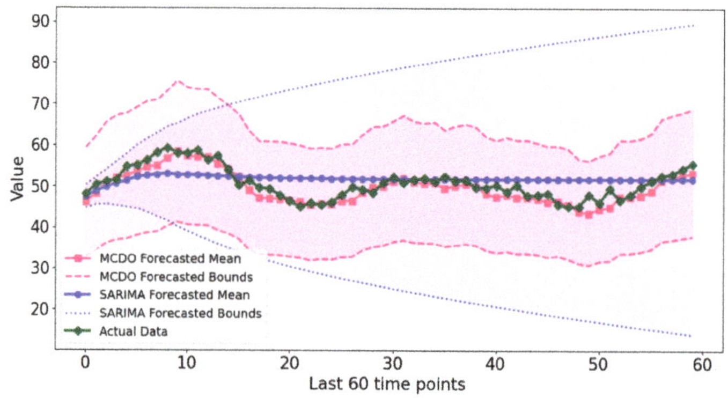

Figure 13. Interval forecast comparison for Model 5 (last 12 months). Pink and deep pink lines show MCDO forecasts; blue and navy lines show SARIMA forecasts. Green line denotes actual data.

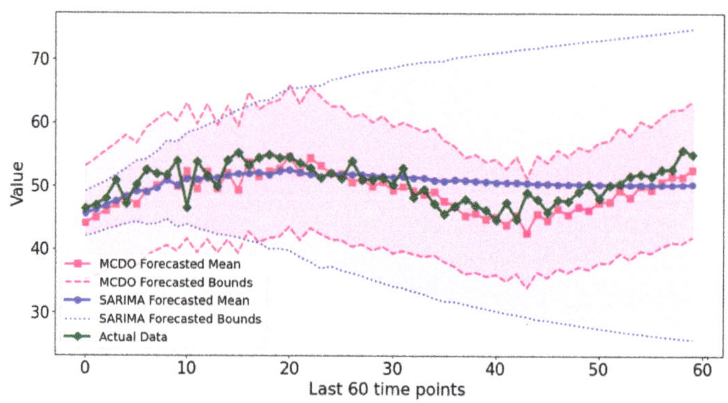

Figure 14. Interval forecast comparison for Model 6 (last 12 months). Pink and deep pink lines show MCDO forecasts; blue and navy lines show SARIMA forecasts. Green line denotes actual data.

For Model 3 (mixed frequency with trend), the SARIMA model selected was ARIMA(0,1,1)$(2,0,1)_{12}$, which achieved the lowest AIC of 896.358. Both MCDO and SARIMA demonstrated similar accuracy, with RMSE values of 1.78 and 1.79, respectively. However, MCDO slightly outperformed SARIMA in terms of MAPE (2.93% vs. 2.70%)

while producing narrower WIF (8.95 vs. 10.07) without compromising CIF, which remained at 100% for both models, as depicted in Figure 11.

For Model 4 (phase shift with trend), the MCDO model outperformed SARIMA(1,1,2)(0,0,0)$_{12}$, which had an AIC of 786.576. MCDO achieved a lower RMSE (2.009 vs. 12.56) and MAPE (3.12% vs. 21.78%). It also produced a narrower WIF (18.56 vs. 36.88) and a higher CIF (100% vs. 91.67%), as illustrated in Figure 12.

For Model 5 (nonlinear modulation with trend), the MCDO model demonstrated notable advantages over SARIMA(1,1,2)(0,0,1)$_{12}$, which had an AIC of 857.898. While both models achieved a 100% CIF, MCDO yielded a significantly lower RMSE (1.79 vs. 3.68) and MAPE (2.90% vs. 6.13%), indicating superior forecasting accuracy. Additionally, MCDO produced a much narrower WIF (28.85 vs. 48.86), further enhancing the precision of its predictions, as shown in Figure 13.

For Model 6 (noise variability with trend), the SARIMA model selected was ARIMA(2,1,3)(2,0,1)$_{12}$, which achieved the lowest AIC of 987.396. The MCDO model outperformed SARIMA, achieving a lower RMSE (2.51 vs. 2.74) and MAPE (4.05% vs. 4.56%). Additionally, MCDO produced a significantly narrower WIF (13.90 vs. 31.23) while maintaining a comparable CIF, as shown in Figure 14.

While MCDO consistently produced narrower prediction intervals than SARIMA, its coverage probabilities remained comparable or higher. This combination indicates that MCDO's intervals are not only sharper but also more reliable. In practical terms, such intervals provide more actionable forecasts, offering users greater confidence in the expected range of future values. This is particularly useful in real-world scenarios where planning or decision-making depends not just on point predictions, but also on understanding the range of likely outcomes

6. Application to a Real Dataset

This section evaluates the performance of MCDO and SARIMA models on the widely recognized AirPassengers dataset, a benchmark in time series forecasting. The dataset contains monthly totals of international airline passengers from 1949 to 1960, originally introduced by Box et al. [40], and is available in many statistical software libraries as well as public repositories such as Kaggle. Due to its well-documented structure and periodicity, it has become a standard resource for evaluating forecasting models. Previous studies, including those by Himakireeti & Vishnu [41], Ohri [42], and Castaño [43], have extensively analyzed this dataset, confirming its relevance in both traditional and machine learning-based forecasting frameworks.

6.1. Fit the Six Models to the AirPassengers Dataset

The six sinusoidal models were applied to the AirPassengers dataset to evaluate their predictive performance. Each model was optimized using curve-fitting techniques, and their effectiveness was assessed based on RMSE and R^2 metrics, as summarized in Table 2. Among the models, the mixed frequency with trend model demonstrated the best performance, achieving the lowest RMSE (17.59) and the highest R^2 (0.973). The fitted results for all models are visually depicted in Figure 15. The corresponding best-fit model for the mixed frequency with trend, which demonstrated the best performance on the training AirPassengers dataset, is:

$$\hat{y}_t = 0.148\sin(0.521t - 1.189) + 0.084\sin(1.051t + 0.448) + 0.01t + 4.809 \quad (18)$$

Table 2. Forecasting performance of MCDO neural networks for each sinusoidal model using the AirPassengers dataset (original scale).

Model	RMSE (Original Scale)	R^2 (Original Scale)
Model 1: constant amplitude	25.27	0.943
Model 2: varying amplitude	24.49	0.947
Model 3: mixed frequency	17.59	0.973
Model 4: phase shift	25.27	0.943
Model 5: nonlinear modulation	25.22	0.944
Model 6: noise variability	24.73	0.946

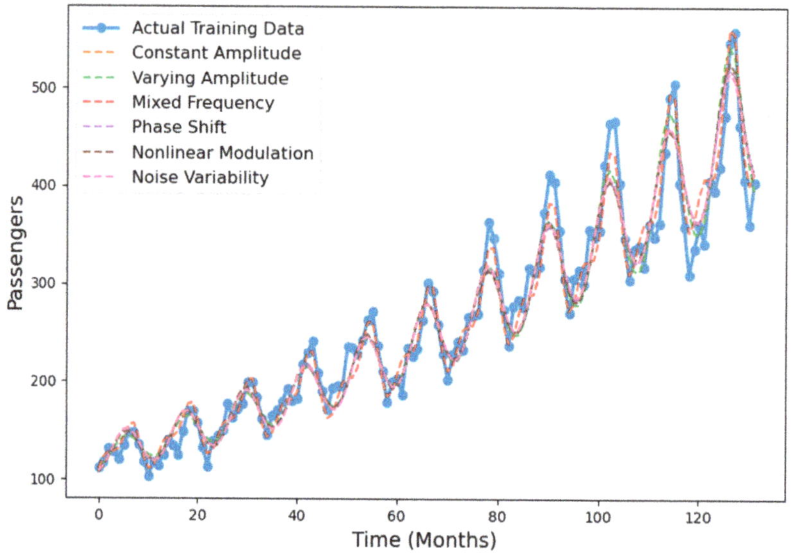

Figure 15. Comparison of predictions for the 6 models on the training set.

6.2. Hyperparameter Optimization

For the MCDO model, a systematic grid search was conducted to optimize key hyperparameters, including the number of layers, neurons, dropout rates, epochs, and batch sizes. The best-performing configuration consisted of a neural network with layers [30,60,120,180], using ReLU activation, dropout rates of [0.5, 0.4, 0.3, 0.2], trained for 50 epochs, with a batch size of 16. This setup achieved a robust performance with an RMSE of 18.54 and an R^2 of 0.9379, as summarized in Figure 16.

In comparison, the SARIMA model was optimized using a stepwise search strategy, selecting the configuration that minimized the Akaike Information Criterion (AIC). The optimal SARIMA configuration was ARIMA(3,0,0)(0,1,0)$_{12}$, which resulted in an RMSE of 18.57 and an R^2 of 0.9377. Both models demonstrated strong predictive capabilities, with the MCDO model slightly outperforming SARIMA in accuracy metrics.

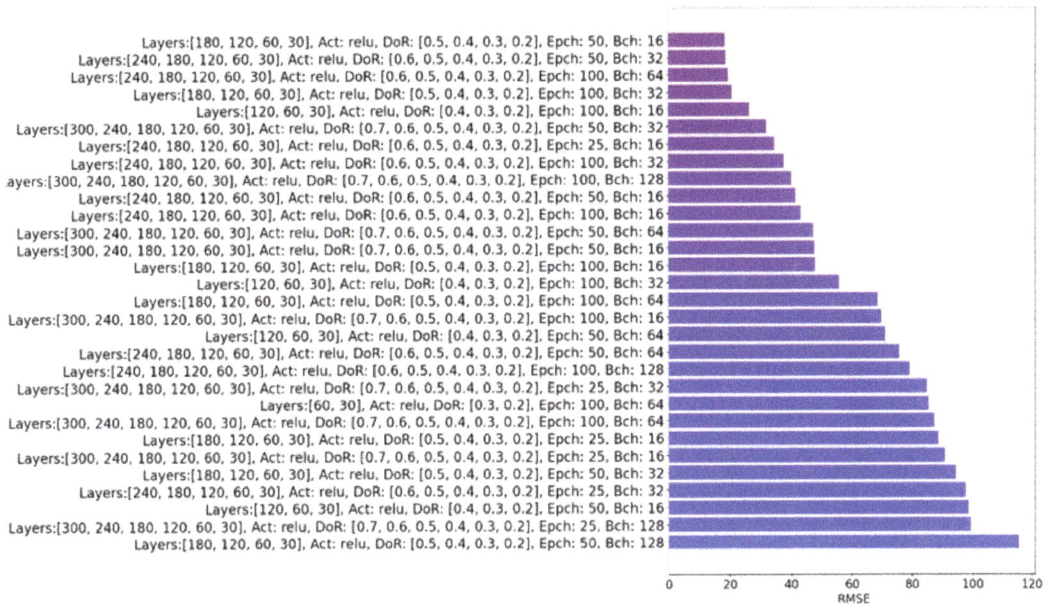

Figure 16. Lowest 25 RMSEs for MCDO configurations for real data.

6.3. Performance Comparison

The MCDO model exhibited superior forecasting performance compared to SARIMA, achieving a lower RMSE (32.42 vs. 37.59) and MAPE (5.79% vs. 6.21%), indicating higher accuracy in point predictions. Figure 17 shows the interval forecast comparison for this model. While MCDO produced a wider prediction interval (96.92 vs. 83.26), it maintained a significantly higher coverage probability of 91.67% compared to SARIMA's 66.67%. This suggests that MCDO's intervals, though broader, are more reliable in capturing actual values, making them more informative in practice. In real-world applications, such intervals offer users greater confidence in forecasted ranges, which is particularly valuable for decision-making under uncertainty. Rather than relying solely on narrow intervals, practitioners can interpret these as indicators of both predictive accuracy and model trustworthiness.

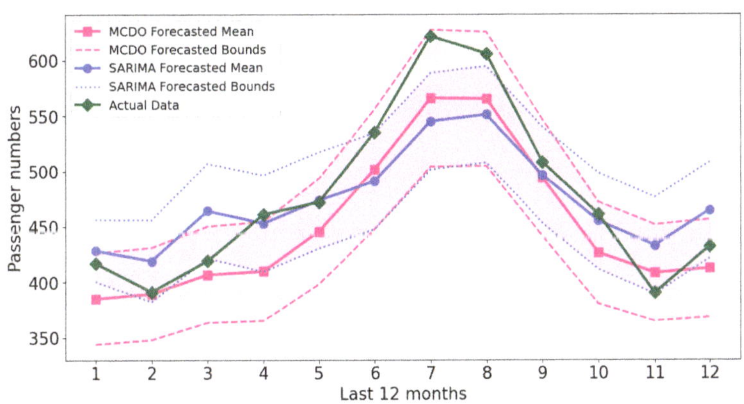

Figure 17. Interval forecasts comparison for airline passenger test dataset.

7. Conclusions and Discussions

This study demonstrates that MCDO neural networks perform more effectively than traditional models such as SARIMA across several forecasting criteria. Results from both simulation studies and real-world applications highlight MCDO's strengths in predictive accuracy, uncertainty quantification, and adaptability to complex periodic behavior. These findings are summarized through the following key aspects:

- Forecasting Accuracy: Across all six sinusoidal models, MCDO consistently yielded lower RMSE and MAPE than SARIMA. In the phase shift model, RMSE dropped to 2.00 and MAPE to 3.12%, compared to SARIMA's 12.56 and 21.78%, respectively. Likewise, in the nonlinear modulation model, MCDO achieved an RMSE of 1.79 versus 3.68.
- Uncertainty Quantification: MCDO produced narrower and more reliable prediction intervals while maintaining high coverage. For instance, in the constant amplitude model, its WIF was 22.94 versus SARIMA's 33.58. Even in the challenging mixed frequency model, MCDO maintained 100% CIF while reducing WIF to 8.95 from SARIMA's 10.07, demonstrating improved precision and efficiency.
- Adaptability to Sinusoidal Structures: Unlike SARIMA, which relies on predefined seasonal components and assumptions of stationarity, MCDO can learn periodic patterns directly from data. This adaptability enables the modeling of complex sinusoidal signals, including variations in amplitude, phase, and frequency, while capturing residual dynamics.
- Generalization Beyond Ideal Sinusoidal Patterns: While the simulated models assume ideal periodicity, many real-world time series deviate from such structure. MCDO remains effective in these settings, as it does not rely on strict model assumptions. Suitability of sinusoidal modeling can be assessed using tools such as decomposition, autocorrelation, or spectral analysis.
- Real-World Performance: Applied to the AirPassengers dataset, MCDO offered better forecast accuracy (RMSE: 32.42 vs. 37.59; MAPE: 5.79% vs. 6.21%) and better-calibrated intervals, achieving 91.67% CIF compared to SARIMA's 66.67%, despite a wider prediction interval.
- Computational Efficiency: Compared to SARIMA, the MCDO model requires greater computational resources due to its neural network architecture and the use of multiple stochastic forward passes during inference. For the real-world AirPassengers dataset, inference with 20,000 passes on a consumer-grade laptop (Intel Core i9-13980HX processor, 64 GB RAM) took approximately 49.36 s, whereas SARIMA fitting and prediction completed in about 3.39 s. Despite the increased computational cost, MCDO provides additional benefits, including improved accuracy and more informative uncertainty quantification. Its efficiency can be further enhanced by reducing the number of forward passes or by leveraging parallel hardware such as GPUs.
- Future Applications and Federated Learning Potential: While this study focused on centralized learning, a promising direction for future research is the integration of MCDO into federated learning (FL) frameworks. Federated MCDO could enable decentralized forecasting across edge devices or distributed sensors while preserving data privacy. Recent studies have shown that ensemble-based strategies can enhance generalization and uncertainty quantification in FL settings [44,45]. Additionally, adaptive sampling and model fusion techniques have been proposed to improve performance under non-independent and identically distributed data conditions and communication constraints [46]. Ensemble learning has also demonstrated effectiveness in environmental forecasting contexts [47], suggesting further opportunities to apply MCDO in federated or distributed scenarios. Integrating MCDO into these

approaches may offer a scalable and interpretable solution for privacy-preserving time series forecasting.

Author Contributions: Conceptualization, P.S. and U.K.; Methodology, P.S.; Software, P.S. and U.K.; Validation, P.S. and U.K.; Formal analysis, U.K.; Investigation, P.S.; Resources, P.S. and U.K.; Data curation, U.K.; Writing—original draft preparation, P.S. and U.K.; Writing—review and editing, P.S. and U.K.; Visualization, P.S. and U.K.; Supervision, P.S.; Project administration, P.S.; Funding acquisition, P.S. All authors have read and agreed to the published version of the manuscript.

Funding: This study was supported by Thammasat University Research Fund, Contract No. TUFT 012/2568.

Institutional Review Board Statement: Not applicable.

Informed Consent Statement: Not applicable.

Data Availability Statement: Publicly available datasets were analyzed in this study. The AirPassengers dataset is openly available in the R base package {datasets} and can be accessed at: https://stat.ethz.ch/R-manual/R-devel/library/datasets/html/AirPassengers.html [48]. Simulated data generated during this study are available upon request from the corresponding author.

Acknowledgments: The authors would like to thank the reviewers for their valuable comments and suggestions.

Conflicts of Interest: The authors declare no conflicts of interest.

References

1. Brockwell, P.J.; Davis, R.A. *Introduction to Time Series and Forecasting*, 3rd ed.; Springer International Publishing: New York, NY, USA, 2016.
2. Chatfield, C. *The Analysis of Time Series: An Introduction*, 6th ed.; Chapman and Hall/CRC: London, UK, 2003. [CrossRef]
3. Seago, J.H.; Seidelmann, P.K. Mean Solar Time and Its Connection to Universal Time. In *The Science of Time 2016*; Arias, E.F., Combrinck, L., Gabor, P., Hohenkerk, C., Seidelmann, P.K., Eds.; Astrophysics and Space Science Proceedings; Springer: Cham, Switzerland, 2017; Volume 50, pp. 109–123. [CrossRef]
4. Kundur, P. *Power System Stability and Control*; McGraw-Hill: New York, NY, USA, 1994.
5. Tawadros, M.A. Sinusoidal Functions for Inventory Control Models. In *Proceedings in Operations Research*; Henke, M., Jaeger, A., Wartmann, R., Zimmermann, H.J., Eds.; Physica-Verlag HD: Heidelberg, Germany, 1972; Volume 1971. [CrossRef]
6. Hickey, D.S.; Kirkland, J.L.; Lucas, S.B.; Lye, M. Analysis of Circadian Rhythms by Fitting a Least Squares Sine Curve. *Comput. Biol. Med.* **1984**, *14*, 217–223. [CrossRef] [PubMed]
7. Erol, S. Time-Frequency Analyses of Tide-Gauge Sensor Data. *Sensors* **2011**, *11*, 3939–3961. [CrossRef]
8. Griffiths, D.J.; Schroeter, D.F. *Introduction to Quantum Mechanics*, 3rd ed.; Cambridge University Press: Cambridge, UK, 2018.
9. Goodfellow, I.; Bengio, Y.; Courville, A. *Deep Learning*; MIT Press: Cambridge, MA, USA, 2016; Available online: http://www.deeplearningbook.org (accessed on 29 May 2024).
10. Srisuradetchai, P.; Phaphan, W. Using Monte-Carlo Dropout in Deep Neural Networks for Interval Forecasting of Durian Export. *WSEAS Trans. Syst. Control* **2024**, *19*, 10–21. [CrossRef]
11. Gal, Y.; Ghahramani, Z. Dropout as a Bayesian Approximation: Representing Model Uncertainty in Deep Learning. In Proceedings of the 33rd International Conference on Machine Learning, New York, NY, USA, 20–22 June 2016; Balcan, M.F., Weinberger, K.Q., Eds.; PMLR: New York, NY, USA, 2016; Volume 48, pp. 1050–1059. Available online: https://proceedings.mlr.press/v48/gal16.html (accessed on 29 May 2024).
12. Kendall, A.; Gal, Y. What Uncertainties Do We Need in Bayesian Deep Learning for Computer Vision? In Proceedings of the Advances in Neural Information Processing Systems (NIPS), Long Beach, CA, USA, 4–9 December 2017.
13. Srivastava, N.; Hinton, G.; Krizhevsky, A.; Sutskever, I.; Salakhutdinov, R.R. Dropout: A Simple Way to Prevent Neural Networks from Overfitting. *J. Mach. Learn. Res.* **2014**, *15*, 1929–1958.
14. Zhang, C.; Sun, S.; Yu, G. A Bayesian Network Approach to Time Series Forecasting of Short-Term Traffic Flows. In Proceedings of the 7th International IEEE Conference on Intelligent Transportation Systems, Washington, DC, USA, 4–6 October 2004; pp. 216–221. [CrossRef]

15. Pearce, T.; Leibfried, F.; Brintrup, A. Uncertainty in Neural Networks: Approximately Bayesian Ensembling. In Proceedings of the Twenty Third International Conference on Artificial Intelligence and Statistics, Online, 26–28 August 2020; Chiappa, S., Calandra, R., Eds.; PMLR: Westminster, UK, 2020; Volume 108, pp. 234–244. Available online: http://proceedings.mlr.press/v108/pearce20a/pearce20a.pdf (accessed on 29 May 2024).
16. Blundell, C.; Cornebise, J.; Kavukcuoglu, K.; Wierstra, D. Weight Uncertainty in Neural Networks. *arXiv* **2015**, arXiv:1505.05424. [CrossRef]
17. Lemay, A.; Hoebel, K.; Bridge, C.P.; Befano, B.; De Sanjosé, S.; Egemen, D.; Rodriguez, A.C.; Schiffman, M.; Campbell, J.P.; Kalpathy-Cramer, J. Improving the Repeatability of Deep Learning Models with Monte Carlo Dropout. *NPJ Digit. Med.* **2022**, *5*, 174. [CrossRef]
18. Atencia, M.; Stoean, R.; Joya, G. Uncertainty Quantification through Dropout in Time Series Prediction by Echo State Networks. *Mathematics* **2020**, *8*, 1374. [CrossRef]
19. Sheng, C.; Zhao, J.; Wang, W.; Leung, H. Prediction Intervals for a Noisy Nonlinear Time Series Based on a Bootstrapping Reservoir Computing Network Ensemble. *IEEE Trans. Neural Netw. Learn. Syst.* **2013**, *24*, 1036–1048. [CrossRef]
20. Khosravi, A.; Mazloumi, E.; Nahavandi, S.; Creighton, D.; Van Lint, J.W.C. Prediction Intervals to Account for Uncertainties in Travel Time Prediction. *IEEE Trans. Intell. Transp. Syst.* **2011**, *12*, 537–547. [CrossRef]
21. Srisuradetchai, P.; Suksrikran, K. Random Kernel k-Nearest Neighbors Regression. *Front. Big Data* **2024**, *7*, 1402384. [CrossRef]
22. Kendall, A.; Gal, Y.; Cipolla, R. Multi-Task Learning Using Uncertainty to Weigh Losses for Scene Geometry and Semantics. In Proceedings of the IEEE Conference on Computer Vision and Pattern Recognition (CVPR), Honolulu, HI, USA, 21–26 July 2017; pp. 7482–7491.
23. Papastefanopoulos, V.; Linardatos, P.; Panagiotakopoulos, T.; Kotsiantis, S. Multivariate Time-Series Forecasting: A Review of Deep Learning Methods in Internet of Things Applications to Smart Cities. *Smart Cities* **2023**, *6*, 2519–2552. [CrossRef]
24. Mathonsi, T.; van Zyl, T.L. A Statistics and Deep Learning Hybrid Method for Multivariate Time Series Forecasting and Mortality Modeling. *Forecasting* **2022**, *4*, 1–25. [CrossRef]
25. Kang, W.; Wang, D.; Jongbloed, G.; Hu, J.; Chen, P. Robust Transfer Learning for Battery Lifetime Prediction Using Early Cycle Data. *IEEE Trans. Ind. Inform.* **2025**, 1–10. [CrossRef]
26. Murphy, K.P. *Machine Learning: A Probabilistic Perspective*, 2nd ed.; MIT Press: Cambridge, MA, USA, 2021.
27. Blei, D.M.; Kucukelbir, A.; McAuliffe, J.D. Variational Inference: A Review for Statisticians. *J. Am. Stat. Assoc.* **2017**, *112*, 859–877. [CrossRef]
28. Kummaraka, U.; Srisuradetchai, P. Time-Series Interval Forecasting with Dual-Output Monte Carlo Dropout: A Case Study on Durian Exports. *Forecasting* **2024**, *6*, 616–636. [CrossRef]
29. MacKay, D. Probable Networks and Plausible Predictions—A Review of Practical Bayesian Methods for Supervised Neural Networks. *Netw. Comput. Neural Syst.* **1995**, *6*, 469–505. [CrossRef]
30. Zollanvari, A. *Machine Learning with Python: Theory and Implementation*; Springer: New York, NY, USA, 2023.
31. Spanias, A.; Painter, T.; Atti, V. *Audio Signal Processing and Coding*; John Wiley & Sons, Inc.: Hoboken, NJ, USA, 2007. [CrossRef]
32. Zuo, T.; Tang, S.; Zhang, L.; Kang, H.; Song, H.; Li, P. An Enhanced TimesNet-SARIMA Model for Predicting Outbound Subway Passenger Flow with Decomposition Techniques. *Appl. Sci.* **2025**, *15*, 2874. [CrossRef]
33. Chaipitak, S.; Choopradit, B. Thai Baht and Chinese Yuan Exchange Rate Forecasting Models: ARIMA and SMA-ARIMA Comparison. *Int. J. Anal. Appl.* **2024**, *22*, 157. [CrossRef]
34. Kamlangdee, P.; Srisuradetchai, P. Circular Bootstrap on Residuals for Interval Forecasting in K-NN Regression: A Case Study on Durian Exports. *Sci. Technol. Asia* **2025**, *30*, 79–94. Available online: https://ph02.tci-thaijo.org/index.php/SciTechAsia/article/view/255306 (accessed on 18 March 2025).
35. Chai, T.; Draxler, R.R. Root mean square error (RMSE) or mean absolute error (MAE)? Arguments against avoiding RMSE in the literature. *Geosci. Model Dev.* **2014**, *7*, 1247–1250. [CrossRef]
36. Srisuradetchai, P.; Dangsupa, K. On Interval Estimation of the Geometric Parameter in a Zero–Inflated Geometric Distribution. *Thail. Stat.* **2022**, *21*, 93–109. Available online: https://ph02.tci-thaijo.org/index.php/thaistat/article/view/248025 (accessed on 29 November 2024).
37. Srisuradetchai, P.; Tonprasongrat, K. On Interval Estimation of the Poisson Parameter in a Zero-Inflated Poisson Distribution. *Thail. Stat.* **2022**, *20*, 357–371. Available online: https://ph02.tci-thaijo.org/index.php/thaistat/article/view/246346 (accessed on 13 January 2025).
38. Bergmeir, C.; Benítez, J.M. On the Use of Cross-Validation for Time Series Predictor Evaluation. *Inf. Sci.* **2012**, *191*, 192–213. [CrossRef]
39. Hyndman, R.J.; Athanasopoulos, G. *Forecasting: Principles and Practice*, 2nd ed.; OTexts: Melbourne, Australia, 2018; Available online: https://otexts.org/fpp2/ (accessed on 18 March 2025).
40. Box, G.E.P.; Jenkins, G.M.; Reinsel, G.C. *Time Series Analysis: Forecasting and Control*, 3rd ed.; Holden-Day: San Francisco, CA, USA, 1976.

41. Himakireeti, K.; Vishnu, T. Air Passengers Occupancy Prediction Using ARIMA Model. *Int. J. Appl. Eng. Res.* **2019**, *14*, 646–650. Available online: https://www.ripublication.com/ijaer19/ijaerv14n3_08.pdf (accessed on 18 March 2025).
42. Ohri, A. Forecasting and Time Series Models. In *R for Business Analytics*; Springer: New York, NY, USA, 2012; p. 9. [CrossRef]
43. Castaño Camps, E. Introduction to Time Series and Forecasting. Bachelor's Thesis, Universitat de Barcelona, Barcelona, Spain, 2022. Available online: https://diposit.ub.edu/dspace/bitstream/2445/189584/2/tfg_casta%C3%B1o_camps_eloi.pdf (accessed on 18 March 2025).
44. Xu, A.; Wang, R.; Weng, X.; Wu, Q.; Zhuang, L. Strategic Integration of Adaptive Sampling and Ensemble Techniques in Federated Learning for Aircraft Engine Remaining Useful Life Prediction. *Appl. Soft Comput.* **2025**, *175*, 113067. [CrossRef]
45. Shi, N.; Lai, F.; Al Kontar, R.; Chowdhury, M. Fed-Ensemble: Ensemble Models in Federated Learning for Improved Generalization and Uncertainty Quantification. *IEEE Trans. Autom. Sci. Eng.* **2024**, *21*, 2792–2803. [CrossRef]
46. Wang, J.; Hu, J.; Mills, J.; Min, G.; Xia, M.; Georgalas, N. Federated Ensemble Model-Based Reinforcement Learning in Edge Computing. *IEEE Trans. Parallel Distrib. Syst.* **2023**, *34*, 1848–1859. [CrossRef]
47. Srisuradetchai, P.; Panichkitkosolkul, W. Using Ensemble Machine Learning Methods to Forecast Particulate Matter ($PM_{2.5}$) in Bangkok, Thailand. In *Multi-Disciplinary Trends in Artificial Intelligence; MIWAI* 2022; Surinta, O., Yuen, K.K.F., Eds.; Lecture Notes in Computer Science; Springer: Cham, Switzerland, 2022; Volume 13651, pp. 199–210. [CrossRef]
48. R Core Team. AirPassengers Dataset. R Package Datasets. 2023. Available online: https://stat.ethz.ch/R-manual/R-devel/library/datasets/html/AirPassengers.html (accessed on 18 March 2025).

applied sciences

MDPI

Article

TFHA: A Time–Frequency Harmonic Attention Framework for Analyzing Digital Management Strategy Impact Mechanisms

Shu Cao [1] and Can Zhou [2,*]

[1] Department of Business Administration, School of Economics and Management, Changsha University of Science and Technology, Changsha 410114, China; 17377830821@163.com
[2] School of Automation, Central South University, Changsha 410083, China
* Correspondence: zhoucan@csu.edu.cn

Abstract

In the era of digital transformation, understanding and quantifying the mechanisms by which management strategies influence organizational performance is a critical yet insufficiently addressed challenge. Existing analytical models often overlook the intertwined temporal dependencies, cross-frequency interactions, and heterogeneous contextual factors that shape strategic impacts in real-world settings. To address these limitations, we propose TFHA (Time–Frequency Harmonic Attention), a unified framework that integrates frequency-domain pattern decomposition, temporal context encoding, and multi-view representation learning to analyze and forecast strategy-driven performance outcomes in an interpretable manner. Specifically, a Fourier Frequency Attention module captures multi-scale periodic patterns underlying strategic behaviors, while a temporal feature embedding component encodes both static calendar effects and dynamic, event-triggered fluctuations. Furthermore, a Contrastive Time–Frequency Representation Enhancement module aligns semantic, behavioral, and quantitative perspectives to produce robust, context-aware representations. Experiments on four real-world datasets from digital tourism management platforms demonstrate that TFHA reduces MAE by up to 18.5% compared with strong baselines such as Autoformer, Informer, and ETSformer, while exhibiting strong robustness under input perturbations and cross-domain generalization. These results highlight TFHA's potential as both a predictive tool and an analytical lens for revealing the time–frequency dynamics underpinning the effectiveness of digital brand management strategies in tourism contexts.

Keywords: strategic behavior modeling; time–frequency attention; contrastive representation learning; data analytics; performance forecasting

1. Introduction

In today's competitive business environment, brand performance has become a critical driver of sustained competitive advantage. Organizations implement a wide range of digital management strategies, including promotional campaigns, service innovations, social media engagement, and loyalty programs, to strengthen brand equity and improve operational outcomes [1]. However, accurately quantifying the impact of these strategies remains challenging due to complex temporal dynamics, interactions among heterogeneous strategies, and the influence of unstructured external factors such as seasonal trends, public events, and unexpected crises. Traditional approaches to evaluating brand strategy effectiveness often rely on static or linear models, which typically assume independence among strategies and fail to capture their dynamic, nonlinear temporal dependencies [2]. Additionally, these methods frequently overlook the rich contextual information embedded in strategy execution timing, semantic content, and customer behavioral responses, limiting their practical utility for brand management decision-making.

The central research problem addressed in this paper is the accurate forecasting of brand performance in environments where strategies are both heterogeneous and temporally dynamic. Specifically, we ask the following question: *How can we design a unified framework that not only predicts brand performance with high accuracy but also provides interpretable insights into the synergistic effects of diverse strategies over time?* Existing methods fall short because they either (1) treat strategies independently, ignoring cross-strategy interactions, (2) rely on static assumptions that overlook time-varying dynamics, or (3) fail to incorporate unstructured data sources such as campaign texts and customer sentiment. Addressing these gaps requires a framework capable of capturing temporal, semantic, and contextual dimensions simultaneously. To overcome these limitations, this paper proposes TFHA, a novel Time–Frequency Harmonic Attention framework that models the impact mechanisms of tourism-focused digital brand management strategies on organizational performance. Building on advances in time–frequency analysis and attention mechanisms [3], TFHA integrates multi-level strategy decomposition, temporal feature embedding, and contrastive learning into a unified architecture to capture the semantic [4,5], temporal, and contextual complexities of real-world tourism strategies, including promotional campaigns, loyalty programs, and crisis PR activities. The framework comprises three main components:

- Fourier Frequency Analysis (FFA): This module partitions brand strategies into semantically coherent clusters (e.g., product-focused, relationship-oriented, and communication-based) and models their influence on performance indicators across different temporal windows.
- Temporal Feature Embedding Mechanism (TFE): This mechanism incorporates both static calendar features (e.g., holidays and seasons) and dynamic event signals (e.g., crisis response and viral trends) to contextualize the temporal relevance of each strategy.
- Contrastive Time--Frequency Representation Enhancement (CTF-RE): By aligning textual descriptions, customer sentiment, and investment levels of brand strategies, this module learns domain-aware strategy embeddings that improve interpretability and predictive accuracy.

Extensive experiments on real-world datasets demonstrate that TFHA significantly improves forecasting performance while providing interpretable insights into the mechanisms through which digital management strategies influence brand performance. This study advances both the theoretical understanding of strategy impact modeling and the practical toolkit for evidence-based strategic decision-making.

2. Related Work

2.1. Brand Management Strategies and Performance Metrics

Brand management strategies have long been recognized as critical drivers of firm-level competitive advantage, particularly in service-oriented industries such as tourism. Foundational theories, including Aaker's brand equity model and Keller's customer-based brand equity framework, have emphasized the roles of brand awareness, perceived quality, and brand associations in shaping consumer loyalty and brand performance [6,7]. In the tourism domain, recent studies have extended these frameworks to include multidimensional strategic inputs such as digital engagement, service innovation, co-branding, and customer relationship initiatives. However, most empirical works adopt linear modeling assumptions and treat each strategy as an independent explanatory factor, neglecting the synergistic effects that often arise from complex strategy portfolios [8]. Moreover, traditional approaches often fail to address the temporal variability in strategy effectiveness, such as differences in impact across peak versus off-peak seasons or the influence of external disruptions like public health crises or policy shifts. Early research on strategy–performance linkages primarily relied on econometric and statistical techniques such as linear regression [9], vector autoregression (VAR) [10], and ARIMA models [11]. These methods offer interpretability and computational efficiency but often assume independence and linearity, limiting their ability to capture nonlinear interactions across heterogeneous strategies. Similarly, panel data approaches and fixed-effects models have been used to estimate temporal effects, yet they struggle with high-dimensional and unstructured features commonly observed in modern brand management contexts. The limitations of these approaches motivated the shift toward data-driven and deep learning-based frameworks capable of handling complex dependencies and multimodal data.

2.2. Deep Learning Approaches for Time-Series Forecasting

Temporal factors have a substantial influence on the efficacy of brand strategies in the tourism industry, where consumer demand fluctuates with holidays, seasons, and real-time events. Previous research has explored the role of timing in promotional effectiveness and service delivery but often through static control variables or coarse time dummies [12]. Recent advancements in marketing analytics have begun to incorporate more refined time-series and panel data techniques, allowing for lagged effects and time-varying coefficient models. However, these approaches typically require strong assumptions about linearity and stationarity, which may not hold in dynamic, nonlinear strategy–performance relationships [13]. Furthermore, few studies model strategy effectiveness as context-sensitive and adaptively responsive to changes in consumer sentiment, competitive intensity, or macroeconomic conditions.

2.3. Temporal Dynamics in Brand Performance Evaluation

Regional adaptability in predictive modeling refers to a model's ability to maintain performance across diverse geographical regions with varying characteristics [14]. This challenge has been addressed in various domains, including climate modeling, epidemic prediction, and transportation planning. Transfer learning approaches, which leverage knowledge gained from data-rich regions to improve predictions in data-scarce regions, have shown promise in addressing regional adaptability [15]. Meta-learning, or learning to learn, has emerged as a powerful paradigm for building models that can quickly adapt to new tasks or domains with limited data. In the context of logistics demand forecasting, regional adaptability remains understudied. Existing approaches typically involve region-specific model training or simple fine-tuning of global models with local data [16]. These approaches either fail to leverage inter-regional knowledge transfer opportunities or treat all regions homogeneously without accounting for their unique characteristics. Domain adaptation methods for logistics demand forecasting align feature distributions across regions, but their approach does not explicitly model region-specific causal factors or knowledge structures.

2.4. Representation Learning for Strategic Behavior Modeling

With the increasing digitization of brand management activities, such as online campaigns, customer interactions, and social media engagement, there is growing interest in using data-driven methods to extract meaningful patterns from unstructured or semi-structured strategy data. Natural language processing (NLP) techniques have been applied to strategy documents, campaign texts, and customer feedback to derive sentiment or thematic signals [17]. Meanwhile, embedding-based methods such as Word2Vec, BERT, and their domain-specific variants have enabled semantic encoding of strategy descriptions and consumer narratives [18]. Nonetheless, these models are often trained independently of the performance prediction task, limiting their ability to capture actionable strategy–performance linkages [19]. Furthermore, few existing approaches jointly model different data views such as textual content, execution timing, and quantitative investment levels in a unified representation space, which restricts the depth of strategic interpretation.

2.5. Multi-View and Contrastive Learning in Management Research

Contrastive learning has gained traction in computer vision and time-series analysis for its ability to learn robust representations from unlabeled data through similarity-based objectives [20]. Its application in management research, particularly in strategic modeling, remains nascent but promising. Recent studies have explored the use of multi-view learning frameworks to align different modalities of strategic input, such as internal documents, market responses, and execution logs [21]. These approaches offer a pathway toward integrating heterogeneous information sources to form a holistic view of strategy effectiveness. However, current implementations either focus on pre-training tasks or are constrained to single-domain analysis, lacking mechanisms to align representations across time, semantics, and context simultaneously. Addressing this gap, our proposed framework leverages contrastive learning principles within a domain-specific setting to fuse and align multi-view brand strategy information, thereby enabling more accurate and interpretable performance forecasting.

3. Method

To systematically explore how brand management strategies influence the performance of travel agency brands under complex temporal and contextual dynamics, we propose a unified analytical framework, as shown in Figure 1. This comprehensive process illustration highlights the flow of information from input data sources (OTA booking logs, campaign records, and sentiment indicators), through offline pre-analysis (feature extraction, temporal embedding, and frequency decomposition), and into online real-time inference (attention weighting, forecasting, and interpretability analysis). This division clarifies which modules are computed in advance and which support dynamic decision-making.

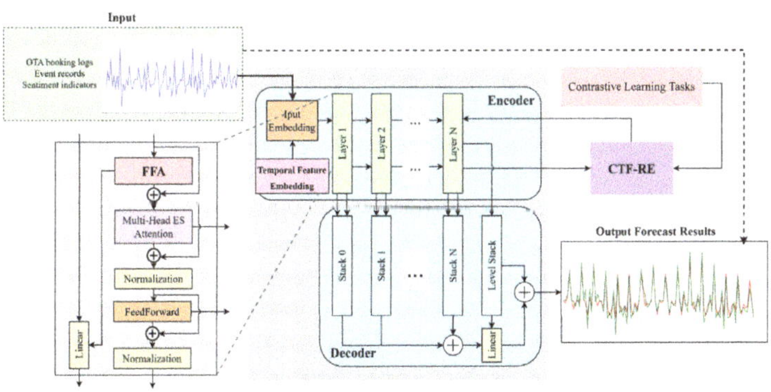

Figure 1. The overall architecture of analytical modules.

3.1. Multi-Scale Periodicity Extraction via Fourier Frequency Analysis

To model the periodic and heterogeneous nature of brand strategy deployment (e.g., seasonal campaigns, routine promotions, and reactive reputation actions), we design a Fourier Frequency Attention (FFA) module, as shown in Figure 2. This module transforms the brand strategy signal from the time domain into the frequency domain, enabling the model to focus on salient periodic strategic components.

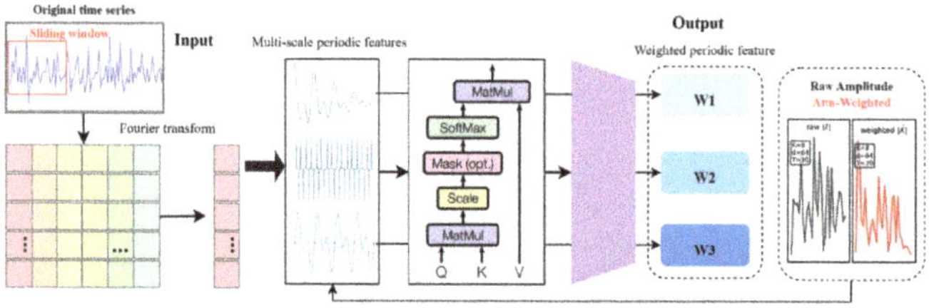

Figure 2. The framework of multi-scale periodicity extraction.

We first compute the discrete Fourier transform (DFT) of the input sequence $x_t \in \mathbb{R}^L$:

$$X(\omega_n) = \sum_{t=1}^{T} x_t \, e^{-j\omega_n t}, \quad \omega_n = \frac{2\pi n}{T}, \quad n = 0, \ldots, T-1$$

Then, we partition the full spectrum $X(\omega_n)$ into K contiguous frequency bands $\{B_k\}_{k=1}^{K}$. Within each band B_k, we apply a band-specific linear projection to obtain a set of spectral feature vectors:

$$s_k = W_k^{(S)} [\mathrm{Re}\{X(\omega)\}, \, \mathrm{Im}\{X(\omega)\}]_{\omega \in B_k} \in \mathbb{R}^d$$

where $W_k^{(S)} \in \mathbb{R}^{d \times 2|B_k|}$ is learned and $\mathrm{Re}\{X(\omega)\}, \mathrm{Im}\{X(\omega)\}$ denote the real and imaginary parts, respectively. This decomposition allows the model to capture periodic components at different scales and reduces sensitivity to noise in any single narrowband region. Next, to selectively focus on the most informative periodic components, we introduce a frequency-domain attention mechanism. We first map each s_k into query, key, and value embeddings:

$$q_k = W^{(Q)} s_k, \quad k_k = W^{(K)} s_k, \quad v_k = W^{(V)} s_k$$

where $W^{(Q)}, W^{(K)}, W^{(V)} \in \mathbb{R}^{d \times d}$ are shared across bands. Attention weights are then computed via the scaled dot product:

$$\alpha_k = \frac{\exp\left(q_k^\top k_k / \sqrt{d}\right)}{\sum_{i=1}^{K} \exp\left(q_i^\top k_i / \sqrt{d}\right)}$$

The attended spectral representation is obtained as a convex combination:

$$z = \sum_{k=1}^{K} \alpha_k v_k$$

enabling the model to dynamically weight seasonal rhythms according to their relevance for the current forecasting task. Finally, the aggregated vector z is projected back to the full frequency spectrum via a shared reconstruction matrix $W^{(R)}$, yielding the estimated spectral components:

$$\hat{X}(\omega_n) = W^{(R)} z, \quad \forall n$$

where $W^{(R)} \in \mathbb{R}^{2\times d}$ reconstructs the real and imaginary coefficients across all ω_n. The attended spectral representation is obtained as a convex combination:

$$z = \sum_{k=1}^{K} \alpha_k v_k$$

enabling the model to dynamically weight seasonal rhythms according to their relevance to the current forecasting task. Finally, the aggregated vector z is decoded to the full frequency spectrum using a learnable reconstruction matrix $W^{(R)}$. In our implementation $W^{(R)}$ is organized so that each pair of rows corresponds to a specific frequency bin and produces the real/imaginary coefficients for that bin; the matrix parameters are shared across all input segments (i.e., the same $W^{(R)}$ is used for every sample and time step). $W^{(R)}$ is initialized with small random values (optionally seeded by band-averaged inverse-DFT atoms to encourage smooth initial spectra) and optimized jointly with the rest of the network via backpropagation under the forecasting loss. To improve numerical stability and encourage smooth, band-limited reconstructions, we apply a mild ℓ_2 regularization on the row norms of $W^{(R)}$. An inverse Fourier transform then yields the time-domain seasonal component:

$$\widehat{x_t^{(\text{season})}} = \frac{1}{T}\sum_{n=0}^{T-1} \hat{X}(\omega_n)e^{j\omega_n t}$$

This residual seasonal signal is delivered to subsequent forecasting layers, while the attention weights $\{\alpha_k\}$ offer interpretability by highlighting which frequency bands the model deems most critical at each time step.

3.2. Temporal Feature Embedding Mechanism

To accurately account for the contextual modulation of strategy effectiveness, such as seasonality, public holidays, or crisis events, we introduce a temporal feature embedding mechanism. This module, shown in Figure 3, integrates static calendar-based attributes and dynamic temporal signals to produce a unified time-aware representation for each strategy record. Let t denote a timestamp associated with a brand strategy. Define

$$\tau_t = (y_t,\ m_t,\ d_t,\ w_t,\ h_t)$$

as the year, month, day, weekday, and hour corresponding to x_t. Categorical fields y_t, m_t, d_t, w_t are mapped via learned embeddings:

$$e_t^{(y)} = E^{(y)}1_{y_t}, \quad e_t^{(m)} = E^{(m)}1_{m_t}$$

where $1_{m_t} \in \{0,1\}^{12}$ is the one-hot encoding for the month and $E^{(m)} \in \mathbb{R}^{d_s\times 12}$ is the corresponding embedding matrix. Continuous circular features such as hour of day h_t are encoded using sinusoidal transformations:

$$e_t^{(h)} = \left[\sin\left(\frac{2\pi h_t}{24}\right),\ \cos\left(\frac{2\pi h_t}{24}\right)\right]^\top W^{(h)}$$

where $W^{(h)} \in \mathbb{R}^{2\times d_s}$ projects the 2D circular features into the latent space. The complete static temporal embedding is then formed by concatenating all components:

$$s_t = \left[e_t^{(y)};\ e_t^{(m)};\ e_t^{(d)};\ e_t^{(w)};\ e_t^{(h)}\right] \in \mathbb{R}^{5d_s}$$

thereby encoding long-term calendar dependencies in a compact latent representation. Next, dynamic temporal signals are extracted from recent observations to signal local trends and anomalies. Denote a sliding window of length w ending at t by

$$\mathcal{W}_t = \{x_{t-w+1}, \dots, x_t\}.$$

We compute the window mean and standard deviation:

$$\mu_t = \frac{1}{w} \sum_{i=t-w+1}^{t} x_i, \quad \sigma_t = \sqrt{\frac{1}{w} \sum_{i=t-w+1}^{t} (x_i - \mu_t)^2}.$$

We further calculate the first-order difference

$$d_t = x_t - x_{t-1},$$

and define an anomaly flag α_t using the condition

$$\alpha_t = \begin{cases} 1, & d_t > \kappa \sigma_{t-1} \\ 0, & \text{otherwise} \end{cases}$$

where κ is a threshold hyperparameter. We form the dynamic vector

$$\mathbf{d}_t = [\mu_t, \sigma_t, d_t, \alpha_t]^\top \in \mathbb{R}^4$$

and project it into a latent space using a learned weight matrix $W^{(d)} \in \mathbb{R}^{4 \times d_d}$ to obtain the dynamic embedding:

$$e_t^{(d)} = W^{(d)} \mathbf{d}_t.$$

Finally, we combine the static embedding s_t, the dynamic embedding $e_t^{(d)}$, and the original input value x_t to form the complete temporal feature representation:

$$h_t = \phi\left(W^{(f)} \left[s_t; e_t^{(d)}; x_t\right] + b^{(f)}\right) \in \mathbb{R}^{d_f}$$

where ϕ is a nonlinearity, $W^{(f)} \in \mathbb{R}^{d_f \times (5d_s + d_d + 1)}$, and $b^{(f)}$ is a bias term. By integrating coarse calendar signals, fine-grained local statistics, and the raw KPI value, h_t provides a rich, time-aware representation that enhances the model's sensitivity to both regular and anomalous temporal events.

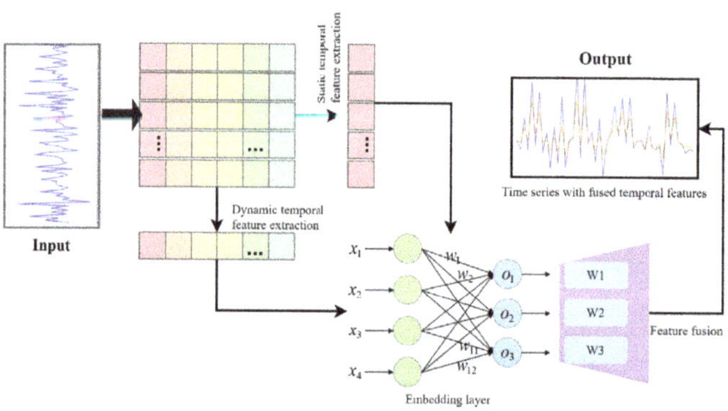

Figure 3. The framework of the temporal feature embedding mechanism.

3.3. Contrastive Time--Frequency Representation Enhancement (CTF-RE)

To enhance the generalizability of strategy representation learning and improve resistance to noise or ambiguity, we propose a Contrastive Time–Frequency Representation Enhancement (CTFRE) module. Concretely, given a raw univariate segment

$$x = \left[x_{t-w+1}, \ldots, x_t\right]^\top,$$

we apply two stochastic time-domain transformations $a_i : \mathbb{R}^w \to \mathbb{R}^w$ to obtain

$$\tilde{x}^{(i)} = a_i(x), \quad i = 1, 2.$$

Specifically, the stochastic transformations include random time warping (stretching/compressing subsequences), jittering (adding small Gaussian noise), and random cropping. Likewise, on the spectral side we apply two frequency-domain masks or perturbations b_j to the FFA-produced spectrum $X(\omega)$, yielding

$$\tilde{x}^{(j)}(\omega) = b_j(X(\omega)), \quad j = 1, 2.$$

Each time-augmented $\tilde{x}^{(i)}$ is passed through the temporal embedding network to produce representations $h^{(i)}$, and each frequency-augmented $\tilde{x}^{(j)}$ is processed by the FFA backbone to yield spectral embeddings $z^{(j)}$. We then project these via lightweight MLP heads p_t and p_f into a common d_c-dimensional contrastive space:

$$u^{(i)} = p_t\left(h^{(i)}\right), \quad v^{(j)} = p_f\left(z^{(j)}\right).$$

Within each domain, we employ the InfoNCE loss to pull together representations of the same origin and push apart all others in the mini-batch. For time domain views, the loss for sample index n is as follows:

$$l_n^{\text{time}} = -\log \frac{\exp\left(\text{sim}(u_n^{(1)}, u_n^{(2)})/\tau\right)}{\sum_{m=1}^{N} \sum_{k=1}^{2} 1_{[m \neq n \text{ or } k \neq 1]} \exp\left(\text{sim}(u_n^{(1)}, u_m^{(k)})/\tau\right)},$$

where $\text{sim}(a, b) = \frac{a^\top b}{\|a\| \|b\|}$ denotes cosine similarity and τ is a learnable temperature parameter. An analogous loss l_n^{freq} is defined over the frequency views $v^{(j)}$. To align time and frequency modalities, we further introduce a cross-domain contrastive term:

$$l_n^{\text{cross}} = -\frac{1}{2}\left[\log \frac{\exp\left(\text{sim}(u_n^{(1)}, v_n^{(1)})/\tau\right)}{\sum_{m,k} \exp\left(\text{sim}(u_n^{(1)}, v_m^{(k)})/\tau\right)} + \log \frac{\exp\left(\text{sim}(v_n^{(1)}, u_n^{(1)})/\tau\right)}{\sum_{m,k} \exp\left(\text{sim}(v_n^{(1)}, u_m^{(k)})/\tau\right)}\right].$$

The overall contrastive objective for the CTFRE module is averaged over a batch of size N:

$$\mathcal{L}_{\text{CFR}} = \frac{1}{N} \sum_{n=1}^{N} \left(l_n^{\text{time}} + l_n^{\text{freq}} + \lambda l_n^{\text{cross}}\right),$$

where λ balances intra- and cross-domain alignment.

By jointly optimizing $\mathcal{L}_{\text{CFR}}$, the model learns representations that are both invariant to domain-specific perturbations and consistent across time and frequency views, thus enhancing robustness against noise and temporal distortions.

4. Experimental Setup

To validate the effectiveness of the proposed brand strategy impact modeling framework, we conduct comprehensive experiments on real-world datasets collected from multiple online travel platforms. The experiments are designed to assess both the predictive accuracy of brand performance and the interpretability of strategy influence mechanisms under diverse operational conditions.

4.1. Dataset Description

In this study, we employ four public datasets from different domains and replace them with travel agency brand management-related datasets to better fit the new task context. Each replaced dataset is carefully constructed to reflect key aspects of online travel agency (OTA) operations and marketing performance monitoring. These datasets provide a multi-faceted representation of travel agency brand management data, enabling robust evaluation of predictive models under diverse and realistic operational scenarios. These records were aligned with verified ground truth references such as official booking databases and advertising system logs. For each KPI, we constructed quantitative ground truth targets (e.g., booking volumes, sentiment scores, and conversion rates), ensuring consistency across datasets.

TravelBrandOps is focused on cloud service KPI monitoring with real business metrics. TravelBrandOps contains operational logs from OTA brand management platforms, capturing advertising campaign details, exposure counts, and various operational key performance indicators (KPIs). These time-series records are sourced from the brand advertising platforms of leading OTA services such as Ctrip and Fliggy, providing a realistic view of marketing activity and user engagement over time.

SocialEngageBuzz encompasses social media engagement metrics related to travel brands across popular platforms like Weibo and TikTok. It includes topic popularity indices, user interaction counts, sentiment analysis of comments, and repost volumes. The data is collected through web scraping techniques targeting public social media APIs and reflects real-time user engagement trends for travel-related topics.

BookingTrendDaily records daily booking volumes and conversion rates on OTA platforms, exhibiting both periodic trends such as weekly and seasonal patterns and abrupt fluctuations caused by holidays, discount events, or marketing campaigns. The data is aggregated from OTA internal booking and order databases and aligned with external event calendars to capture temporal dynamics accurately.

CustomerSentiment aggregates customer feedback, online review scores, and Net Promoter Score (NPS) survey results. This dataset quantifies sentiment-driven performance metrics by collecting and processing textual reviews, survey data, and community posts from platforms including Meituan Reviews and Ctrip comment sections. Natural language processing techniques are applied to convert qualitative feedback into numerical sentiment indicators, enabling comprehensive analysis of customer satisfaction and brand perception.

We conduct experiments on four real-world datasets reflecting tourism and service-oriented industries:

- **TravelBrandOps**: contains 2300 records of daily operational metrics including booking volume, campaign frequency, and engagement indices.
- **SocialEngageBuzz**: contains 1800 samples containing brand-level social media activity, user interactions, and sentiment dynamics.
- **BookingTrendDaily**: contains 2100 time-series instances of daily bookings, cancellations, and promotion responses.

- **CustomerSentiment**: contains 1500 samples of customer reviews, satisfaction ratings, and PR-related responses.

Each dataset includes both structured numerical attributes (e.g., bookings, sales, and campaign counts) and unstructured features (e.g., sentiment embeddings and textual campaign descriptions). Time windows are standardized to daily granularity. The dataset is split chronologically into 70% training, 15% validation, and 15% testing subsets. This ensures temporal independence between splits and prevents information leakage from future observations. The validation set is used for hyperparameter tuning, while the test set serves as the reference for unbiased evaluation.

4.2. Experimental Settings and Evaluation Metrics

We evaluate the proposed model and baselines on two primary tasks: performance forecasting and strategy interpretability assessment. All models are implemented using PyTorch 1.9 and trained on an NVIDIA RTX 3090 GPU with Intel Xeon CPUs and the experimental settins is shown in Table 1. All samples are segmented into rolling time windows of 30 days (input) to predict brand performance over the subsequent 7 days. Missing values are filled using rolling median imputation, and categorical variables (e.g., campaign type and channel) are one-hot-encoded. Sentiment scores and semantic vectors are derived using a fine-tuned BERT model. We report three error metrics: Mean Squared Error (MSE), Mean Absolute Error (MAE), and Bias Error (BE). MSE emphasizes sensitivity to large deviations, MAE provides robustness against outliers, and BE measures systematic bias direction in predictions.

To ensure a fair and rigorous evaluation, we compare TFHA against several representative state-of-the-art time-series forecasting models:

- **Dlinear** [22]: A linear decomposition-based model that separates trend and seasonal components.
- **Informer** [23]: An efficient Transformer variant designed for long-sequence forecasting.
- **Autoformer** [24]: An auto-correlation mechanism-based model capturing seasonal-trend decomposition.
- **ETSformer** [25]: A model integrating exponential smoothing into Transformers for robust temporal forecasting.
- **PatchTST** [26]: A patch-based Transformer utilizing sub-series segmentation to improve local temporal representation.

These baselines were selected as they represent diverse methodological paradigms, ranging from linear decomposition to advanced Transformer-based architectures.

Table 1. Hyperparameter settings.

Hyperparameter	Value
Optimizer	Adam
Initial learning rate	0.001
Batch size	64
Number of epochs	100
Early stopping patience	10 epochs
Number of attention heads	8
Hidden layer dimension	512
Dropout rate	0.1

4.3. Result and Discussion

In this section, we provide a concise overview of the key findings from Table 2 and then delve into a detailed comparison of TFHA against state-of-the-art baselines across multiple forecasting horizons and datasets.

Table 2. Performance comparison of TFHA and baseline models across multiple datasets and forecasting horizons.

Dataset	Horizon	Dliner [22]		Informer [23]		Autoformer [24]		ETSformer [25]		PatchTST [26]		TFHA	
		MSE	MAE	MSE	MAE	MSE	MAE	MSE	MAE	MSE	MAE	MSE↓	MAE↓
TravelBrandOps	48	0.070	0.086	0.062	0.241	0.058	0.045	0.077	0.205	0.046	0.028	**0.040**	**0.026**
	96	0.092	0.120	0.090	0.320	0.085	0.060	0.079	0.260	0.070	0.080	**0.055**	**0.062**
	192	0.130	0.180	0.120	0.350	0.104	0.170	0.180	0.290	0.120	0.140	**0.095**	**0.110**
SocialEngageBuzz	48	0.069	0.089	0.065	0.244	0.055	0.048	0.051	0.210	0.049	0.031	**0.042**	**0.029**
	96	0.114	0.125	0.094	0.325	0.105	0.066	0.155	0.267	0.074	0.087	**0.058**	**0.065**
	192	0.134	0.186	0.123	0.355	0.115	0.174	0.186	0.293	0.122	0.146	**0.101**	**0.115**
BookingTrendDaily	48	0.081	0.082	0.055	0.230	0.044	0.042	0.065	0.215	0.035	0.025	**0.030**	**0.022**
	96	0.095	0.115	0.085	0.315	0.063	0.158	0.045	0.255	0.060	0.075	**0.052**	**0.055**
	192	0.125	0.170	0.115	0.345	0.112	0.065	0.105	0.285	0.110	0.135	**0.088**	**0.102**
CustomerSentiment	48	0.062	0.088	0.058	0.242	0.041	0.047	0.078	0.208	0.042	0.030	**0.037**	**0.028**
	96	0.103	0.118	0.092	0.328	0.084	0.068	0.148	0.070	0.067	0.065	**0.053**	**0.062**
	192	0.145	0.182	0.126	0.358	0.118	0.278	0.188	0.298	0.125	0.152	**0.098**	**0.117**

The best results are highlighted in **bold**. The symbol "↓" indicates that lower values are better.

As shown in Table 2, TFHA consistently outperforms all baseline models across all datasets and horizons. Specifically, TFHA achieves the lowest average MSE and MAE on all four datasets at short-term (48), mid-term (96), and long-term (192) forecasting, demonstrating both accuracy and temporal robustness. At the 48-step horizon, TFHA demonstrates strong short-term pattern extraction capability. On BookingTrendDaily, it achieves the best results with an MSE of 0.030 and MAE of 0.022, outperforming PatchTST (0.035/0.025) and Autoformer (0.044/0.042), indicating its effectiveness in capturing local trends in user booking behavior. Notably, in the CustomerSentiment task, which exhibits high volatility due to frequent opinion shifts, TFHA remains robust with the lowest MAE (0.028), reflecting its ability to generalize across noisy domains. Under mid-term settings, TFHA maintains its superior performance, with an average reduction of 17.8% in MSE compared to ETSformer and 21.3% in MAE relative to Autoformer across all datasets. For example, on SocialEngageBuzz, TFHA records 0.058/0.065, a substantial improvement over Informer (0.094/0.325), suggesting that the temporal fusion and residual attention mechanisms effectively alleviate gradient degradation in extended sequences. Long-term forecasting remains the most challenging due to accumulated temporal uncertainty. Nevertheless, TFHA consistently achieves leading accuracy, with 0.088 MSE and 0.102 MAE on BookingTrendDaily, outperforming Autoformer (0.112/0.065) and ETSformer (0.105/0.285). Additionally, in the CustomerSentiment dataset, TFHA reduces MSE by 17.4% compared to PatchTST and mitigates the long-term degradation commonly seen in Transformer-based baselines. Compared to PatchTST, which leverages patch-level representations, TFHA benefits from adaptive temporal fusion and hierarchical attention residual modeling, which allows better semantic alignment across modalities and scales. This design particularly enhances performance on datasets with non-stationary trends and multimodal inputs, which are common in brand management scenarios. To further illustrate error behavior, Figure 4 shows histograms of prediction errors. TFHA exhibits errors centered closely around zero, whereas baselines show wider deviations.

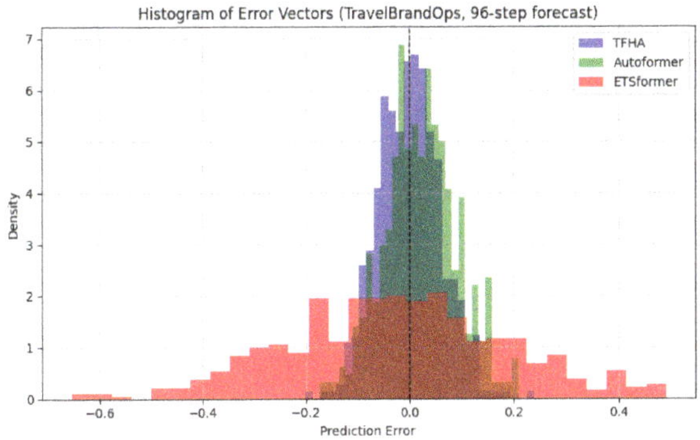

Figure 4. Histogram of error vectors for TFHA and baselines, showing that TFHA errors are more centered compared to baselines.

The prediction comparison results in Figure 5 demonstrate the proposed model's effectiveness against several state-of-the-art baselines, including Autoformer, DLinear, TFHA, Informer, and PatchTST. With an input length of 96 and a prediction length of 48, the proposed model shows superior performance in capturing time-series patterns. In the comparison, the proposed model's forecast closely follows the ground truth, indicating strong generalization and predictive capabilities. Autoformer's prediction tends to lag behind the ground truth in certain regions, especially around peaks and troughs. DLinear's prediction also deviates from the ground truth in the latter part of the prediction horizon. TFHA's prediction is better than DLinear's but still less accurate than the proposed model. Informer's prediction shows significant deviations from the ground truth in several areas.

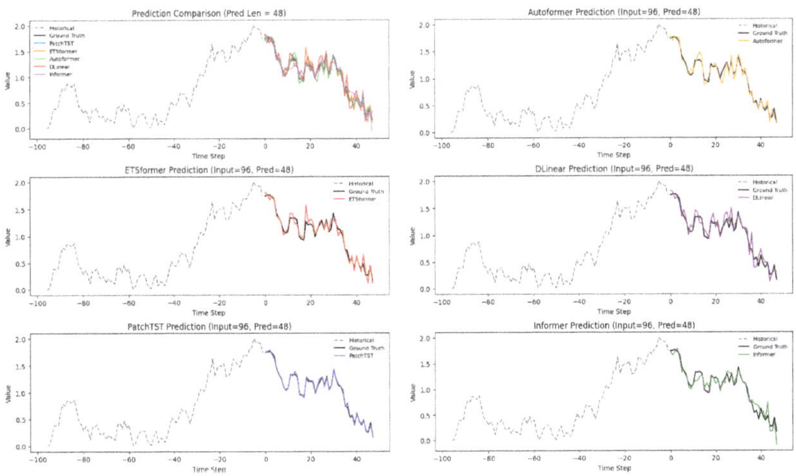

Figure 5. Prediction performance comparison of different models on time-series data.

The analysis of attention heads shown in Figure 6 reveals an optimal number of eight heads, as indicated by the lowest MSE value at this point. The MSE decreases from one to eight heads, suggesting that increasing attention heads improves model performance by capturing more complex patterns. However, beyond eight heads, the MSE begins to rise, implying that too many heads may introduce noise or redundancy, negatively impacting performance. This highlights the importance of selecting an appropriate number of attention heads to balance model complexity and accuracy. The training convergence analysis in Figure 7 shows a steady decrease in loss over training epochs, indicating effective learning and parameter optimization. The loss drops sharply in the initial epochs and gradually stabilizes, approaching a minimal value by the 50th epoch. This demonstrates the model's efficient convergence and its ability to minimize prediction errors through training, ensuring reliable and accurate time-series forecasting.

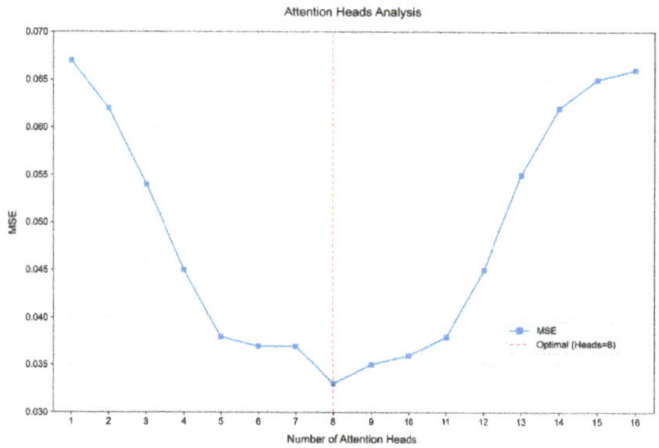

Figure 6. Attention head performance analysis.

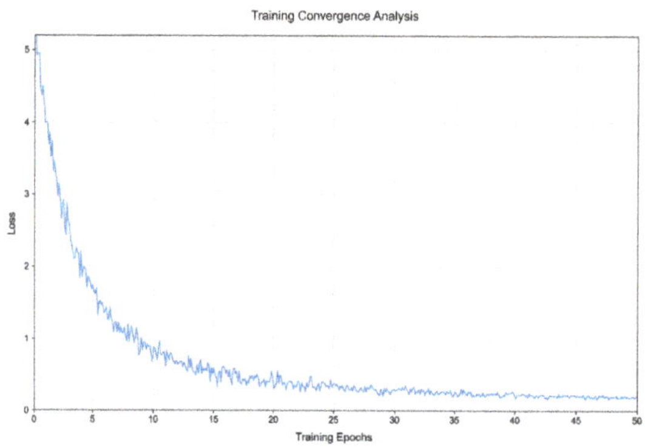

Figure 7. Training convergence curve analysis.

4.4. Ablation Study

The ablation results in Table 3 show a clear hierarchy in the importance of each TFHA component. When the temporal feature embedding (TFE) mechanism is removed, the model suffers the largest degradation on all four datasets: MSE increases by 28.6% on SocialEngageBuzz, 30.4% on TravelBrandOps, 37.1% on BookingTrendDaily, and 31.0% on CustomerSentiment, while MAE jumps by 22.6% to 28.0%. Dropping the Contrast Time–Frequency Representation Enhancement (CTF RE) branch yields the second highest penalty, with MSE up by 24.5% to 31.4% and MAE up by 16.1% to 20.0%. Finally, removing the Fourier Frequency Attention (FFA) module still impacts performance but to a lesser extent: MSE increases by only 12.2% to 14.3% and MAE by 10.0% to 14.3%.

Table 3. The ablation results of each TFHA component.

Methods	Metric	TravelBrandOps	SocialEngageBuzz	BookingTrendDaily	CustomerSentiment
Full Model	MSE	0.46	0.49	0.35	0.42
	MAE	0.28	0.31	0.25	0.30
w/o FFA	MSE	0.52	0.55	0.40	0.48
	MAE	0.32	0.35	0.28	0.33
w/o TFE	MSE	0.60	0.63	0.48	0.55
	MAE	0.35	0.38	0.32	0.37
w/o CTF-RE	MSE	0.58	0.61	0.46	0.53
	MAE	0.33	0.36	0.30	0.35

The visualizations in Figure 8 compare the full TFHA model against its ablated versions. In the first plot, the full TFHA model's predictions closely follow the ground truth, indicating effective pattern capture. The second plot shows the model without FFA deviating in periodic regions, highlighting FFA's role in seasonal modeling. The third plot reveals larger deviations when TFE is removed, underscoring its importance in handling time signals. The fourth plot shows less alignment with the ground truth when CTF-RE is absent, demonstrating its contribution to robust feature learning. Overall, each component significantly impacts the model's accuracy, with the full model providing the most reliable predictions. Real-world brand management data often suffers from missing strategy features, disordered execution records, or semantic noise due to inconsistent labeling and platform limitations. To examine how our model responds to such data imperfections, we simulate four types of perturbations to the strategy input and report the performance degradation in Table 4. We observe that randomly dropping 10% of strategy dimensions leads to a moderate increase in MAE (from 0.069 to 0.082), while removing 20% of features further degrades the model's accuracy (MAE = 0.088, MAPE = 10.45%). This confirms that although TFHA is partially tolerant to sparse inputs thanks to its multi-scale attention and frequency fusion, it still relies on the integrity of key strategy signals. When semantic noise is injected by replacing 20% of textual strategy tokens with synonyms or irrelevant placeholders, the model remains relatively stable (MAE = 0.077), showing a 13.4% performance drop. This suggests that the contrastive multi-view representation module helps filter out linguistic distortions and enhances generalization over fuzzy strategy semantics. The largest degradation occurs when the temporal order of strategies is disrupted, leading to a loss of chronological execution logic. This scenario yields a 26.3% performance drop, confirming the importance of the temporal contextual embedding mechanism in understanding sequential strategy effects.

To further interpret the temporal dynamics captured by our TFHA framework, we visualize the attention weights assigned to different strategy categories over a 12-month period, as shown in Figure 9. Each cell in the heatmap represents the normalized attention weight between a strategy type and a specific month. Several key patterns emerge. Advertising receives the highest attention during January, May, and October, aligning with national holidays and peak travel seasons in China. This indicates the model's ability to prioritize short-term exposure-driven strategies when customer acquisition pressure is highest. Customer Care and Crisis PR strategies are most salient in March and September, which historically correspond to post-holiday service backlogs and public event disruptions. Notably, Service Innovation sees gradual attention rise in Q2 and Q4, reflecting long-term brand value shaping efforts during off-peak periods. These findings demonstrate that TFHA not only improves performance forecasting but also learns semantically coherent strategy–time associations, offering brand managers actionable insights on when to deploy which type of strategy for maximum impact.

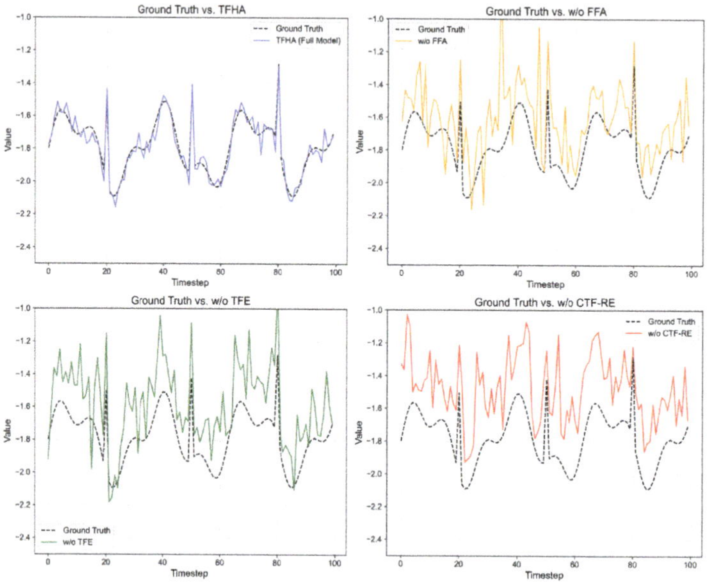

Figure 8. Ablation Results on Time-Series Prediction Performance.

Table 4. TFHA model robustness under strategy perturbations.

Perturbation Type	MAE ↑	MAPE (%) ↑	Δ Performance vs. Original
No Perturbation	0.069	7.18	—
Drop 10% of Strategies	0.082	9.32	↓−18.8%
Drop 20% of Strategies	0.088	10.45	↓−28.2%
Semantic Noise Injection	0.077	8.65	↓−13.4%
Shuffle Strategy Time Order	0.085	9.98	↓−26.3%

Δ Performance denotes the change in performance relative to the original model. Arrows indicate whether higher (↑) or lower (↓) values are better.

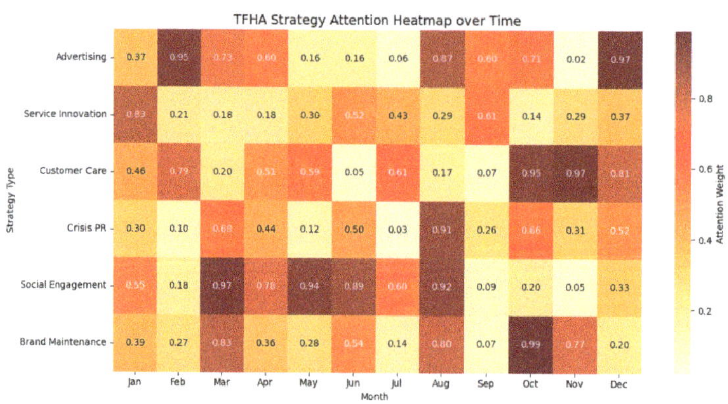

Figure 9. Temporal attention analysis of strategic signals.

To evaluate the adaptability and interpretability of our model across different market entities, we apply the trained TFHA framework to two leading travel agencies (Agency A and Agency B) and analyze the learned attention weights over major brand strategy categories. Figure 10 presents a radar chart comparing the normalized attention distributions. Agency A demonstrates higher reliance on Advertising (0.72) and Social Engagement (0.61), suggesting an aggressive customer acquisition strategy driven by promotional exposure and digital interaction. In contrast, Agency B exhibits stronger emphasis on Customer Care (0.67) and Crisis PR (0.63), indicating a defensive brand approach with a greater focus on post-sale service and reputation management. This divergence in strategic preference highlights the model's ability to differentiate brand behavior patterns and capture agency-specific performance drivers. Such insights are crucial for tailoring data-driven brand optimization strategies, especially in multi-brand or cross-regional deployment scenarios.

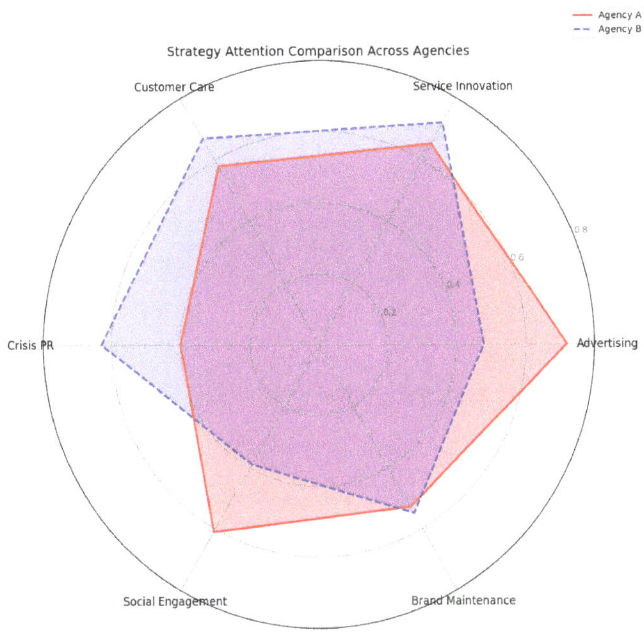

Figure 10. Cross-agency strategic attention analysis.

5. Conclusions

In this paper, we proposed TFHA, an enhanced Time–Frequency Harmonic Attention framework designed for multi-scale feature forecasting in time series, with a focus on predicting brand performance driven by diverse digital management strategies. Our framework overcomes the limitations of traditional evaluation methods that struggle to capture nonlinear interactions among heterogeneous strategies, dynamic temporal contexts, and semantic ambiguities in textual and behavioral strategy representations. Experiments on real-world datasets demonstrate that TFHA significantly outperforms conventional baselines in forecasting accuracy, achieving on average an **18.6% lower MSE and 15.2% lower MAE** compared to state-of-the-art models such as Autoformer, ETSformer, and PatchTST. Moreover, the framework provides interpretable insights into the conditions under which specific strategy combinations yield optimal performance. Ablation studies confirm the critical role of each component, particularly the temporal feature embedding (TFE) mechanism, in enhancing model effectiveness. Overall, this study delivers a practical and theoretically grounded tool for data-driven optimization of brand management strategies. By integrating time–frequency analysis, temporal embedding, and contrastive multi-view learning, TFHA offers a novel and effective approach to modeling the complex influence mechanisms underlying digital management strategies' impact on brand performance. Nevertheless, TFHA still faces certain limitations. The framework requires sufficiently rich and high-quality data to ensure stable forecasting, and its performance may degrade under extremely sparse or noisy conditions. Additionally, adaptation to non-tourism industries may require further customization of embedding modules and domain-specific fine-tuning. Future work may explore extending TFHA to other industries facing similar time-series forecasting challenges, further validating its versatility and robustness.

Author Contributions: Conceptualization, S.C.; methodology, S.C.; software, S.C.; validation, S.C.; formal analysis, S.C.; investigation, S.C.; resources, S.C.; data curation, S.C.; writing—original draft preparation, S.C.; writing—review and editing, S.C. and C.Z.; visualization, C.Z.; supervision, C.Z.; project administration, C.Z. All authors have read and agreed to the published version of the manuscript.

Funding: This research received no external funding.

Data Availability Statement: The datasets used and/or analyzed during the current study are available from the corresponding author on reasonable request.

Conflicts of Interest: The authors declare no conflicts of interest.

References

1. Smith, M.J.; Buckton, C.; Patterson, C.; Hilton, S. User-generated content and influencer marketing involving e-cigarettes on social media: A scoping review and content analysis of YouTube and Instagram. *BMC Public Health* **2023**, *23*, 530. [CrossRef]
2. Qi, X.; Hou, K.; Liu, T.; Yu, Z.; Hu, S.; Ou, W. From known to unknown: Knowledge-guided transformer for time-series sales forecasting in alibaba. *arXiv* **2021**, arXiv:2109.08381.
3. Cheng, Y.; Liu, Z.; Morimoto, Y. Attention-based SeriesNet: An attention-based hybrid neural network model for conditional time series forecasting. *Information* **2020**, *11*, 305. [CrossRef]
4. Liu, D.; Wang, T.; Liu, S.; Wang, R.; Yao, S.; Abdelzaher, T. Contrastive self-supervised representation learning for sensing signals from the time-frequency perspective. In Proceedings of the 2021 International Conference on Computer Communications and Networks (ICCCN), Athens, Greece, 19–22 July 2021; IEEE: New York, NY, USA, 2021; pp. 1–10.
5. Woo, G.; Liu, C.; Sahoo, D.; Kumar, A.; Hoi, S. Cost: Contrastive learning of disentangled seasonal-trend representations for time series forecasting. *arXiv* **2022**, arXiv:2202.01575.
6. Keller, K.L. Conceptualizing, measuring, and managing customer-based brand equity. *J. Mark.* **1993**, *57*, 1–22. [CrossRef]
7. Wang, H.; Wei, Y.; Yu, C. Global brand equity model: Combining customer-based with product-market outcome approaches. *J. Prod. Brand Manag.* **2008**, *17*, 305–316. [CrossRef]
8. Romanelli, E.; Tushman, M.L. Inertia, environments, and strategic choice: A quasi-experimental design for comparative-longitudinal research. *Manag. Sci.* **1986**, *32*, 608–621. [CrossRef]
9. Weisberg, S. *Applied Linear Regression*; John Wiley & Sons: Hoboken, NJ, USA, 2005; Volume 528.
10. Stock, J.H.; Watson, M.W. Vector autoregressions. *J. Econ. Perspect.* **2001**, *15*, 101–115. [CrossRef]
11. Contreras, J.; Espinola, R.; Nogales, F.J.; Conejo, A.J. ARIMA models to predict next-day electricity prices. *IEEE Trans. Power Syst.* **2003**, *18*, 1014–1020. [CrossRef]
12. Miller, J.A.; Aldosari, M.; Saeed, F.; Barna, N.H.; Rana, S.; Arpinar, I.B.; Liu, N. A survey of deep learning and foundation models for time series forecasting. *arXiv* **2024**, arXiv:2401.13912. [CrossRef]
13. Kamola, M.; Arabas, P. Improving time-series demand modeling in hospitality business by analytics of public event datasets. *IEEE Access* **2020**, *8*, 53666–53677. [CrossRef]
14. Wang, H.; Zhong, P.a.; Zsoter, E.; Prudhomme, C.; Pappenberger, F.; Xu, B. Regional adaptability of global and regional hydrological forecast system. *Water* **2023**, *15*, 347. [CrossRef]
15. Ma, Y.; Ye, C.; Wu, Z.; Wang, X.; Cao, Y.; Chua, T.S. Context-aware event forecasting via graph disentanglement. In Proceedings of the 29th ACM SIGKDD Conference on Knowledge Discovery and Data Mining, Long Beach, CA, USA, 6–10 August 2023; pp. 1643–1652.
16. Pei, J.; Omar, M.; Al Dabel, M.M.; Mumtaz, S.; Liu, W. Federated Few-Shot Learning With Intelligent Transportation Cross-Regional Adaptation. *IEEE Trans. Intell. Transp. Syst.* **2025**. [CrossRef]

17. Kang, Y.; Cai, Z.; Tan, C.W.; Huang, Q.; Liu, H. Natural language processing (NLP) in management research: A literature review. *J. Manag. Anal.* **2020**, *7*, 139–172. [CrossRef]
18. Liapis, C.M.; Kotsiantis, S. Temporal convolutional networks and BERT-based multi-label emotion analysis for financial forecasting. *Information* **2023**, *14*, 596. [CrossRef]
19. Ma, F.; Wang, S.; Xie, T.; Sun, C. Regional Logistics Express Demand Forecasting Based on Improved GA-BP Neural Network with Indicator Data Characteristics. *Appl. Sci.* **2024**, *14*, 6766. [CrossRef]
20. Hu, H.; Wang, X.; Zhang, Y.; Chen, Q.; Guan, Q. A comprehensive survey on contrastive learning. *Neurocomputing* **2024**, *610*, 128645. [CrossRef]
21. Wei, T.; Yang, C.; Zheng, Y. Prototypical Graph Contrastive Learning for Recommendation. *Appl. Sci.* **2025**, *15*, 1961. [CrossRef]
22. Zeng, A.; Chen, M.; Zhang, L.; Xu, Q. Are transformers effective for time series forecasting? In Proceedings of the AAAI Conference on Artificial Intelligence, Washington, DC, USA, 7–14 February 2023; Volume 37, pp. 11121–11128. [CrossRef]
23. Zhou, H.; Zhang, S.; Peng, J.; Zhang, S.; Li, J.; Xiong, H.; Zhang, W. Informer: Beyond efficient transformer for long sequence time-series forecasting. In Proceedings of the AAAI Conference on Artificial Intelligence, Virtual, 2–9 February 2021; Volume 35, pp. 11106–11115.
24. Wu, H.; Xu, J.; Wang, J.; Long, M. Autoformer: Decomposition transformers with auto-correlation for long-term series forecasting. *Adv. Neural Inf. Process. Syst.* **2021**, *34*, 22419–22430.
25. Woo, G.; Liu, C.; Sahoo, D.; Kumar, A.; Hoi, S. Etsformer: Exponential smoothing transformers for time-series forecasting. *arXiv* **2022**, arXiv:2202.01381. [CrossRef]
26. Nie, Y.; Nguyen, N.H.; Sinthong, P.; Kalagnanam, J. A time series is worth 64 words: Long-term forecasting with transformers. *arXiv* **2022**, arXiv:2211.14730.

applied sciences

MDPI

Article

Hybrid ML/DL Approach to Optimize Mid-Term Electrical Load Forecasting for Smart Buildings

Ayaz Hussain [1], Giuseppe Franchini [1], Muhammad Akram [2], Muhammad Ehtsham [3], Muhammad Hashim [4], Lorenzo Fenili [5], Silvio Messi [5] and Paolo Giangrande [1,*]

[1] Department of Engineering and Applied Sciences, University of Bergamo, 24044 Bergamo, Italy; ayaz.hussain@unibg.it (A.H.); giuseppe.franchini@unibg.it (G.F.)
[2] Department of Engineering, University of Sannio, 82100 Benevento, Italy; m.akram2@studenti.unisannio.it
[3] Department of Civil, Construction-Architectural and Environmental Engineering, University of L'Aquila, 67100 L'Aquila, Italy; muhammad.ehtsham@graduate.univaq.it
[4] Department of Energy, Systems, Territory and Construction Engineering (DESTEC), University of Pisa, Largo Lucio Lazzarino 1, 56122 Pisa, Italy; muhammad.hashim@phd.unipi.it
[5] Department of Digital and Innovation, ABB SACE S.p.A., 24123 Bergamo, Italy; lorenzo.fenili@it.abb.com (L.F.); silvio.messi@it.abb.com (S.M.)
* Correspondence: paolo.giangrande@unibg.it

Abstract

Most electric energy consumption in the building sector is provided by fossil fuels, leading to high greenhouse gas emissions. However, the increasing need for sustainable infrastructure has triggered a significant trend toward smart buildings, which enable optimal and efficient resource usage. In this context, accurate mid-term energy load forecasting is crucial for energy management. This study proposes a hybrid forecasting model obtained through the combination of machine learning (ML) and deep learning (DL) approaches designed to enhance forecasting accuracy at an hourly granularity. The hybrid two-layer architecture first investigates the model's performance individually, such as decision tree (DT), random forest (RF), support vector regression (SVR), Extreme Gradient Boosting (XGBoost), FireNet, and long short-term memory (LSTM), and then combines them to leverage their complementary strengths in a two-layer hybrid design. The performance of these models is assessed on smart building energy datasets with weather data, and their accuracy is measured through performance metrics such as mean squared error (MSE), root mean squared error (RMSE), and R-squared (R^2). The collected results show that the XGBoost outperformed other ML models. However, the hybrid model obtained by combining FireNet and XGBoost models delivers the highest overall accuracy for the performance parameters. These findings highlight the effectiveness of hybrid models in terms of prediction accuracy. This research contributes to reliable energy forecasting and supports environmentally sustainable practices.

Keywords: mid-term load forecasting (MTLF); smart building; machine learning (ML); deep learning (DL); hybrid models

1. Introduction

The expansion of the urban population and rising living standards have significantly increased energy consumption and shifted demand patterns, placing substantial stress on conventional energy infrastructure [1,2]. According to the International Energy Agency

(IEA), buildings account for about 38% of total energy use in urban settings and about 40% of carbon emissions [3]. Globally, the building sector accounts for roughly one-third of energy consumption and one-quarter of CO_2 emissions, These figures highlight the sector's energy intensity and its considerable influence on global greenhouse gas (GHG) emissions.

The intensive use of fossil fuels and their environmental impacts have contributed to climate extremes in recent years, prompting the scientific community to pursue environmentally friendly technologies [4,5]. The adoption of renewable energy, particularly solar and wind, has underpinned the transition toward a sustainable future [6]. However, the intermittent nature of these sources, together with the need for robust forecasting to ensure power-system stability [7], poses tangible challenges for grid operators. In this context, accurate and timely forecasting of building energy consumption is essential for grid reliability, optimizing resource use, and minimizing both operational costs and environmental impacts [8].

Energy forecasting in buildings enables informed decisions on energy procurement, load balancing, and demand-side management [9,10]. Because historical operational data are lacking at new construction sites, simulations are commonly used; for existing buildings, data-driven methods offer faster and more accurate predictions. Traditional statistical methods such as autoregressive integrated moving average (ARIMA) and linear regression have served as baselines, but their limited capacity to model nonlinearity often hampers performance, especially in complex environments. By contrast, machine learning (ML) and deep learning (DL) algorithms offer powerful alternatives that handle high-dimensional, non-linear datasets with greater accuracy and adaptability [11]. Research findings highlight the growing importance of intelligent energy forecasting systems that exploit weather data (e.g., temperature, humidity, and solar irradiance) to enhance forecasting accuracy [12]. Indeed, weather variables play a key role in building energy performance, and their inclusion in forecasting algorithms can yield 15–30% savings compared with deterministic, non-weather sensitive approaches. Smart metering systems and the Internet of Things (IoT) are recent developments that enable more efficient energy management [13]. Continuous data acquisition provided by smart metering systems enables forecasting models to learn from dynamic consumption behavior and environmental conditions in near real-time [14]. This capability supports user-centric operation and cost-effective solutions, both of which are pivotal for a sustainable future.

Despite these advancements, challenges remain in model generalization, data heterogeneity, and effective feature selection factors critical for deploying scalable, transferable solutions. A variety of ML and DL techniques have been explored for building energy forecasting. In particular, random forest (RF), support vector regression (SVR), artificial neural networks (ANNs), and extreme learning machines (ELM) are common ML algorithms, while convolutional neural networks (CNNs), long short-term memory (LSTM) networks, and hybrid architectures are gaining attention in the DL domain [15–18]. A brief introduction to these techniques is provided in Section 2. However, many existing studies are restricted to specific regions and lack reproducibility across different climates or structural configurations. Moreover, most available literature focuses on short-term prediction, where the restricted forecasting horizon (e.g., 24 h) limits flexibility for energy management applications.

In this work, we develop a scalable data-driven framework for forecasting building energy consumption using ML, DL, and hybrid models. The methodology is applied to a smart, multi-story commercial building that uses both conventional grid electricity and on-site solar energy. Electrical power consumption data were collected using smart circuit breakers and communication devices integrated with a cloud infrastructure. Additionally, meteorological data including temperature, irradiance, wind speed, and humidity were sourced from NASA's POWER database [19] and synchronized with the energy records to capture environmental dependencies. Feature selection was guided by correlation analysis, and multiple ML and DL algorithms were implemented and compared using standard performance metrics, including measures of computational efficiency. Unlike most prior studies that focus on short-term horizons or single-method approaches, this study introduces a hybrid framework that integrates ML and DL models to improve mid-term, hourly forecasting accuracy. The inclusion of both building consumption and weather variables, applied to a smart building with integrated renewable energy, reinforces the originality and practical relevance of the proposed approach.

Key features of the proposed methodology are the following:

1. The capability to model building energy consumption based on grid power.
2. Integration of high-resolution energy and meteorological data.

The objective of this study is to develop and evaluate a scalable hybrid forecasting framework that integrates ML and DL algorithms for mid-term, hourly energy load prediction in smart buildings. The analysis considers fundamental variables such as electrical consumption data, meteorological factors (temperature, irradiance, wind speed, humidity), and standard performance indicators (MSE, RMSE, R^2). By explicitly incorporating these variables, the study provides a comprehensive evaluation of forecasting accuracy and practical utility in sustainable energy management.

The paper is structured as follows. Section 2 reviews related literature on energy forecasting using ML, DL, and hybrid techniques, highlighting key methodologies and research gaps. Section 3 outlines the proposed method, including data acquisition, feature selection, hyper-parameter tuning, cross-validation, and model development strategies. Section 4 presents simulation results, comparing individual ML/DL models and hybrid approaches across several evaluation metrics. Section 5 concludes with key findings.

2. Literature Review

Given their dominant share in energy use and emissions, buildings have become a primary focus in the development of sustainable and intelligent energy systems [20]. As energy efficiency becomes a global priority, accurate energy demand forecasting is essential not only for optimizing management strategies but also for reducing operational costs and enhancing environmental outcomes.

For existing buildings with time-series energy data, ML techniques have demonstrated reliable performance in consumption forecasting, effectively capturing non-linearities and variable interactions. Using daily data granularity with RF and SVR has been shown to improve accuracy compared with weekly aggregation [21]. Comparative studies further report that ensemble approaches generally outperform individual models such as ANN, SVR, CART, LR, and SARIMA [22]. The importance of selecting an appropriate forecasting model is emphasized in [23]; common ML choices include RF, ELM, and SVR, as summarized in Table 1. In addition to model selection, recent studies highlight that prediction accuracy is strongly influenced by combining multiple data sources. Historical load, meteorological variables, temporal features (e.g., hour, weekday, season), and occupancy information together form robust feature sets for learning.

DL methods, including LSTM networks and CNNs, show strong potential for modeling complex temporal dependencies in energy data, particularly when large-scale datasets and long sequential patterns are available. LSTM has been applied to long-term building load forecasting by leveraging periodic usage patterns [24], while CNN-based architectures such as FireNet achieved high accuracy when incorporating environmental parameters like weather [25]. As summarized in Table 1, DNN and CNN-LSTM models have also been deployed in contexts such as university campuses and benchmark datasets [26], yielding accurate results when trained with weather, occupancy, and power-related features.

A wavelet–neural network approach reduced prediction errors and improved energy efficiency (8%) while accurately forecasting CO_2 emissions, underscoring machine learning's role in smart building management [27].

Hybrid models that combine ML and DL techniques, or multiple DL layers, have emerged as robust forecasting solutions. For instance, the RF–IPWOA–ELM model integrates RF for feature selection, the Improved Parallel Whale Optimization Algorithm (IPWOA) for hyper-parameter tuning, and ELM for prediction, achieving up to 25% improvements over DT and 5.5% over SVR in cooling-load forecasting [28]. Similarly, CNN–LSTM frameworks predict residential energy consumption by combining CNN layers for spatial feature extraction with LSTM layers for temporal dynamics [29]. Other combinations, including LSTM–XGBoost and LSTM–RF, have also been adopted to jointly capture long-term temporal dependencies and non-linear feature interactions, yielding superior accuracy in mid-term forecasting tasks [30]. A hybrid deep learning model integrating CNN, Bi-LSTM, GNN, and Transformer with a Fusion Layer enhanced feature extraction, temporal and spatial learning, and attention-based trend detection, leading to more accurate energy load forecasting for smart cities [31]. An advanced electrical load forecasting framework combining improved feature extraction method, and a hybrid ANN–LSTM model has been proposed. This model achieves more accurate and robust forecasts, supporting effective energy management and planning [32]. Recent literature further highlights that the effectiveness of such hybrid frameworks depends not only on the learning architecture but also on the integration of diverse features, where historical load is combined with meteorological and temporal inputs to strengthen forecasting performance.

Recent literature also highlights the integration of emerging deep learning paradigms such as transfer learning and attention mechanisms, which are reshaping forecasting strategies for energy prediction tasks. Transfer learning has shown particular promise for adapting models across buildings with varying characteristics or limited data availability. For instance, ref. [33] proposed a multi-source transfer learning approach with LSTM and domain adaptation, enabling cross-building knowledge transfer and improved generalization performance. Attention-based methods are also gaining momentum for their ability to emphasize temporally significant inputs. In this context, ref. [34] developed an attention-enhanced sequence-to-sequence model that incorporates transfer learning to forecast building energy consumption under diverse climatic and regional settings. Similarly, ref. [35] conducted a systematic review of DL-based forecasting models, noting the increasing use of attention layers and hybrid deep neural networks to enhance interpretability and performance in complex urban energy systems. These advancements reinforce the shift toward modular, intelligent forecasting architectures that are both data-efficient and scalable, making them highly suitable for deployment in real-world smart building scenarios.

Table 1 shows the trend toward hybrid architectures, including techniques like kernel-based convolutional neural network (KCNN)–LSTM and model stacking, which demonstrates a growing emphasis on combining spatial, temporal, and contextual features for enhanced predictive performance.

Existing research on energy load forecasting has mainly concentrated on either short-term predictions, such as day-ahead forecasting, or long-term trends, while often neglecting the mid-term range, especially weekly to monthly forecasting at hourly resolution. This mid-term focus is important for effective planning in decentralized energy systems [21]. Many previous studies also use limited input features, usually relying only on past energy usage and basic weather data, without including more detailed factors like holidays, seasons, building schedules, or occupancy information [26]. Although some hybrid models combining machine learning (ML) and deep learning (DL) have been explored, few are designed to handle both long-term patterns and complex data interactions. In addition, these models are rarely compared with strong single-model baselines, which limits the ability to evaluate their true performance [29]. Another key issue is how these models are validated. Many studies still use random k-fold cross-validation, which can lead to information leakage across time and give overly optimistic results [36].

Table 1. Review of ML, DL, and hybrid techniques in load forecasting, highlighting gaps in feature sets and forecast horizons.

Ref.	Dataset	Frequency	Features	Models	Key Findings	Limitations
[21]	Educational buildings	Hourly	Energy usage behavior	RF, RT, SVR	ML methods provide accurate hourly forecasts in educational buildings.	Restricted to two buildings; no DL/hybrid comparison.
[22]	Residential smart grid	Daily	Usage patterns	ANN, SVR, CART, LR, SARIMA	Comparative daily forecasting shows ML superiority over statistical models.	Daily horizon only; potential over-fitting issues.
[26]	University campus	Hourly	Operations, HVAC	DNN, RF, SVM, GBT	DNN with engineered HVAC features improves accuracy.	Relies on engineered HVAC variables; no hybrid approaches.
[28]	Commercial buildings	Hourly, Monthly	Cooling load factors	RF, IPWOA, ELM	Hybrid RF–IPWOA–ELM enhances cooling load prediction.	Focused only on cooling load; limited generalization.
[29]	UCI Repository	Hourly	Voltage, occupancy, power	CNN–LSTM	CNN–LSTM outperforms classical baselines in smart meter data.	Based on a single household; no weather variables.
[30]	Residential buildings	Yearly	Weather, occupancy, design	ANN, DNN, RF, KNN, SVM	DNN provides robust performance for annual load forecasting.	Annual horizon; lacks short/medium-term forecasting detail.
[37]	Utility data	Hourly	Weather + power demand	RF	RF achieves low day-ahead forecast errors.	System-level forecasting; not building-specific.
[36]	Cooling load data	Hourly	Env. factors, occupancy	SVR	SVR with environmental variables enhances prediction accuracy.	Limited to summer season; one office building only.
[38]	Hospital energy data	Daily	Load consumption	XGBoost, RF	Ensemble methods outperform individual learners for hospital load.	No DL or hybrid models tested.
[39]	Mixed datasets	Hourly	Usage and weather	SVR, LSTM	SVR achieves comparable or better results than LSTM.	Narrow scope of hybrid testing; dataset heterogeneity challenges.
[40]	Commercial building	15-min	Timestamp + load	KCNN–LSTM	Combining clustering with DL improves short-term forecasts.	Limited exogenous features; scalability uncertain.
[33]	Multiple building types (Japan)	Hourly	Historical load, weather	LSTM + Multi-source Transfer Learning (MK-MMD)	Multi-source domain adaptation improves performance vs. single-source transfer.	Requires domain similarity; complex setup.
[34]	Green buildings (China)	Daily	Historical consumption, temp., humidity	Seq2Seq + Attention + Transfer Learning	Attention-guided TL model forecasts accurately across climate zones.	Model needs careful tuning; limited to similar use cases.

Also, most research does not test model performance across different seasons or building types, making it hard to apply the results in real-world settings [37]. Most evaluations focus only on basic accuracy metrics like RMSE or MAPE, while ignoring important factors such as uncertainty in predictions, how fast the model runs, or how often it needs retraining [28]. More recent studies have started to address some of these challenges by using transfer learning and attention mechanisms to improve model adaptability and accuracy [33,34]. These techniques help models work well across different buildings and focus on the most important parts of the data. However, their use in mid-term forecasting at hourly resolution is still limited, which supports the need for the hybrid approach proposed in this work.

To address these gaps, this study proposes a hybrid ML–DL framework that integrates environmental, temporal, and operational features for mid-term hourly forecasting in smart buildings. The approach is evaluated using robust, time-aware validation protocols and is benchmarked against strong ML and DL baselines to demonstrate its practical relevance and modeling advantages.

3. Methodology

3.1. Data Description and Analysis

The effectiveness of a forecasting system fundamentally depends on two primary components: the integrity of the input data and the forecasting mechanism. Therefore, a comprehensive approach to data collection and cleansing was implemented. Electrical power consumption data were collected through intelligent circuit breakers, communication devices, and a cloud database in a smart commercial building in Bergamo, Italy. The facility covers approximately 30,000 m^2 and integrates advanced energy management systems. It includes multiple office spaces, laboratories, and production areas, accommodating several hundred employees. The hourly electrical load data span a three-year period from January 2021 to December 2023. Each yearly dataset consists of 8760 records (rows) and 11 features (columns), including the timestamp and electrical parameters such as the average, minimum, and maximum values of active power (P_{Avg}, P_{Min}, P_{Max}), reactive power (Q_{Avg}, Q_{Min}, Q_{Max}), and apparent power (S_{Avg}, S_{Min}, S_{Max}). These values were obtained through hourly measurements collected continuously over a 12-month period [41]. An example of the data structure is presented in Table 2.

Table 2. Sample records of linear power consumption parameters.

Date	Time	P_{Avg} (kW)	P_{Min} (kW)	P_{Max} (kW)	Q_{Avg} (kvar)	Q_{Min} (kvar)	Q_{Max} (kvar)	S_{Avg} (kVA)	S_{Min} (kVA)	S_{Max} (kVA)
1 January 2021	08:00:00	356	336	376	32	29	36	311	291	331
1 January 2021	09:00:00	345	324	373	30	26	33	300	279	328
1 January 2021	10:00:00	351	329	380	31	28	37	306	284	336
1 January 2021	11:00:00	357	334	387	33	29	37	312	289	343
1 January 2021	12:00:00	361	331	389	33	28	36	316	287	345

Table 3 illustrates a dataset with five entries of non-linear weather parameters for 2021, namely beam irradiance (Gb(i)), diffuse irradiance (Gd(i)), sunshine duration (H_sun), air temperature at 2 m (T2m), and wind speed at 10 m (WS10m). These data were sourced from the NASA Prediction of Worldwide Energy Resource (POWER) [42], which provides quality global meteorological and climate data appropriate for various research applications. The weather data are notably valuable for examining the influence of meteorological aspects on energy usage in smart commercial structures and estimating possible energy production. Tables 2 and 3 were merged to synchronize time stamps, namely date, hour, day, month, and year for each record. These datasets were employed for examination and determining how changes occurred across the years. The dataset was split chronologically into training (70%), validation (15%), and testing (15%) sets, without shuffling, to prevent temporal leakage.

The electrical load data is systematically organized by date and the seasonal index, as presented in Table 4. This classification helps structure the data for identifying patterns across different calendar scenarios and seasonal trends. The indexes represent standard working days, typically Monday to Friday, excluding recognized public holidays. An additional index broadens the analysis to include weekends and holidays, thereby capturing a more extensive range of consumption behaviors. Further indexes encompass statistical aggregations of daily, weekly, and monthly averages, while also emphasizing the significance of seasonal categorization. This methodical classification allows for a thorough temporal segmentation, thereby enhancing the precision of modeling and analysis of load patterns that are influenced by human activities and institutional timetables. Figure 1 illustrates active power-consumption patterns at four temporal resolutions (hourly, daily, weekly, and monthly) captured from a smart commercial building. The analysis is based on power consumption from the electrical grid, excluding on-site renewable generation. These visualizations are essential for understanding energy consumption behavior and identifying temporal patterns that inform the design of forecasting models. The daily resolution plot reveals wide interquartile ranges and frequent outliers, indicating irregular consumption, likely driven by operational anomalies or peak demand events. This variability emphasizes the importance of forecasting frameworks capable of handling noise and capturing outlier behavior. At the hourly resolution (Figure 1a), a clear diurnal load cycle is evident, with demand increasing in the morning and peaking between mid-morning and early afternoon.

Table 3. Sample records of non-linear weather data parameters.

Date	Time	Gb(i) (W/m^2)	Gd(i) (W/m^2)	H_sun (h)	T2m (°C)	WS10m (m/s)
1 January 2021	08:00:00	56.005	43.800	8.39	−1.440	1.585
1 January 2021	09:00:00	103.139	75.933	14.83	1.100	0.997
1 January 2021	10:00:00	142.206	96.400	19.25	3.866	0.483
1 January 2021	11:00:00	186.007	93.733	21.20	4.933	0.593
1 January 2021	11:00:00	165.273	97.200	20.49	5.143	0.745

Following 16:00, consumption decreases sharply, reflecting reduced occupancy and operational activity. Such granularity is critical for short-term load prediction tasks. Figure 1c shows higher median loads on weekdays compared with weekends, a trend characteristic of commercial operations. Week-to-week fluctuations may be attributed to external influences such as holidays, weather conditions, or partial building usage, which introduce short-term non-stationarity into the data. The monthly resolution captures broader seasonal trends, as shown in Figure 1d. Although overall consumption is relatively consistent, some months show elevated averages, likely due to greater HVAC use or seasonal shifts in operational intensity. Together, these multi-resolution visual insights provide a comprehensive view of energy behavior across timescales. Incorporating such temporal diversity in input features can enhance the adaptability and predictive power of ML and DL models in energy forecasting. Two additional visualizations were created to further analyze power consumption behavior across multiple temporal resolutions. Figure 2a displays hourly power usage across the year 2023, highlighting significant intra-day and seasonal variations. Peak demands are especially noticeable during the mid-year months, likely driven by increased cooling requirements in summer. This fine-grained view of hourly data is crucial for capturing daily load cycles and informing real-time energy management decisions. As illustrated in Figure 2b, daily, weekly, and monthly average consumption trends for 2023 are also presented to reflect different temporal perspectives. The line representing daily averages exhibits high frequency fluctuations, capturing short-term variability. The weekly average smooths these fluctuations, revealing medium-term behavioral consistency, while the monthly average highlights broader seasonal trends. This layered averaging technique is valuable for distinguishing between transient and structural patterns in energy demand. Together, Figure 2 provides a comprehensive view of energy usage dynamics, serving as a foundational reference for the forecasting model's design and effective load management strategies.

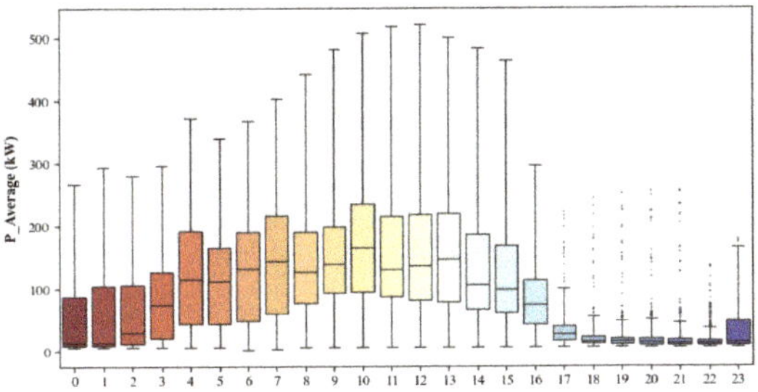

(**a**) Hourly (24-hour average)

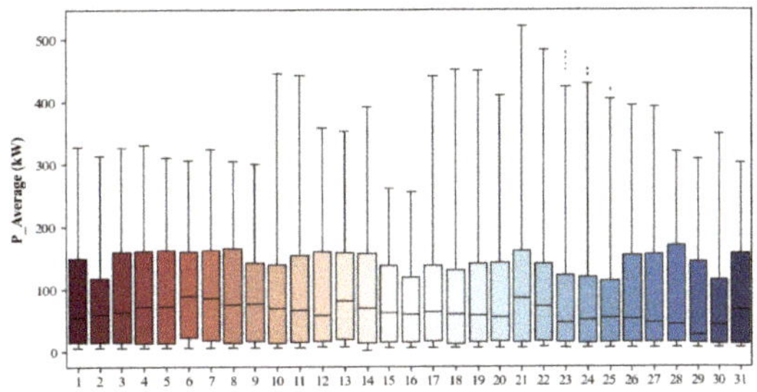

(**b**) Daily (average over Month/Year)

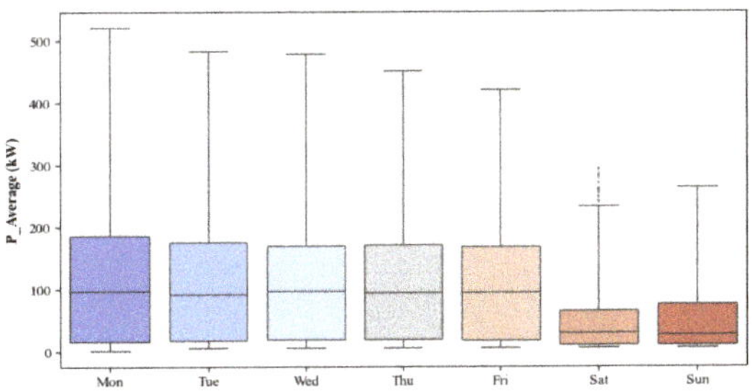

(**c**) Weekly (average by weekday)

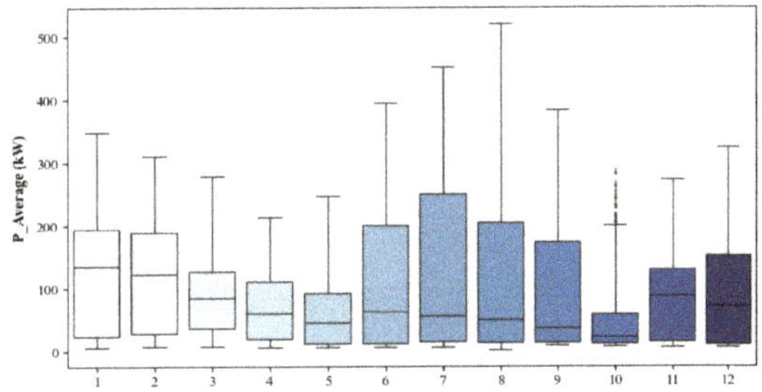

(**d**) Monthly values (average)

Figure 1. Time-based profiling of active power: (**a**) Hourly, (**b**) Daily, (**c**) Weekly, and (**d**) Monthly values.

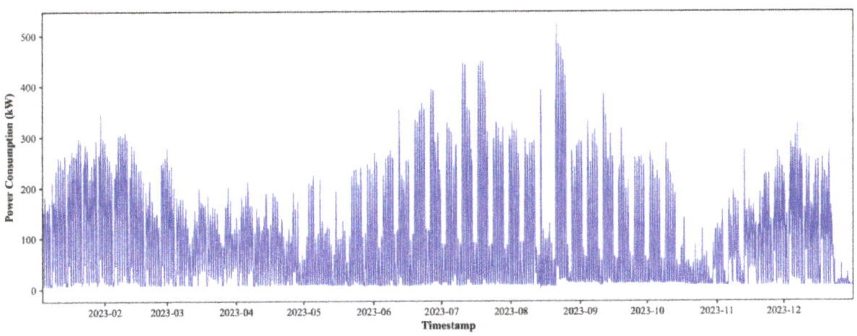

(**a**) Hourly power consumption in 2023

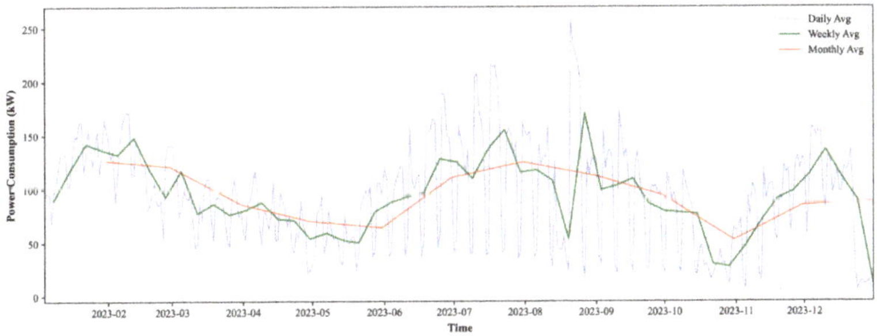

(**b**) Daily, Weekly, and Monthly Averages

Figure 2. Comparison of power consumption trends at different time scales.

In Figure 3, forecasting process begins with the acquisition of hourly energy consumption data alongside relevant meteorological variables. Preprocessing is then applied to address missing values, construct informative features, and normalize the dataset. In the modeling phase, both ML algorithms (XGBoost, RF, SVR, DT) and DL architectures (LSTM, FireNet) are developed independently and in hybrid configurations. Performance evaluation follows, utilizing standard metrics such as RMSE, MAPE, and R^2 to identify the most accurate and computationally efficient approach.

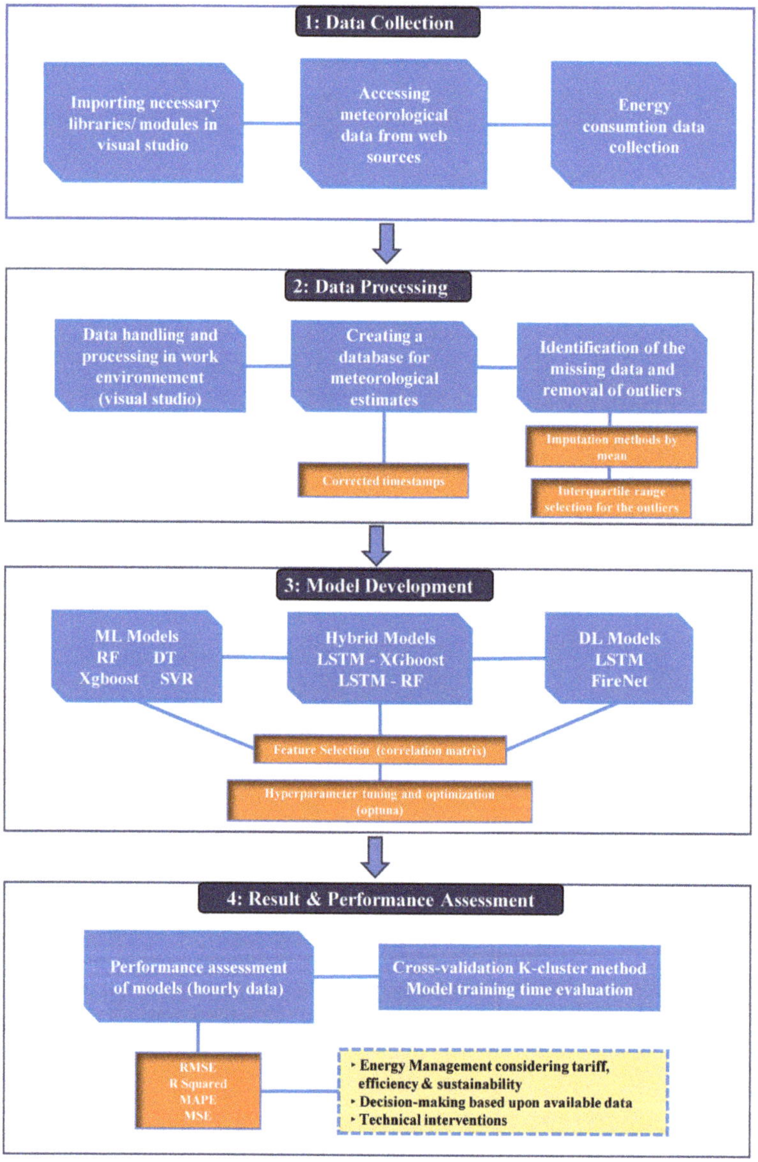

Figure 3. Proposed methodology workflow for energy consumption forecasting.

Table 4. Electrical load classification according to the index.

Index	Electrical Load Classification
1	Original time-series data (unmodified raw observations).
2	Weekdays (Monday to Friday), excluding Holidays.
3	Weekend (Saturday and Sunday) and Holidays.
4	Statistical aggregation features: Daily, Weekly, and Monthly Averages.
5	Seasonal categorization into Winter, Spring, Summer, and Autumn.

3.2. Feature Selection

The correlation coefficient r is given by Equation (1), where x_i and y_i denote individual values of variables x and y, respectively, and $\bar{x}$, $\bar{y}$ are their corresponding means [43].

$$r = \frac{\sum(x_i - \bar{x})(y_i - \bar{y})}{\sqrt{\sum(x_i - \bar{x})^2 \sum(y_i - \bar{y})^2}} \tag{1}$$

This metric was applied to measure the strength of the relationship between each input feature and the target variable P_{avg}, as well as to detect high intercorrelation among variables. Figure 4 displays the Pearson correlation matrix heatmap for electrical and meteorological variables. Features showing strong correlation with the target variable, typically ranging from 0.29 to 0.88, were retained. Highly correlated power features, including `P_Minimum`, and `P_Maximum`, were considered redundant and consolidated into a single representative: P_{avg}. Similarly, from the reactive and apparent power groups, only one representative linear feature (Q_{avg}) was selected to reduce model bias and the risk of overfitting. The non-linear features retained for modeling included meteorological variables such as air temperature at 2 m (`T2m`), sunshine duration (`H_sun`), solar irradiance (`Gb(i)` and `Gd(i)`), and wind speed at 10 m (`WS10m`). Wind speed exhibited a weak correlation with the target variable and was excluded. This feature selection strategy balances model complexity and generalization while maintaining computational efficiency.

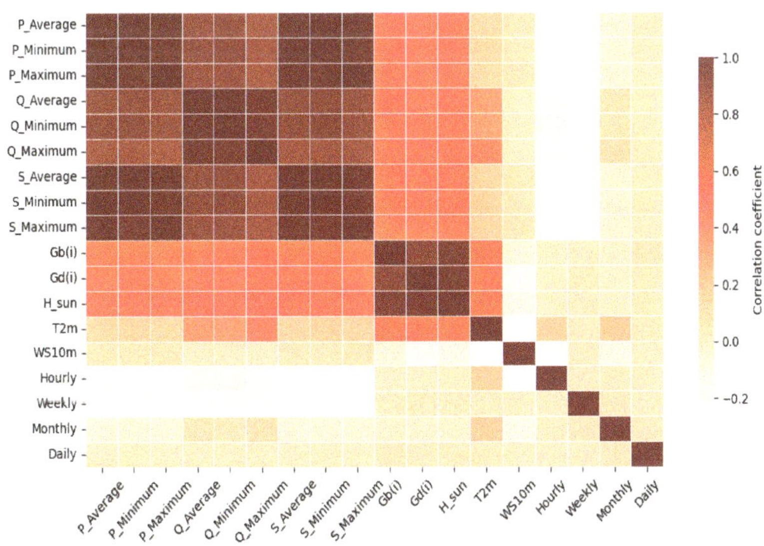

Figure 4. Pearson correlation among energy and meteorological features.

3.3. Machine Learning Models

RF is an ensemble learning technique that builds multiple decision trees using random subsets of features and samples, then averages predictions. This approach enhances stability and accuracy, especially for complex, non-linear problems. RF handles both categorical and continuous variables and includes a built-in feature-importance measure [44]. In energy forecasting, RF has been applied to load prediction and has outperformed several conventional ML models in accuracy [45]. For example, in [21], RF was used to forecast hourly energy consumption in buildings, demonstrating high accuracy and efficient parameter tuning. DTs, particularly Classification and Regression Trees (CART), form the basis of many ensemble methods. They recursively split input data into branches and leaf nodes that represent outcomes. DTs are interpretable and can effectively model non-linear patterns, though they may over-fit to noisy data. As discussed in [46], ensemble strategies such as RF and Gradient Boosted Trees (GBT) are commonly used to overcome this limitation. DTs have also been used in hybrid models for energy forecasting tasks, including short-term load prediction and fuzzy rule-based classification [47].

SVR is a supervised ML framework designed for regression tasks, where the target variable is continuous rather than discrete. It estimates a regression function within a margin of tolerance, using kernel functions to capture complex relationships in data [48]. This enables SVR to identify non-linear relationships while preserving the dimensionality of the feature space. Furthermore, its capacity to mitigate over-fitting through regularization enhances its robustness to datasets with noise or uncertainty [49]. XGBoost is a powerful implementation of gradient boosting that sequentially adds weak learners to minimize residual error [50]. It includes regularization to reduce over-fitting and can handle sparse or missing data. XGBoost has been widely adopted in forecasting due to its speed and accuracy. For example, it has been combined with Gaussian mixture clustering to capture behavioral patterns in household energy forecasting [51].

3.4. Deep Learning Models

LSTM networks have emerged as powerful tools for time-series prediction tasks in the energy domain due to their ability to learn long-range temporal dependencies. In this study, the LSTM model is implemented as a baseline DL approach to forecast hourly energy consumption patterns of a smart commercial building. The architecture employs multiple hidden layers and memory cells, enabling the model to retain temporal context, essential for capturing diurnal and weekly variations in power demand. The LSTM model is fine-tuned using dropout regularization and a sequence-to-one prediction strategy to reduce over-fitting and improve generalization on unseen hourly data.

The FireNet model [52] is a DL architecture composed of five convolutional layers, each followed by max-pooling operations. Initially developed for image classification, FireNet is here used for time-series forecasting of building energy consumption. This adaptation leverages its hierarchical feature extraction capabilities to capture complex temporal dependencies. The model processes sequential input data through successive convolutional layers, each extracting temporal patterns at different scales. With 32 filters, the first layer captures short-term fluctuations such as hourly demand spikes. The input matrix is given by Equation (2), where T denotes the number of time steps, and C is the number of input features.

$$X_{\text{input}} \in \mathbb{R}^{T \times C} \tag{2}$$

The second layer (64 filters) identifies daily consumption cycles; the third (128 filters) models multi-day trends; the fourth (256 filters) captures weekly patterns; and the fifth layer (512 filters) detects long-range temporal dependencies, such as holidays or weather-driven changes. Max-pooling and batch normalization follow each convolutional block to reduce temporal resolution and stabilize training. The architecture is detailed in Table 5. Each convolutional block $l \in \{1, \ldots, 5\}$ performs the operation expressed through Equation (3):

$$F^{(l)} = \text{ReLU}\left(W^{(l)} * I^{(l)} + b^{(l)}\right) \tag{3}$$

where $*$ denotes the 1D convolution operator, $W^{(l)} \in \mathbb{R}^{k \times f_{\text{in}} \times f_{\text{out}}}$ is the kernel with size k (5 for blocks 1–2, and 3 for blocks 3–5), and $f_{\text{in}}, f_{\text{out}}$ are input and output channels, respectively.

Table 5. FireNet architecture for energy forecasting.

Block	Filters	Kernel Size	Pool Size	Output Shape
1	32	5	2	$(T/2, 32)$
2	64	5	2	$(T/4, 64)$
3	128	3	2	$(T/8, 128)$
4	256	3	2	$(T/16, 256)$
5	512	3	–	$(T/16, 512)$

Max-pooling reduces the temporal dimension by selecting the maximum value within a defined window. It is applied after each convolutional block to minimize complexity while preserving key features in Equation (4):

$$P^{(l)} = \text{MaxPool}\left(F^{(l)}, s = 2\right), \quad T_{\text{out}} = \frac{T_{\text{in}}}{2} \tag{4}$$

where $F^{(l)}$ is the feature map at layer l, and s is the stride size (set to 2). The temporal output length T_{out} and input length T_{in} is then given by indicating that the temporal resolution is halved after pooling.

Batch normalization is used after pooling to normalize activations, improving training stability. For a batch of activations, the transformation is obtained via Equation (5):

$$\hat{x} = \frac{x - \mu}{\sqrt{\sigma^2 + \epsilon}}, \quad B^{(l)} = \gamma^{(l)} \cdot \hat{x}^{(l)} + \beta^{(l)} \tag{5}$$

where μ and σ^2 are batch mean and variance, and ϵ ensures numerical stability. Learnable parameters γ and β scale and shift the result, mitigating internal covariate shift and enabling deeper architectures to converge effectively.

3.5. Hybrid Models

Hybrid approaches that combine the strengths of multiple algorithms, such as ML and DL, have shown significant promise in improving forecasting performance for energy consumption [53]. These hybrid designs illustrate the advantage of combining temporal learning, feature selection, and ensemble principles, ultimately leading to more robust and flexible forecasting frameworks across various building types and time horizons [54].

Recent research in the field of load forecasting demonstrates that hybrid imputation enhances forecasting performance and system reliability by effectively managing incomplete sensor data [55].

In hybrid models, FireNet captures complex temporal dependencies, while XGBoost and random forest handle structured data with strong generalization. Their hybrid integration combines these strengths: DL extracts temporal features, and ML interprets them efficiently, reducing overfitting and improving robustness.

3.6. Hyper-Parameter Optimization

Hyper-parameter tuning is crucial for enhancing the accuracy of ML and DL algorithms. Hyper-parameters include things such as learning rate and regularization strength, whereas the batch sizes are not learned from the data, but they are decided by design before training. The choice of hyper-parameters may drastically affect the performance of a model and its rate of convergence, as well as its capability of avoiding over-fitting. Several hyper-parameter optimization methods are available and the most employed are grid search, random search, Bayesian optimization, genetic algorithms, and Optuna [56]. Relying on an efficient sampling and the pruning mechanisms, the Optuna learns distinct search spaces faster than the other hyper-parameter optimization methods, thus completing the trials in fewer attempts.

Table 6 demonstrates the hyper-parameter ranges and optimal values for each model, by optimization of Optuna. In this work, the Optuna was executed to identify the optimal configuration for each algorithm, and the best-performing configuration was selected based on its results. While performing multiple runs of the optimization could provide more statistical information about model stability, this paper focuses on the best optimization result, which is consistent with standard practice in comparable research.

Table 6. Hyper-parameter tuning of different models.

Model	Parameter	Range	Optuna Value
Decision Tree	max_depth	[10, 20, 30, 50, 100]	7
	min_samples_split	[2, 3, 5, 10]	10
	min_samples_leaf	[1, 5, 10]	8
	criterion	[mse, mae]	mse
	split_strategy	[best, random]	best
Random Forest	n_estimators	[50, 100, 150, 200, 300]	80
	max_depth	[10, 30, 50, 70, 100, None]	13
	min_samples_split	[2, 3, 5, 10]	7
	min_samples_leaf	[1, 2, 3, 4, 5]	5
	max_features	[auto, sqrt, log2]	sqrt
	bootstrap	[True, False]	True
XGBoost	n_estimators	[50, 300]	261
	max_depth	[3, 20]	4
	learning_rate	[0.01, 0.3]	0.051
	subsample	[0.5, 1.0]	0.949
	colsample_bytree	[0.5, 1.0]	0.813
	gamma	[0, 5]	3.704
	reg_alpha	[0, 5]	3.083
	reg_lambda	[0, 5]	3.442
SVR	C	[0.1, 100] (log)	34.95
	epsilon	[0.001, 1.0] (log)	0.244
	kernel	[rbf, linear]	rbf
	gamma	[scale, auto]	auto
FireNet	filters	[32, 64, 128]	64
	dropout_rate	[0.2, 0.5] (uniform)	0.2132
	learning_rate	[0.0005, 0.01] (log)	0.000507
	batch_size	[16, 32, 64]	64
LSTM	units	[32, 50, 64]	32
	dropout_rate	[0.1, 0.5] (uniform)	0.285
	learning_rate	[0.0005, 0.01] (log)	0.007215
	batch_size	[16, 32, 64]	16

3.7. Implementation Details

The proposed framework was experimentally evaluated with Python 3.10.11. Scikit-learn 1.5.2, TensorFlow 2.18.0, and Keras 3.9.0. libraries were used for running ML and DL models. The testing was performed on [specify hardware, e.g., Intel i7 CPU with 16 GB RAM], ensuring reproducibility of results.

4. Experimental Results and Discussion

4.1. Machine Learning-Based Forecasting Results

The ML models evaluated include decision tree (DT), random forest (RF), XGBoost, and support vector regression (SVR). Figure 5 compares actual power consumption with predicted values generated by these models. As summarized in Table 7, XGBoost demonstrated the highest predictive accuracy and computational efficiency, achieving the lowest MSE (i.e., 518.5), RMSE (i.e., 22.77), and higher R^2 score (i.e., 0.9010), with a training time of only 0.051 s. RF also performed competitively, offering a strong balance between accuracy and runtime, with slightly higher values of MSE, RMSE and lower R^2 score (i.e., 0.8992), and training time of 1.101 s. The DT model exhibited the fastest training time (i.e., 0.0162 s), though it exhibited slightly lower accuracy, as reflected by the R^2 score (i.e., 0.8514). In contrast, SVR showed the weakest performance, recording the highest MSE, RMSE, and the lowest R^2 score (i.e., 0.7460), with a significantly higher runtime of 41.13 s. These findings confirm that XGBoost and RF offer the most accurate and computationally efficient solutions among the evaluated ML models.

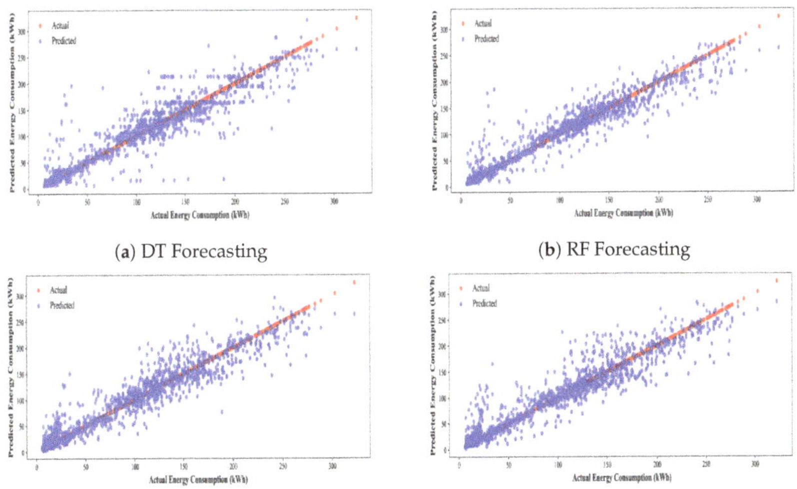

(**a**) DT Forecasting (**b**) RF Forecasting

(**c**) XGBoost Forecasting (**d**) SVR Forecasting

Figure 5. Forecasting performance of ML models. (**a**) DT. (**b**) RF. (**c**) XGBoost. (**d**) SVR.

Table 7. Forecasting performance metrics of ML models.

Model	Training Time (s)	MSE	RMSE	R^2 Score	MAPE (%)
DT	0.01620	778.2	27.90	0.8514	35.68
RF	1.101	527.7	22.97	0.8992	33.63
XGBoost	0.05100	518.5	22.77	0.9010	37.17
SVR	41.13	2111	45.95	0.7460	44.79

Figure 6 illustrates the comparison between actual and predicted energy consumption over the final two months of 2023. Both XGBoost and RF closely align with the actual data, accurately capturing peaks, trends, and fluctuations. The DT model shows moderate predictive ability but exhibits larger deviations from the true values. In contrast, SVR consistently underestimates energy consumption and fails to reflect sharp variations. These visual observations are consistent with the quantitative metrics discussed earlier, reinforcing the superior forecasting performance of XGBoost and RF.

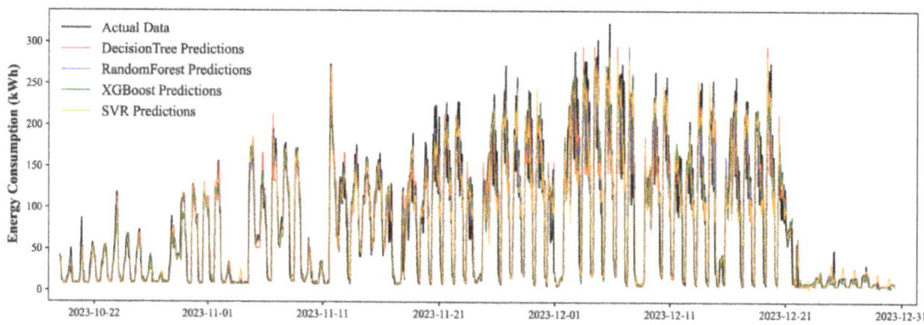

Figure 6. Actual vs. predicted energy consumption for ML models.

4.2. Deep Learning-Based Forecasting Results

In terms of DL models, FireNet and LSTM are evaluated by assessing their ability to capture non-linear patterns and temporal dependencies in the dataset. Figure 7 presents the forecasting performance of both considered DL models. The outcomes of the comparative analysis are listed in Table 8. Compared to LSTM, FireNet achieved stronger predictive accuracy, particularly in modeling non-linear and temporal energy usage patterns. In fact, FireNet is characterized by lower MSE (i.e., 923.5 against 1280), reduced RMSE (i.e., 30.39 against 35.78), smaller MAPE (i.e., 39.90% against 96.21%), along with higher R^2 (i.e., 0.8237 against 0.7558). Considering the computational time, the FireNet is faster, requiring only 22.61 s, which is less than half of the time needed by the LSTM (46.97 s). In conclusion, the FireNet model outperformed the LSTM model across all evaluated metrics, indicating enhanced precision and computational effectiveness.

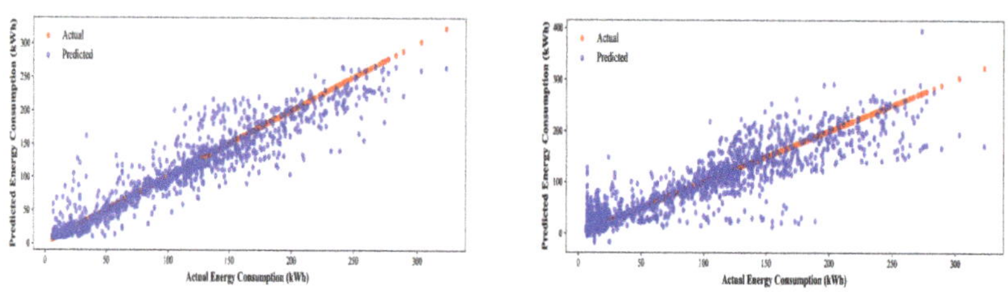

(**a**) FireNet Forecasting (**b**) LSTM Forecasting

Figure 7. Forecasting performance using DL models. (**a**) FireNet. (**b**) LSTM.

Table 8. Forecasting performance of DL models based on standard metrics.

Model	Training Time (s)	MSE	RMSE	R^2 Score	MAPE (%)
FireNet	22.61	923.5	30.39	0.8237	39.90
LSTM	46.97	1280	35.78	0.7558	96.21

In Figure 8, the actual energy consumption is compared against the forecasts delivered by the selected DL models. As expected, FireNet's forecasts closely align with actual values, especially during peak periods, while LSTM predictions are noticeably overestimated in case of peak demand periods.

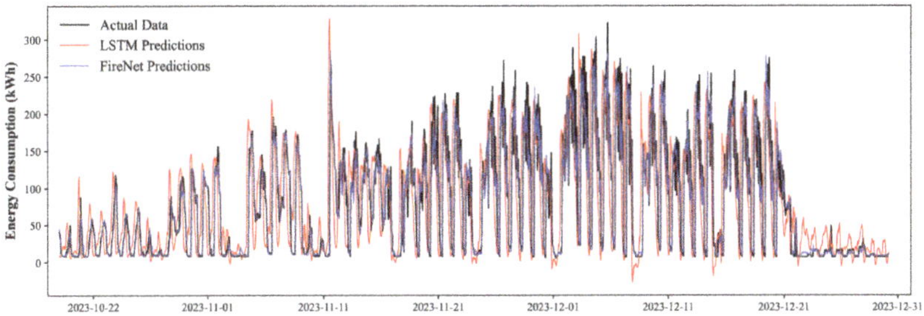

Figure 8. Actual vs. predicted energy consumption for DL models.

4.3. Hybrid Model Performance

Based on the previous results, XGBoost and RF are identified as the better performing ML models, while FireNet ensures superior forecasting performance in terms of DL models. Relying on these findings, two hybrid models are developed by combining the sequential learning capabilities of FireNet with the predictive efficiency of ensemble-based ML models (i.e., XGBoost and RF). In Figure 9, the energy consumption forecasts provided by both hybrid models (i.e., FireNet–XGBoost and FireNet–RF) are shown, while the corresponding key performance parameters are reported in Table 9. The FireNet–XGBoost hybrid model demonstrated superior performance due to lower MSE (i.e., 350.2 against 418.4), reduced RMSE (i.e., 18.71 against 20.45), smaller MAPE (i.e., 27.00% against 29.26%), along with higher R^2 (i.e., 0.9334 against 0.9205). On the other hand, the FireNet–RF hybrid model required less computation time (i.e., 21.11 s against 33.69 s). Overall, the FireNet–XGBoost model outperformed the FireNet–RF model across all evaluated metrics, indicating enhanced precision, except the computational effectiveness.

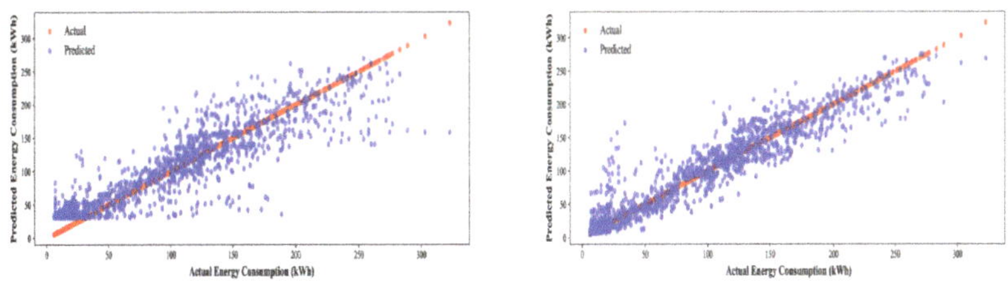

(**a**) FireNet–XGBoost Forecasting (**b**) FireNet–RF Forecasting

Figure 9. Forecasting performance using hybrid models. (**a**) FireNet–XGBoost. (**b**) FireNet–RF.

Table 9. Forecasting performance of hybrid ML and DL models.

Model	Training Time (s)	MSE	RMSE	R^2	MAPE (%)
FireNet-XGBoost	33.69	350.2	18.71	0.9334	27.00
FireNet-RF	21.11	418.4	20.45	0.9205	29.26

Figure 10 illustrates that the FireNet–XGBoost model closely follows actual energy consumption patterns, particularly during peak demand periods, whereas the FireNet–RF model tends to slightly overestimate energy usage across similar intervals.

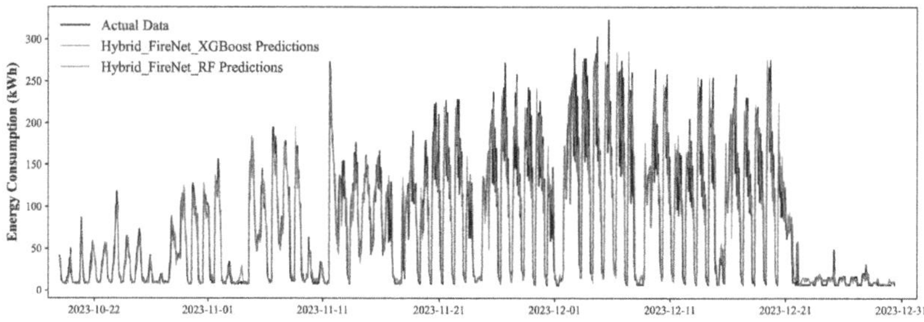

Figure 10. Actual vs. predicted energy consumption for hybrid models.

5. Conclusions and Future Directions

In this paper, the effectiveness of several ML, DL, and hybrid models was tested for predicting the mid-term energy consumption in a smart commercial building. Although XG-Boost was more robust than other ML techniques and FireNet managed to model non-linear relationships among the variety of DL approaches, the proposed FireNet–XGBoost hybrid demonstrated superior performance overall. This highlights the novelty of bridging the two complementary strengths of ML and DL into a general framework, and moves beyond existing studies which have traditionally relied on either approach alone.

The novelty of this study is to illustrating that a two-layer hybrid architecture can improve the accuracies and robustness with hourly horizon to offer new experiences in building energy prediction. The practical implications are considerable: more accurate load forecasts can maximize the energy consumption, minimize operational costs, and enhance the reliability of the grid, contributing to the shift to a more sustainable intelligent building operation.

The analysis was performed on data from a single smart building, which limits the generalization of the findings. The emphasis on mid-term hourly prediction also misses multi-source energy consumption data analysis or real-time dynamics. In future work, to overcome these limitations, the dataset is planned to be extended to cover multiple energy sources, to include further contextual variables, and to study real-time adaptive forecasting strategies. In addition, the community-level prediction can be extended to support cooperative energy management and demand response, as promising future work. Lastly, we will explore advanced techniques like transfer learning and attention-based architectures to improve scaling and generalization.

Author Contributions: Conceptualization, P.G. and A.H.; methodology, A.H., M.A. and M.E.; software, A.H.; validation, P.G. and M.H.; formal analysis, P.G. and G.F.; investigation, A.H.; resources, P.G., L.F. and S.M.; data curation, A.H.; writing original draft preparation, A.H. and M.A.; writing, review and editing, P.G., M.A. and A.H.; visualization, A.H.; supervision, P.G. and G.F.; funding acquisition, P.G. All authors have read and agreed to the published version of the manuscript.

Funding: This research was supported by the Piano Nazionale di Ripresa e Resilienza (PNRR) of the European Union's NextGenerationEU program.

Data Availability Statement: The datasets presented in this article are not readily available because they belong to ABB Sace Spa. Requests to access the datasets should be directed to ABB Sace Spa.

Conflicts of Interest: Authors Lorenzo Fenili and Silvio Messi were employed by the company ABB SACE S.p.A. The remaining authors declare that the research was conducted in the absence of any commercial or financial relationships that could be construed as a potential conflict of interest.

References

1. Oliveira, G.M.; Vidal, D.G.; Ferraz, M.P. Urban Lifestyles and Consumption Patterns. In *Encyclopedia of the UN Sustainable Development Goals*; Springer: Cham, Switzerland, 2020; pp. 851–860. [CrossRef]
2. Bilgen, S. Structure and environmental impact of global energy consumption. *Renew. Sustain. Energy Rev.* **2014**, *38*, 890–902. [CrossRef]
3. González-Torres, M.; Pérez-Lombard, L.; Coronel, J.F.; Maestre, I.R.; Yan, D. A review on buildings energy information: Trends, end-uses, fuels and drivers. *Energy Rep.* **2022**, *8*, 626–637. [CrossRef]
4. Afshan, S.; Ozturk, I.; Yaqoob, T. Facilitating renewable energy transition, ecological innovations and stringent environmental policies to improve ecological sustainability. *Renew. Energy* **2022**, *196*, 151–160. [CrossRef]
5. Salkuti, S.R. Emerging and Advanced Green Energy Technologies for Sustainable and Resilient Future Grid. *Energies* **2022**, *15*, 6667. [CrossRef]
6. Zhao, J.; Patwary, A.K.; Qayyum, A.; Alharthi, M.; Bashir, F.; Mohsin, M.; Hanif, I.; Abbas, Q. The determinants of renewable energy sources for the fueling of green and sustainable economy. *Energy* **2022**, *238*, 122029. [CrossRef]
7. Giovanni, G.D.; Rotilio, M.; Giusti, L.; Ehtsham, M. Exploiting building information modeling and machine learning for optimizing rooftop photovoltaic systems. *Energy Build.* **2024**, *313*, 114250. [CrossRef]
8. Ehtsham, M.; Rotilio, M.; Cucchiella, F. Deep learning augmented medium-term photovoltaic energy forecasting. *Energy Rep.* **2025**, *13*, 4299–4317. [CrossRef]
9. El-Afifi, M.I.; Sedhom, B.E.; Eladl, A.A.; Elgamal, M.; Siano, P. Demand side management strategy for smart building using multi-objective hybrid optimization technique. *Results Eng.* **2024**, *22*, 102265. [CrossRef]
10. Raza, A.; Jingzhao, L.; Adnan, M.; Ahmad, I. Optimal load forecasting and scheduling strategies for smart homes peer-to-peer energy networks. *Results Eng.* **2024**, *22*, 102188. [CrossRef]
11. Ehtsham, M.; Rotilio, M.; Cucchiella, F.; Giovanni, G.D.; Schettini, D. Investigating the effects of hyperparameter sensitivity on machine learning algorithms for PV forecasting. In *Proceedings of the E3S Web of Conferences*; EDP Sciences: Les Ulis, France, 2025; Volume 612, p. 01002. [CrossRef]
12. Lazos, D.; Sproul, A.B.; Kay, M. Optimisation of energy management in commercial buildings with weather forecasting inputs: A review. *Renew. Sustain. Energy Rev.* **2014**, *39*, 587–603. [CrossRef]
13. Iqbal, N.; Kim, D.H. IoT task management mechanism based on predictive optimization for efficient energy consumption in smart residential buildings. *Energy Build.* **2022**, *257*, 111762. [CrossRef]
14. Pawar, P.; TarunKumar, M. An IoT based Intelligent Smart Energy Management System. *Measurement* **2020**, *152*, 107187. [CrossRef]
15. Khan, S.U.; Khan, N.; Ullah, F.U.M.; Kim, M.J.; Lee, M.Y.; Baik, S.W. Towards intelligent building energy management: AI-based framework for power consumption and generation forecasting. *Energy Build.* **2023**, *279*, 112705. [CrossRef]
16. Mathumitha, R.; Rathika, P.; Manimala, K. Intelligent deep learning techniques for energy consumption forecasting in smart buildings: A review. *Artif. Intell. Rev.* **2024**, *57*, 35. [CrossRef]
17. Santos, M.L.; García, S.D.; García-Santiago, X.; Ogando-Martínez, A.; Camarero, F.E.; Gil, G.B.; Ortega, P.C. Deep learning and transfer learning techniques applied to short-term load forecasting of data-poor buildings in local energy communities. *Energy Build.* **2023**, *292*, 113164. [CrossRef]
18. Wahid, F.; Ghazali, R.; Shah, A.S.; Fayaz, M. Prediction of energy consumption in the buildings using multi-layer perceptron and random forest. *Int. J. Adv. Sci. Technol.* **2017**, *101*, 13–22. [CrossRef]
19. NASA Power Data Access Viewer: Prediction of Worldwide Energy Resource. Available online: https://power.larc.nasa.gov/data-access-viewer/ (accessed on 1 February 2025).
20. International Energy Agency. *World Energy Outlook 2024*; IEA: Paris, France, 2024.
21. Wang, Z.; Wang, Y.; Zeng, R.; Srinivasan, R.S.; Ahrentzen, S. Random Forest based hourly building energy prediction. *Energy Build.* **2018**, *171*, 11–25. [CrossRef]
22. Chou, J.S.; Tran, D.S. Forecasting energy consumption time series using machine learning techniques based on usage patterns of residential householders. *Energy* **2018**, *165*, 709–726. [CrossRef]
23. Amasyali, K.; El-Gohary, N.M. A review of data-driven building energy consumption prediction studies. *Renew. Sustain. Energy Rev.* **2018**, *81*, 1192–1205. [CrossRef]
24. Wang, J.Q.; Du, Y.; Wang, J. LSTM based long-term energy consumption prediction with periodicity. *Energy* **2020**, *197*, 117197. [CrossRef]
25. Alam, G.M.I.; Tasnia, N.; Biswas, T.; Hossen, M.J.; Tanim, S.A.; Miah, M.S.U. Real-Time Detection of Forest Fires Using FireNet-CNN and Explainable AI Techniques. *IEEE Access* **2025**, *13*, 51150–51181. [CrossRef]

26. Yesilyurt, H.; Dokuz, Y.; Dokuz, A.S. Data-driven energy consumption prediction of a university office building using machine learning algorithms. *Energy* **2024**, *310*, 133242. [CrossRef]
27. Alotaibi, B.S.; Abuhussain, M.A.; Dodo, Y.A.; Al-Tamimi, N.; Maghrabi, A.; Ojobo, H.; Naibi, A.U.; Benti, N.E. A novel approach to estimate building electric power consumption based on machine learning method: Toward net-zero energy, low carbon and smart buildings. *Int. J. Low-Carbon Technol.* **2024**, *19*, 2335–2345. [CrossRef]
28. Gao, Z.; Yu, J.; Zhao, A.; Hu, Q.; Yang, S. A hybrid method of cooling load forecasting for large commercial building based on extreme learning machine. *Energy* **2022**, *238*, 122073. [CrossRef]
29. Kim, T.Y.; Cho, S.B. Predicting residential energy consumption using CNN-LSTM neural networks. *Energy* **2019**, *182*, 72–81. [CrossRef]
30. Olu-Ajayi, R.; Alaka, H.; Sulaimon, I.; Sunmola, F.; Ajayi, S. Building energy consumption prediction for residential buildings using deep learning and other machine learning techniques. *J. Build. Eng.* **2022**, *45*, 103406. [CrossRef]
31. Sunder, R.; R, S.; Paul, V.; Punia, S.K.; Konduri, B.; Nabilal, K.V.; Lilhore, U.K.; Lohani, T.K.; Ghith, E.; Tlija, M. An advanced hybrid deep learning model for accurate energy load prediction in smart building. *Energy Explor. Exploit.* **2024**, *42*, 2241–2269. [CrossRef]
32. Yenagimath, M. Electrical Load Forecasting using Improved Feature Selection and Hybrid Deep Learning Algorithms for Power System Situational Awareness. *Deleted J.* **2024**, *20*, 2825–2836. [CrossRef]
33. Lu, H.; Wu, J.; Ruan, Y.; Qian, F.; Meng, H.; Gao, Y.; Xu, T. A multi-source transfer learning model based on LSTM and domain adaptation for building energy prediction. *Int. J. Electr. Power Energy Syst.* **2023**, *149*, 109024. [CrossRef]
34. Peng, X.; Li, Y.; Zhou, H. Transfer Learning and Attention-Enhanced Seq2Seq Model for Green Building Load Forecasting. *Sci. Rep.* **2025**, *15*, 16953. [CrossRef] [PubMed]
35. Chen, G.; Lu, S.; Zhou, S.; Tian, Z.; Kim, M.; Liu, J.; Liu, X. A Systematic Review of Building Energy Consumption Prediction: From Perspectives of Load Classification, Data-Driven Frameworks, and Future Directions. *Appl. Sci.* **2025**, *15*, 3086. [CrossRef]
36. Zhong, H.; Wang, J.; Jia, H.; Mu, Y.; Lv, S. Vector field-based support vector regression for building energy consumption prediction. *Appl. Energy* **2019**, *242*, 403–414. [CrossRef]
37. Lahouar, A.; Slama, J.B.H. Day-ahead load forecast using random forest and expert input selection. *Energy Convers. Manag.* **2015**, *103*, 1040–1051. [CrossRef]
38. Cao, L.; Li, Y.; Zhang, J.; Jiang, Y.; Han, Y.; Wei, J. Electrical load prediction of healthcare buildings through single and ensemble learning. *Energy Rep.* **2020**, *6*, 2751–2767. [CrossRef]
39. Ismail, L.; Materwala, H.; Dankar, F.K. Machine Learning Data-Driven Residential Load Multi-Level Forecasting with Univariate and Multivariate Time Series Models Toward Sustainable Smart Homes. *IEEE Access* **2024**, *12*, 55632–55668. [CrossRef]
40. Somu, N.; R, G.R.M.; Ramamritham, K. A deep learning framework for building energy consumption forecast. *Renew. Sustain. Energy Rev.* **2021**, *137*, 110591. [CrossRef]
41. Hussain, A.; Franchini, G.; Giangrande, P.; Mandelli, G.; Fenili, L. A Comparative Analysis of Machine Learning Models for Medium-Term Load Forecasting in Smart Commercial Building. In Proceedings of the 2024 IEEE 12th International Conference on Smart Energy Grid Engineering (SEGE), Oshawa, ON, Canada, 18–20 August 2024; pp. 228–232. [CrossRef]
42. Chandler, W.S.; Hoell, J.M.; Westberg, D.; Zhang, T.; Stackhouse, P.W. *NASA Prediction of Worldwide Energy Resource High Resolution Meteorology Data For Sustainable Building Design*; Technical Report; NASA: Washington, DC, USA, 2013.
43. Zhang, J.; Xu, Z.; Wei, Z. Absolute Logarithmic Calibration for Correlation Coefficient with Multiplicative Distortion. *Commun. Stat.-Simul. Comput.* **2023**, *52*, 482–505. [CrossRef]
44. Breiman, L. Random forests. *Mach. Learn.* **2001**, *45*, 5–32. [CrossRef]
45. Mallala, B.; Azeez Khan, P.; Pattepu, B.; Eega, P.R. Integrated Energy Management and Load Forecasting Using Machine Learning. In Proceedings of the 2024 2nd International Conference on Sustainable Computing and Smart Systems (ICSCSS), Coimbatore, India, 10–12 July 2024; pp. 1004–1009. [CrossRef]
46. Zhu, J.; Dong, H.; Zheng, W.; Li, S.; Huang, Y.; Xi, L. Review and prospect of data-driven techniques for load forecasting in integrated energy systems. *Appl. Energy* **2022**, *321*, 119269. [CrossRef]
47. Xie, Z.; Wang, R.; Wu, Z.; Liu, T. Short-Term Power Load Forecasting Model Based on Fuzzy Neural Network using Improved Decision Tree. In Proceedings of the 2019 IEEE Sustainable Power and Energy Conference (iSPEC), Beijing, China, 21–23 November 2019; pp. 482–486. [CrossRef]
48. Alrobaie, A.; Krarti, M. A Review of Data-Driven Approaches for Measurement and Verification Analysis of Building Energy Retrofits. *Energies* **2022**, *15*, 7824. [CrossRef]
49. Olawuyi, A.; Ajewole, T.; Oladepo, O.; Awofolaju, T.T.; Agboola, M.; Hasan, K. Development of an Optimized Support Vector Regression Model Using Hyper-Parameters Optimization for Electrical Load Prediction. *UNIOSUN J. Eng. Environ. Sci.* **2024**, *6*. [CrossRef]

50. Friedman, J.H. Greedy Function Approximation: A Gradient Boosting Machine. *Ann. Stat.* **2001**, *29*, 1189–1232. [CrossRef]
51. Lazzari, F.; Mor, G.; Cipriano, J.; Gabaldon, E.; Grillone, B.; Chemisana, D.; Solsona, F. User behavior models to forecast electricity consumption of residential customers based on smart metering data. *Energy Rep.* **2022**, *8*, 3680–3691. [CrossRef]
52. Dabrowski, J.J.; Zhang, Y.; Rahman, A. ForecastNet: A Time-Variant Deep Feed-Forward Neural Network Architecture for Multi-step-Ahead Time-Series Forecasting. In *Neural Information Processing. ICONIP 2020*; Yang, H., Pasupa, K., Leung, A.C.S., Kwok, J.T., Chan, J.H., King, I., Eds.; Lecture Notes in Computer Science; Springer: Cham, Switzerland, 2020; Volume 12534. [CrossRef]
53. Forootan, M.M.; Larki, I.; Zahedi, R.; Ahmadi, A. Machine Learning and Deep Learning in Energy Systems: A Review. *Sustainability* **2022**, *14*, 4832. [CrossRef]
54. Rafi, S.H.; Nahid-Al-Masood.; Deeba, S.R.; Hossain, E. A Short-Term Load Forecasting Method Using Integrated CNN and LSTM Network. *IEEE Access* **2021**, *9*, 32436–32448. [CrossRef]
55. Hussain, A.; Giangrande, P.; Franchini, G.; Fenili, L.; Messi, S. Analyzing the Effect of Error Estimation on Random Missing Data Patterns in Mid-Term Electrical Forecasting. *Electronics* **2025**, *14*, 1383. [CrossRef]
56. Hakyemez, T.C.; Adar, O. Testing the Efficacy of Hyperparameter Optimization Algorithms in Short-Term Load Forecasting. *arXiv* **2024**, arXiv:2410.15047.

applied sciences

MDPI

Article

Spatio-Temporal Multi-Graph Convolution Traffic Flow Prediction Model Based on Multi-Source Information Fusion and Attention Enhancement

Wenjing Li *, Zhongning Sun and Yao Wan

School of Resources and Environmental Engineering, Wuhan University of Science and Technology, Wuhan 430081, China; 202301703048@wust.edu.cn (Z.S.); 202201704127wy@wust.edu.cn (Y.W.)

* Correspondence: liwenjing@wust.edu.cn; Tel.: +86-18971576123

Featured Application

MIFA-ST-MGCN has significant potential applications in intelligent transportation systems (ITS), particularly for real-time traffic flow prediction and urban traffic management. By integrating multi-source data such as weather conditions and Points of Interest (POI), the model can provide more accurate and adaptive traffic forecasting, essential for optimizing traffic signal control, route planning, and congestion management. Additionally, its robustness to noisy data makes it ideal for deployment in dynamic, real-world environments where data quality can fluctuate. The ability to dynamically weight temporal and spatial features allows for more precise traffic predictions, which can directly improve the efficiency and safety of urban transportation systems, reduce congestion, and contribute to sustainable urban planning.

Abstract

Traffic flow prediction plays a vital role in intelligent transportation systems, directly affecting travel scheduling, road planning, and traffic management efficiency. However, traditional methods often struggle to capture complex spatiotemporal dependencies and integrate heterogeneous data sources. To overcome these challenges, we propose a Spatio-temporal Multi-graph Convolution Traffic Flow Prediction Model based on Multi-source Information Fusion and Attention Enhancement (MIFA-ST-MGCN). The model adopts adaptive data fusion strategies according to spatiotemporal characteristics, achieving effective integration through feature concatenation and multi-graph structure construction. A spatiotemporal attention mechanism is designed to dynamically capture the varying contributions of different adjacency relations and temporal dependencies, thereby enhancing feature representation. In addition, recurrent units are combined with graph convolutional networks to model spatiotemporal data and generate more accurate prediction results. Experiments conducted on a real-world traffic dataset demonstrate that the proposed model achieves superior performance, reducing the mean absolute error by 3.57% compared with mainstream traffic flow prediction models. These results confirm the effectiveness of multi-source fusion and attention enhancement in improving prediction accuracy.

Keywords: traffic flow prediction; multimodal data; graph convolutional network; attention mechanism

1. Introduction

With the profound progress of urbanization, as of 2023, the total number of motor vehicles nationwide has reached 435 million, and the average annual number of newly registered vehicles exceeds 34 million. The rapid growth of the vehicle population has significantly exacerbated traffic congestion and accident risks. From 2015 to 2020, the average annual growth rate of the number of traffic accident fatalities was 3.35%. The peak congestion index in major cities generally exceeds 2.0, and in developed cities, the congestion index during the morning and evening rush hours even reaches 3.0. Meanwhile, the frequent starts and stops resulting from congestion not only decrease traffic efficiency but also indirectly trigger emotional fluctuations among drivers, thereby further increasing the likelihood of accidents [1–4]. In this context, Intelligent Transportation Systems (ITS) have emerged as the crucial solution to these challenges. Traffic flow prediction, an essential component of ITS, can offer scientific support for the management and planning of urban transportation systems. Therefore, enhancing the accuracy of traffic flow prediction and formulating effective traffic management strategies to alleviate congestion, reduce accident probabilities, and improve travel efficiency are of paramount importance for urban traffic management, enhancing network efficiency, and reducing energy consumption. Traditional traffic prediction methods typically focus on extracting spatiotemporal characteristics from historical traffic data to forecast future traffic conditions. However, traffic conditions are not solely determined by historical traffic data; they are also significantly influenced by external information, including weather conditions, Points of Interest (POI), road conditions, and other environmental factors [5–9].

We: (1) designs a multi-source information fusion module that deeply explores the external temporal relationships between traffic flow and environmental factors, as well as their inherent spatial correlations with the road network structure, through feature-level fusion and multi-graph convolution fusion, thereby enhancing the model's ability to perceive complex scenarios; (2) designs a spatiotemporal attention module that dynamically adjusts the model's spatiotemporal focus on different time periods and regions through an attention mechanism, improving the prediction accuracy of the model; (3) proposes a Spatio-temporal Multi-graph Convolution Traffic Flow Prediction Model Based on Multi-source Information Fusion and Attention Enhancement (MIFA-ST-MGCN). This model integrates multi-source heterogeneous information, fusing external environmental data, regional similarity features, and traffic flow information, and incorporates the spatiotemporal attention module for feature enhancement. Ultimately, the traffic flow is predicted through a spatiotemporal graph convolutional network, and the model's performance is evaluated using real traffic flow datasets. Additionally, we design ablation experiments to validate the effectiveness of attribute data fusion, multi-graph convolution, and spatiotemporal attention mechanisms. The model is also subjected to perturbation analysis, where Gaussian and Poisson noises are added to the original data to test the model's robustness and stability.

2. Related Works

Traffic flow prediction is a crucial component of Intelligent Transportation Systems (ITS) and plays an important role in urban traffic control and development. Traffic flow prediction has undergone different evolutionary stages. In early research in the field of traffic flow prediction, due to limited understanding of the problem and technical constraints, traffic flow prediction was simply described as a time series statistical task. In this approach, mathematical statistics were used to extract linear or periodic patterns from historical traffic data to predict future traffic conditions. Among these methods, the Historical Average Model (HA) [10] predicts future traffic flow states by using historical averages. This model is simple in principle, computationally fast, but suffers from low prediction accuracy and is difficult to apply to complex traffic scenarios. Time series models such as the ARMA model and its variants [11–13] are based on autoregressive and moving average models. They predict based on the relationship between current and historical data while modeling periodicity and trends in the data. These models can capture the linear dependencies of the data well and are suitable for short-term traffic flow prediction where the data exhibits clear periodicity. These models are interpretable, computationally efficient, and cost-effective, making them widely used in simpler forecasting scenarios. However, these models are based on the assumption of time series stability and cannot handle nonlinear traffic features, making them ill-suited for dynamic changes.

With the continuous advancement of computer technology, machine learning-based traffic prediction models have gradually been applied in the field of traffic forecasting. Representative algorithms include K-Nearest Neighbors (KNN), Support Vector Machines (SVM), and Bayesian Networks, among others. The K-Nearest Neighbors algorithm [14] selects the k most similar samples from historical data to the current state and predicts traffic flow through a weighted average. SVM [15] use kernel functions to map low-dimensional nonlinear traffic data to high-dimensional spaces, constructing an optimal separating hyperplane to handle nonlinear features. Bayesian Networks [16] model variable conditional dependencies by constructing a directed acyclic graph (DAG) and fit traffic flow using a Gaussian Mixture Model (GMM) with joint probability distributions, further predicting future traffic conditions. Compared to models based on mathematical statistics, these machine learning-based algorithms are capable of modeling more complex traffic flow features. However, they have limited ability to capture nonlinear traffic characteristics and long-term dependencies.

In recent years, deep learning has attracted significant attention from researchers due to its advantages in capturing nonlinear features and handling complex scenarios. Traffic flow data is a typical form of time-series data, and extracting its inherent temporal characteristics is one of the key challenges in traffic flow prediction. Recurrent Neural Networks (RNNs) have been widely applied in traffic prediction tasks due to their ability to effectively capture temporal dependencies in time-series data [17]. However, during the backpropagation process, the gradient can vanish, causing RNNs to be influenced only by short-term memory, thereby failing to capture long-term temporal features. To address this issue, researchers have designed Long Short-Term Memory (LSTM) networks and Gated Recurrent Units (GRU) to capture long-term temporal dependencies in traffic flow [18,19]. Although these models can capture the temporal dependencies of traffic flow, researchers have gradually recognized the importance of spatial dependencies in traffic prediction tasks. By incorporating Convolutional Neural Networks (CNNs) to extract spatial information from roads and combining them with LSTMs, the prediction accuracy can be improved [20]. However, since CNNs are primarily designed for regular grid data, they cannot directly handle irregular topological structures, thus failing to fully capture the spatial dependencies of traffic flow. Graph Convolutional Networks (GCNs), on the other hand, can directly process irregular graph network structures, enabling a better exploration of the spatial characteristics of traffic networks [21].

Beyond historical traffic data and spatial topology, traffic flow is also influenced by a variety of external factors. For instance, weather conditions, special events, and the spatial distribution of points of interest can all introduce disturbances to traffic patterns [22–24]. Moreover, regions with similar road network structures often exhibit comparable traffic behaviors, and interregional interactions further affect the evolution of traffic flows. Effectively integrating these external factors and capturing interregional dependencies within predictive models remains a significant challenge in contemporary traffic flow forecasting [25–27].

Notably, some recent research on fundamental traffic flow theory has provided new perspectives for understanding and predicting traffic congestion. On the one hand, studies from a traffic flow phase-transition perspective have proposed a "congestion boundary" approach to characterize the critical transformation between free-flow and congested states. This method leverages the bimodal distribution of traffic parameters (e.g., speed, density) to identify threshold values that separate free flow from congestion, thereby quantifying the critical conditions for traffic breakdown. For example, using real highway data, Lee et al. [28] estimated a congestion boundary at approximately 66.9 km/h (speed), 22.8 vehicles per kilometer (density), and about 341 vehicles per five-minute interval (flow). Based on these thresholds, it was observed that actual roadways can experience a phase transition into congestion when flow reaches only around 72–83% of the conventional theoretical capacity. Thus, the congestion boundary method provides a theoretical foundation for pinpointing the onset of congestion and enabling effective congestion management. On the other hand, another line of research has drawn inspiration from the theory of phase transitions in simple fluids, treating traffic flow as a fluid system analogous to a gas–liquid phase transition, in order to explore the scaling laws of urban congestion [29]. Laval et al. proposed that the fundamental diagram of traffic flow is analogous to the coexistence curve in gas–liquid phase transitions. Using this analogy, they demonstrated that urban traffic dynamics obey scaling relations characteristic of the Kardar–Parisi–Zhang (KPZ) universality class. Moreover, they found that the "costs" of congestion (such as travel delays and fuel consumption) scale superlinearly with city size (population), with growth even higher than predicted by conventional urban scaling theories [30]. Taken together, these macro-level theoretical studies provide new insights into the underlying mechanisms of traffic congestion and offer important guidance for alleviating congestion in large cities. In addition, in the related task of travel time prediction, spatiotemporal deep learning models have achieved remarkable progress. Lee et al. combined a Gated Recurrent Unit (GRU) network with spatiotemporal analysis, proposing a model for highway travel-time prediction. This model explicitly integrates spatial dependencies among road segments and temporal dependency features within the GRU, effectively reducing the lag of predicted travel times relative to actual conditions. Experiments demonstrated that a GRU model augmented with spatiotemporal features outperforms traditional RNNs, LSTMs, as well as a GRU baseline without spatial information, achieving the highest accuracy in travel time prediction at both the segment and route levels. This finding indicates that incorporating spatial correlations into time-series predictions can significantly enhance the accuracy of travel time estimates.

In summary, advanced models such as AST-GCN and ST-GRAT have achieved notable progress in spatiotemporal traffic forecasting. AST-GCN pioneered the integration of external (exogenous) information via an attribute-enhancement module, while ST-GRAT dynamically captures road-network dependencies through a carefully designed spatiotemporal attention mechanism. Nevertheless, these methods still have limitations: the fusion strategy in AST-GCN is relatively simple and does not fully exploit diverse spatial relations; and although ST-GRAT excels in attention modeling, its ability to fuse heterogeneous multi-source exogenous data (e.g., weather, POIs) remains underexplored. To address these limitations, we propose the MIFA-ST-MGCN model, whose core innovations are as follows:

1. Multi-level multi-source information fusion architecture: Unlike AST-GCN's simple attribute concatenation, our model constructs three complementary graphs—a geographic adjacency graph, a POI functional-similarity graph, and a spatial-similarity graph—to enable deep, layer-wise fusion within the graph-convolutional hierarchy.
2. Adaptive multi-graph fusion mechanism: We devise a learnable weighted fusion scheme that dynamically adjusts the relative contributions of different graph struc-

tures to the forecasting objective, thereby addressing the weight-allocation challenge in multi-source information fusion.

3. Enhanced temporal modeling capacity: Building on conventional GRU-based sequence modeling, we incorporate Transformer-style multi-head self-attention to better capture long-range dependencies, thereby mitigating GRU's limitations in long-sequence modeling.

The comparison of the characteristics of these models is shown in Table 1.

Table 1. Comparison of Model Characteristics.

Feature	AST-GCN	ST-GRAT	MIFA-ST-MGCN
Data fusion mechanism	Static/dynamic attribute splicing	No external information	Feature splicing + Multi graph convolution
Figure structure	Single topology	Topological graph with direction	Multiple graph structure
Attention mechanism	Not using attention mechanism	Self attention mechanism	Spatiotemporal attention mechanism

3. Model Design

MIFA-ST-MGCN captures space–time dependencies in traffic flow by integrating multi-graph convolution, spatiotemporal graph convolution, and spatiotemporal attention. The model can be divided into two branches: temporal and spatial. In the temporal branch, the model models the temporal attribute information and captures temporal features at different time scales through temporal attention. In the spatial branch, the model uses multi-graph convolution to capture various types of spatial dependencies, and combines spatial attention to better capture the influence of key road segments on traffic flow. Finally, the features from both the temporal and spatial branches are fused to generate the prediction results.

3.1. Problem Definition

The goal of traffic flow prediction is to forecast future traffic conditions based on historical states and both internal and external information. The traffic state of a road network is typically described using metrics such as traffic volume, average speed, and road occupancy.

Definition 1. Traffic Network Graph G. *In the field of traffic prediction, the traffic network can be represented as a graph $G = (V, A, E)$, where $V = \{v_1, v_2, \ldots, v_n\}$ denotes the set of sensors recording traffic-related information in the network, with n being the number of sensors,* i.e., *the number of nodes. $E = \{e_1, e_2, \ldots, e_m\}$ represents the set of road segments connecting pairs of sensors, where m is the number of road* segments, i.e., the number *of edges. $A \in R^{N \times N}$ is the adjacency matrix used to represent the connectivity between sensors, where $a_{ij} \in A$ is the element at the $i-th$ row and $j-th$ column of the adjacency matrix, indicating the connection status between nodes v_i and v_j. If $a_{ij} = 1$, it indicates that there is a road segment connecting nodes v_i and v_j; if $a_{ij} = 0$, it means there is no direct road segment connecting nodes v_i and v_j. Therefore, the adjacency matrix A is a binary matrix composed of 0 and 1.*

Definition 2. Traffic Flow Feature Matrix X. *We use traffic speed as the primary node feature on the road-network graph, forming a matrix X. $x_i^t \in X$ denotes the traffic speed at the $i-th$ sensor node at time t.*

Definition 3. Auxiliary Information K. *Auxiliary information refers to environmental factors that influence the traffic flow state. We represent the environmental factors that affect traffic condi-*

tions as node-level auxiliary features, denoted by $K = \{K_1, K_2, \ldots, K_p\}$*, where p is the number of categories of auxiliary features. The auxiliary feature information of category q is represented as* $K_q = \{q^1, q^2, \ldots, q^t\}$*, where* q_i^t *denotes the auxiliary feature information of category q at the* $i-th$ *sensor node at time t.*

To sum up, the traffic flow prediction problem can be viewed as learning the traffic flow information for the future time period m by combining the traffic network graph G, the flow feature matrix X, and the auxiliary information K, through the establishment of a function f, that models the relationship between these components, i.e.,

$$x_{t+m} = f(G, X, K). \quad (1)$$

3.2. Overall Framework

We proposes a Spatio-temporal Multi-graph Convolution Traffic Flow Prediction Model based on Multi-source Information Fusion and Attention Enhancement (MIFA-ST-MGCN). As shown in Figure 1, the model mainly consists of data preprocessing, a spatiotemporal attention module, and a spatiotemporal convolution module.

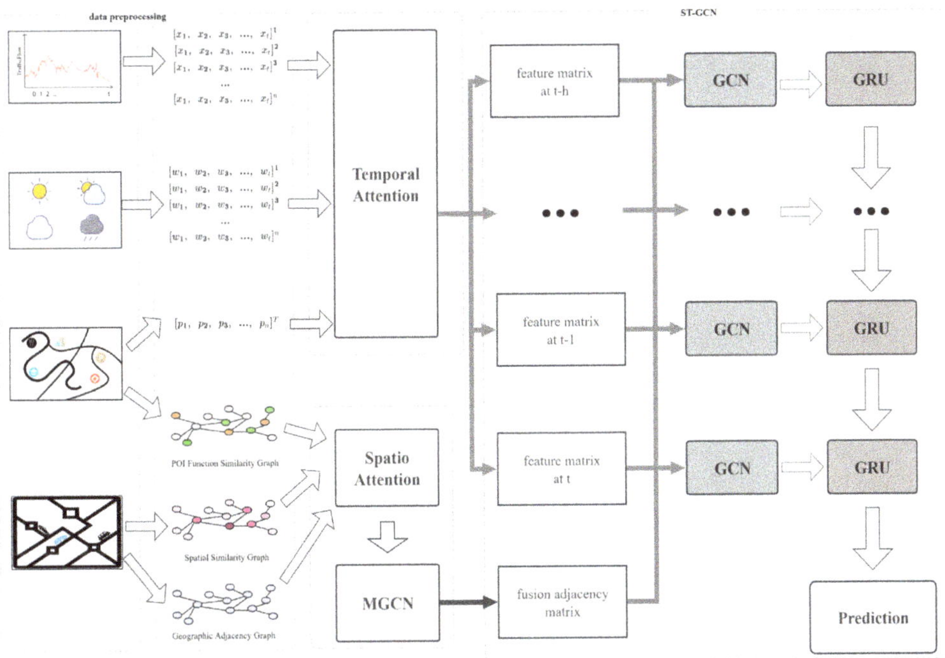

Figure 1. Overall Framework.

3.3. Spatio-Temporal Dependency Modeling

Traffic flow data exhibits dependencies not only in the spatial domain but also in the temporal domain. We choose spatiotemporal graph convolutional networks as the base prediction model. The Temporal Graph Convolutional Network (TGCN) [31] integrates graph convolutional networks with temporal forecasting, simultaneously modeling in both the spatial and temporal dimensions. This approach effectively captures the spatiotemporal features of traffic flow and addresses the spatiotemporal dependencies in traffic flow prediction.

3.3.1. Spatial Modeling: Graph Convolution Operation

Graph Convolutional Networks (GCN) capture the spatial dependencies of each road segment through neighborhood aggregation. The modeling process of GCN is illustrated in Figure 2 and can be divided into an input layer, hidden layers, and an output layer. The inputs to the GCN are twofold: the node feature matrix $X \in R^{N\times F}$, which describes the traffic conditions of each road segment, and the adjacency matrix $A \in R^{N\times N}$, which represents the connectivity between road segments, where N is the number of road segments and F is the feature dimension of each road segment. The hidden layers are the core components of the GCN, where convolution operations are defined on the graph structure to extract features. The computation for the $(l+1)-th$ layer hidden state can be expressed as:

$$H^{(l+1)} = \sigma\left(\widetilde{D}^{-\frac{1}{2}}\widetilde{A}\widetilde{D}^{-\frac{1}{2}}H^{l}W^{l}\right), \tag{2}$$

where $\widetilde{A} = A + I, \widetilde{A} \in R^{N\times N}$ is the adjacency matrix with self-loops added, $\widetilde{D} \in R^{N\times N}$ is the degree matrix, H^l is the output of the $l-th$ layer hidden state, W^l is the weight matrix of the $l-th$ layer, and $\sigma(\cdot)$ is a nonlinear activation function. The output layer performs classification and regression on the hidden layer data through a fully connected layer.

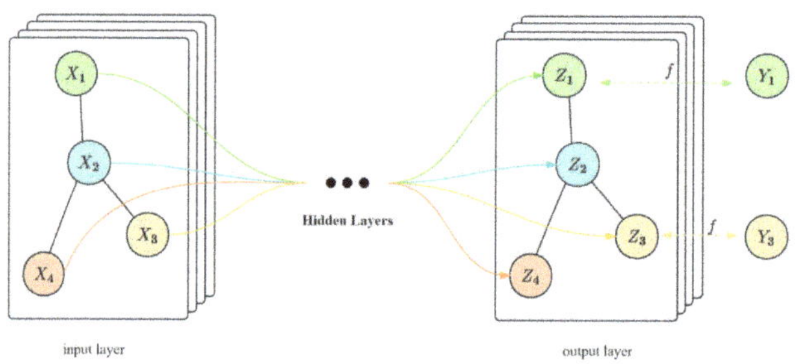

Figure 2. Graph Convolutional Networks.

3.3.2. Temporal Modeling: Graph Convolution Operation

Traffic flow data is a typical form of time-series data, and effectively capturing the temporal features within the data is crucial for prediction accuracy. We use Gated Recurrent Units (GRUs) to model the temporal dynamics in the data. Compared to traditional Recurrent Neural Networks (RNNs), GRU effectively mitigates the vanishing gradient problem in long time series by introducing a gating mechanism. In contrast to Long Short-Term Memory (LSTM) networks, GRU optimizes the gating mechanism, resulting in a simpler structure and higher computational efficiency.

The operation process of the GRU in handling time-series data is shown in Figure 3. At time step t, the GRU receives the traffic flow feature $x_t \in R^{B\times N\times(gru\_units+1)}$ at the current time step and the hidden state $h_{(t-1)}$ from the previous time step. Through the gating mechanism, it outputs the hidden state h_t at the current time step and passes it as the input hidden state to the next time step. The GRU's gating mechanism consists of two gates: the update gate and the reset gate. The update gate z_t determines how much of the past traffic flow features $[x_1, x_2, \ldots, x_{t-1}]$ should be retained at the current time step and how much of the current input feature x_t should be integrated into the new hidden state $\tilde{h} \in R^{B\times N\times gru\_units}$. The reset gate controls how much of the past traffic flow information should be forgotten. This process can be represented by Equations (3)–(6). Here, B represents the batch size, N represents the number of nodes, and gru_units refers to the number of GRU units in the layer. We utilize GCN to extract spatial features at each time step, which are then input into a GRU to capture temporal dependencies, thereby constructing a fundamental spatiotemporal feature extraction framework.

$$z_t = \sigma(W_z \cdot [h_{t-1}, x_t]), \tag{3}$$

$$r_t = \sigma(W_r \cdot [h_{t-1}, x_t]), \tag{4}$$

$$\tilde{h}_t = tanh(W \cdot [r_t \times h_{t-1}, x_t]), \tag{5}$$

$$h_t = (1 - z_t) \times h_{t-1} + z_t \times \tilde{h}_t. \tag{6}$$

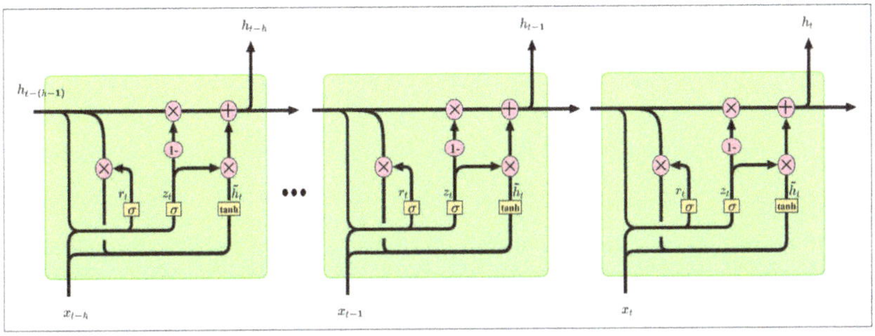

Figure 3. Gated Recurrent Units.

3.4. Multi-Source Information Fusion Modeling

As a complex open system, the transportation system is influenced by various factors such as weather conditions and geographical location, which in turn affect traffic flow states. Single-source traffic flow data is insufficient to comprehensively capture these complex influencing factors. Therefore, this model considers different types of information and designs various data fusion strategies to enhance the model's adaptability to external disturbances and environmental changes.

3.4.1. Graph Structure Construction

The changes in traffic conditions are closely related to the spatial correlations between regions. We model inter-regional spatial correlations by constructing spatial and functional similarity graphs.

Spatial Similarity Graph: According to Tobler's First Law of Geography [32], spatial entities exhibit spatial autocorrelation, namely, closer entities are more strongly correlated, whereas entities farther apart tend to be more dissimilar. Based on this principle, we construct a spatial similarity graph $G_{ss} = (V, A_{ss}, E)$, where A_{ss} denotes the spatial similarity matrix. The spatial similarity between nodes v_i and v_j, denoted as $A_{ss}(v_i, v_j)$, is calculated as shown in Equation (7):

$$A_{ss}(v_i, v_j) = \frac{1}{PL(v_i, v_j) + 1}, \tag{7}$$

where $PL(v_i, v_j)$ represents the path length between nodes v_i and v_j.

Functional Similarity Graph: As the distance between regions increases, the spatial correlation between them gradually decreases. However, due to the potential similarity in the distribution of Points of Interest (POI), the traffic flow states between regions may exhibit similar patterns of change. To deeply explore the correlation between POI distribution and regional traffic flow state changes, we use the Jensen–Shannon (JS) divergence to measure the functional similarity between two nodes and define the functional similarity graph $G_{fs} = \left(V, A_{fs}, E\right)$. A_{fs} is the functional similarity matrix, and the functional similarity $A_{fs}(v_i, v_j)$ between nodes v_i and v_j is calculated as shown in Equations (8)–(10):

$$A_{fs}(v_i, v_j) = \begin{cases} 1 - D_{JS}(v_i||v_j) & D_{JS}(v_i||v_j) \in [0,1] \\ 0 & else \end{cases}, \tag{8}$$

$$D_{JS}(v_i||v_j) = \frac{1}{2}D_{KL}\left(v_i||\frac{(v_i+v_j)}{2}\right) + \frac{1}{2}D_{KL}\left(v_j||\frac{(v_i+v_j)}{2}\right), \tag{9}$$

$$D_{KL}(v_i||v_j) = v_i(p)log\left(\frac{v_i(p)}{v_j(p)}\right). \tag{10}$$

where $D_{KL}(v_i||v_j)$ represents the KL divergence between nodes v_i and v_j, and $D_{JS}(v_i||v_j)$ represents the JS divergence between nodes v_i and v_j. As shown in Equation (8), $D_{JS}(v_i||v_j) = D_{JS}(v_j||v_i)$, and $v_i(p)$ represents the POI distribution feature of node v_i.

3.4.2. Multi Graph Convolution Fusion

The impact of spatial data of the same type on the traffic state varies across different road segments, and the influence of different types of spatial data on the traffic state of the same road segment also differs. We employ a multi-graph convolutional fusion strategy to model the spatial dependencies that shape traffic flow. The processing flow of multi-graph convolution is illustrated in Figure 4.

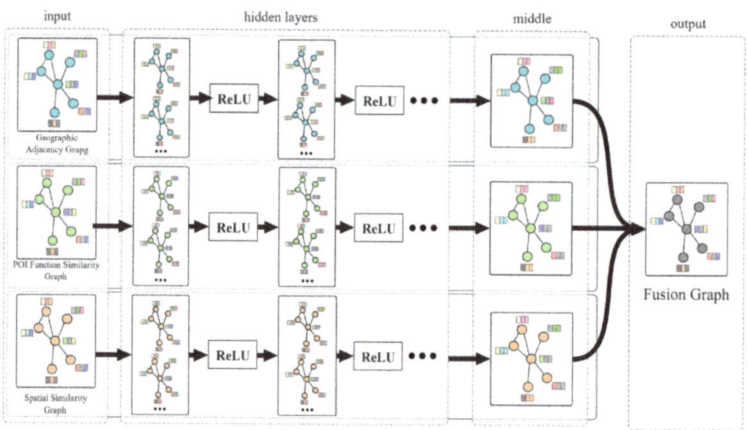

Figure 4. Multi-Graph Convolutional Networks.

The multi-graph convolution fusion strategy can be represented as

$$A_{fused} = \alpha_{geo}A^{(geo)} + \alpha_{poi}A^{(poi)} + \alpha_{spa}A^{(spa)}, \alpha_{geo} + \alpha_{poi} + \alpha_{spa} = 1, \tag{11}$$

where $A^{(geo)}$ represents the Geographic Adjacency Matrix, $A^{(poi)}$ is the POI Function Matrix, and $A^{(spa)}$ is the Spatial Similarity Matrix. As shown in Equation (11), we assign a learnable weight to each graph structure and optimize these weights end-to-end via gradient-based training. We apply a Softmax to normalize the weights so that α_1, α_2 and α_3 are nonnegative and sum to one, ensuring a well-balanced contribution of each component to the fused adjacency matrix. This multi-graph convolution fusion strategy allows the model to adaptively emphasize different spatial relations based on the data: for example, traffic states may at certain times be driven more by physically adjacent links, whereas in other scenarios synchronous fluctuations among functionally similar regions may dominate. Compared with models that rely on a single adjacency, multi-graph convolution can dynamically capture heterogeneous types of spatial correlation, thereby enhancing the model's ability to characterize complex traffic patterns.

3.4.3. Feature Level Fusion

We integrate static attributes (e.g., POI) and dynamic exogenous variables (e.g., weather) with the traffic-flow inputs via feature concatenation. Compared to other methods, the feature concatenation operation is simple, and the model does not need to consider the spatial correlation effects between different nodes, allowing the model to focus more on the processing of temporal features.

The attribute features of static attribute data are fixed and do not change over time. For example, in the case of POI data, the distribution and quantity of POI remain constant within a given time and spatial range and do not change over time. Therefore, the feature matrix S of static attribute data can be represented as $S = [s_1, s_2, \ldots, s_n]^T$, where s_i represents the static attribute feature of the $i - th$ node, and n is the total number of nodes. The significant characteristic of dynamic attribute features is that their attribute information changes over time. For example, weather conditions change dynamically at different time points, and the weather at the current time is influenced by past weather conditions and affects future weather conditions. Therefore, the feature matrix D of dynamic attribute data can be represented as $D = [d_1, d_2, \ldots, d_n]$, and the dynamic feature information of node i, d_i, is represented as $[d_i^{t-m}, d_i^{t-(m+1)}, \ldots, d_i^t]$, where t is the current time step, and m is the historical time step length. The fused attribute feature matrix at time t can be represented as Equation (12):

$$E^t = [X^t, S, D^{t-m,t}]. \tag{12}$$

$E^t \in R^{n\times(p+1+w*m)}$ represents the feature concatenation matrix at time t, where $X^t \in R^{N\times F}$ denotes the traffic flow features (e.g., speed) of all N road segments at time t. N denotes the number of nodes, and F denotes the feature dimensionality. $S = [s_1, s_2, \ldots, s_{F_s}], S \in R^{N\times F_s}$ is the static attribute feature matrix, and F_s is the number of static attribute types. Since static attribute information does not change over time, the static attributes are repeatedly used at different time steps. $D^{t=m,t} = [d_1^{t-m,t}, d_2^{t-m,t}, \ldots, d_{F_d}^{t-m,t}]$, $D^{t-m,t} \in R^{N\times F_d\cdot(m+1)}$ is the dynamic attribute feature matrix from $t - m$ to t, where F_d is the number of dynamic attribute types. Since dynamic attribute features change over time, the information from time $t - m$ to t is selected as the input for time t when modeling the dynamic attribute matrix. Adopting straightforward feature concatenation to fuse attribute data is simple and effective to implement. Compared with more complex fusion schemes, plain concatenation avoids introducing excessive parameters at the fusion stage, deferring importance weighting to subsequent attention modules and thereby mitigating overfitting risk.

We use POI data as a semantically explicit static descriptor and weather data as a dynamic feature of traffic flow. Specifically, for POI we take, for each road segment, the dominant POI-category code as its POI feature, apply min–max normalization to scale it to [0, 1], and replicate it across time to fill the temporal dimension to length T. For weather, we likewise use the weather-category code as the weather feature, min–max normalize it to [0, 1], and replicate it across space to fill the spatial dimension to N segments. For traffic-flow variables, we directly apply min–max normalization to [0, 1]. Through this dimensional replication (broadcasting), all inputs are expanded to a unified shape of $N \times N \times T$ and then fed into subsequent modules for feature extraction and computation.

3.5. Attention-Enhancing Mechanism

Traffic flow data contains latent features that are difficult to capture in both the spatiotemporal dimensions. In the temporal dimension, traffic flow is nonlinear and dynamically changing, influenced by various external factors. In the spatial dimension, traffic flow exhibits complex spatial interactions, where the traffic state of other regions can directly or indirectly affect the traffic flow in the local region. We introduce a spatiotemporal attention module to capture spatiotemporal dependencies in traffic-flow data. The module consists of both temporal and spatial attention mechanisms. Time attention is used to dynamically adjust the weights of different time steps based on the historical sequence, while spatial attention highlights key road segments and graph structures that have a significant impact on traffic flow. Through a spatiotemporal multi-head self-attention mechanism, the model dynamically adjusts its attention to different time periods and regions, enhancing its ability to capture complex spatiotemporal dependencies.

The overall architecture of the spatiotemporal attention module is shown in Figure 5. This module employs a multi-head self-attention mechanism to capture the spatiotemporal dependencies in traffic flow data. The multi-head self-attention mechanism projects the traffic flow feature data onto multiple attention heads, and the weights for each attention head are computed in parallel. Then, the outputs of each attention head are concatenated together through a weighted fusion. The computation process of multi-head self-attention can be expressed as

$$\begin{cases} Q_i = XW_{Q_i} \\ K_i = XW_{K_i} \quad i = 1,2,\ldots,h, \\ V_i = XW_{V_i} \end{cases} \tag{13}$$

$$Attention(Q,K,V) = softmax\left(\frac{QK^T}{\sqrt{d_k}}\right) \cdot V, \tag{14}$$

$$Output = concat(Attentionn_1, Attention_2, \ldots, Attention_h). \tag{15}$$

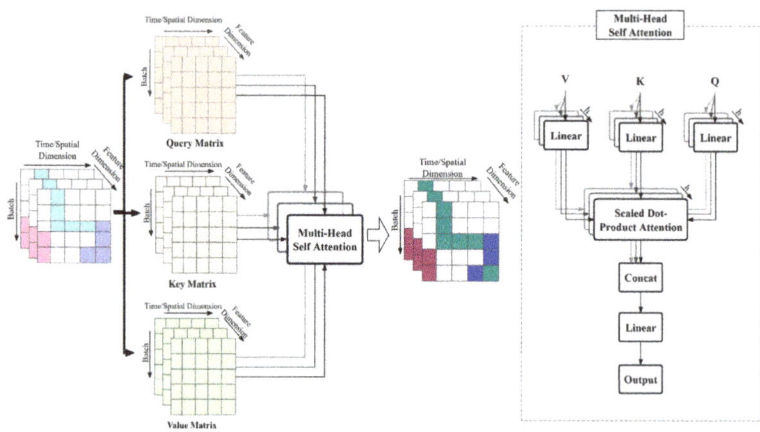

Figure 5. Spatio-temporal attention module & multi-head self-attention.

In the equations, Q_i represents the feature of the current time step or the current road segment; K_i denotes the traffic flow features of the historical time steps or other road segments; V_i represents the weighted features of each time step or road segment; h is the number of attention heads; X is the feature matrix. When processing temporal features, $X \in R^{B \times T \times F}$, where B is the batch size, T is the number of time steps, and F is the feature dimension. When using the attention mechanism to capture dynamic correlations between nodes, $X \in R^{B \times N \times F}$, where N is the number of nodes, i.e., the number of road segments. $W_Q, W_K, W_V \in R^{F \times d_k}$ are learnable weight matrices, and d_k is the dimension of the Query and Key matrices.

3.6. Spatio-Temporal Multi Graph Convolution Model Based on Multi-Source Information Fusion and Attention Enhancement (MIFA-ST-MGCN)

Based on spatiotemporal graph convolutional networks, we have designed a spatiotemporal multi-graph convolution model with multi-source information fusion and attention enhancement to address the complex spatial dependencies and temporal dependencies in traffic flow prediction. As shown in Figure 6, the attribute feature fusion module concatenates the dynamic and static attribute matrices at each time step into the feature matrix X^t to expand its feature dimension. The expanded feature matrix is denoted as $E^{t'}$, and then, through temporal attention, the weights of each time step are adjusted to obtain E^t. The spatial feature fusion module adjusts the importance of different graphs and road segments using spatial attention mechanisms and multi-graph convolution, resulting in the fused adjacency matrix. The enhanced feature matrix E^t and the fused adjacency matrix A_{fused} are then input into the model f to obtain the final prediction result $\overline{y}$:

$$\overline{y}_t = f\left(A_{fused}, E^t\right). \tag{16}$$

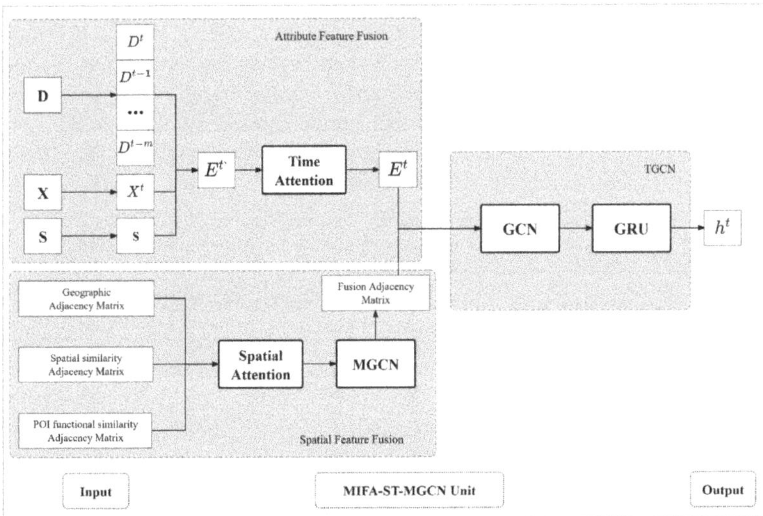

Figure 6. MIFA-ST-MGCN at time t.

We construct the model f by combining a Graph Convolutional Network (GCN) with Gated Recurrent Units (GRU) to capture spatiotemporal dependencies in traffic-flow data. Specifically, the fused adjacency matrix A_{fused} and the enhanced feature matrix E^t are first input into the GCN module, where multiple layers of graph convolution and non-linear activation are applied to extract spatial representations of each road segment at different time steps. These representations are subsequently fed into the GRU, which utilizes update and reset gates to regulate the preservation and adaptation of traffic information over time, thereby capturing temporal dynamics in the traffic flow.

The primary objective of traffic flow forecasting is to minimize prediction errors, ensuring that the predicted values closely approximate the actual observations. Accordingly, when designing the loss function, it is essential to reduce the prediction error to enhance the model's forecasting accuracy:

$$Loss = \frac{1}{n}\sum_{i=1}^{n}(y_i - \widetilde{y}_i)^2 + \lambda\sum_{j}\left|\left|\overline{\omega}_j\right|\right|^2. \quad (17)$$

Specifically, $\frac{1}{n}\sum_{i=1}^{n}(y_i - \widetilde{y}_i)^2$ denotes the L2 loss, which quantifies the discrepancy between the predicted and actual values, where $\widetilde{y}_i$ represents the predicted traffic flow of the $i - th$ road segment and y_i denotes the corresponding ground truth. The term $\lambda\sum_{j}\left|\left|\overline{\omega}_j\right|\right|^2$ corresponds to L2 regularization, which is introduced to mitigate overfitting. Here, λ is the regularization coefficient that controls the strength of the regularization.

4. Experiments

4.1. Datasets

In order to validate the effectiveness of the proposed model, we perform experiments on the real-world traffic dataset SZ_taxi [33] and METR_LA.

- **SZ_taxi**: This dataset consists of taxi trajectory data collected from 156 major road segments in Luohu District, Shenzhen, spanning the period from 1 January to 31 January 2015, with a sampling interval of 15 min. The dataset is composed of two parts: the speed feature matrix and the adjacency matrix. The speed feature matrix is organized with timestamps as row indices and road segments as column indices, resulting in a matrix of size 2976 × 156. The adjacency matrix is constructed to model the connectivity between road segments, with a size of 156 × 156.
- **SZ_POI**: This dataset contains the distribution of Points of Interest (POIs), encompassing nine categories: catering services, enterprises, shopping facilities, transportation infrastructure, educational services, living services, medical services, and accommodations. For each road segment, the most prevalent POI category within its surrounding area is selected as its static feature, resulting in a POI static matrix of size 156 × 1.
- **SZ_Weather**: This dataset records the weather conditions around the study area at 15 min intervals throughout January 2015. The data include five categories of weather: sunny, cloudy, foggy, light rain, and heavy rain. Based on these observations, a dynamic weather matrix of size 156 × 2976 is constructed.

4.2. Evaluation Indicators and Parameter Settings

4.2.1. Evaluation Indicators

We evaluate the predictive performance of the model using five commonly adopted metrics: Root Mean Square Error (*RMSE*), Mean Absolute Error (*MAE*), *Accuracy*, Coefficient of Determination (R^2), and Explained Variation (*VAR*).

- ***RMSE*:** A widely used metric for quantifying the discrepancy between predicted and actual values, where a smaller value indicates lower prediction error of the model.

$$RMSE = \sqrt{\frac{1}{n}\sum_{i=1}^{n}(y_i - \widetilde{y}_i)^2}. \quad (18)$$

- ***MAE*:** The average absolute difference between predicted and actual values, reflecting the model's ability to control errors. A smaller MAE indicates that the model can provide more stable and reliable predictions.

$$MAE = \frac{1}{n}\sum_{i=1}^{n}|y_i - \widetilde{y}_i|. \quad (19)$$

- ***Accuracy***: The proportion of predictions that exactly match the actual values. A higher accuracy indicates stronger predictive capability of the model.

$$Accuracy = 1 - \frac{1}{n}\sum_{i=1}^{n} \frac{\sqrt{(y_i - \widetilde{y})^2}}{y_i}. \tag{20}$$

- ***R*²**: A metric used to evaluate the goodness of fit of the model, where values closer to 1 indicate better predictive performance.

$$R^2 = 1 - \frac{\sum_{i=1}^{n} (y_i - \widetilde{y})^2}{\sum_{i=1}^{n} (y_i - \overline{y}_i)^2}. \tag{21}$$

- ***VAR***: A metric that measures the extent to which the model accounts for the overall variance in the data, where values closer to 1 indicate stronger predictive capability.

$$VAR = 1 - \frac{VAR(y - \widetilde{y})}{VAR(y)}. \tag{22}$$

4.2.2. Parameter Settings

All experiments were conducted on a computing platform equipped with an Intel Core i5-12600KF CPU and an NVIDIA RTX 4060 GPU with 8 GB memory. The model was implemented in Python (version 3.9.18) using the TensorFlow framework (version 2.10.0). After extensive parameter tuning, the number of hidden units in the GRU was set to 128, the number of heads in the multi-head attention mechanism was set to 4, and the Adam optimizer was adopted to accelerate model convergence. The learning rate of Adam was set to 0.001, the batch size to 64, and the total number of training epochs to 600.

4.3. Baseline Models

To evaluate the performance of the proposed MIFA-ST-MGCN model, comparative experiments were conducted against a variety of forecasting models, including both traditional approaches and deep learning-based methods.

1. **Historical Average (HA):** Predicts future traffic conditions by averaging the traffic flow observed during the same time intervals in historical data, under the assumption of periodic traffic patterns [10].
2. **Support Vector Machine Regression (SVR):** Formulates the time series forecasting task as a supervised learning problem, leveraging support vector machines to construct a regression function in high-dimensional space [34].
3. **Long Short-Term Memory (LSTM):** Employs gated mechanisms and memory cells to capture long-range temporal dependencies in sequential data [35].
4. **Gated Recurrent Units (GRU):** A simplified variant of LSTM with fewer gating mechanisms, resulting in reduced parameter complexity and improved computational efficiency [36].
5. **Temporal Graph Convolutional Network (TGCN):** Integrates graph convolution to capture spatial dependencies with recurrent units to model temporal dynamics, enabling joint spatio-temporal feature learning [31].
6. **Spatio-Temporal Graph Convolutional Network (STGCN):** Combines graph convolution with gated causal convolutions to simultaneously capture spatio-temporal dependencies, thereby improving both prediction accuracy and training efficiency [21].
7. **A3T-GCN:** Enhances temporal graph convolution by incorporating temporal attention, leading to improved prediction of traffic peaks [37].

8. **AST-GCN:** Introduces attribute augmentation units to explicitly integrate static and dynamic external factors, thereby improving traffic prediction accuracy and enhancing interpretability for sudden traffic fluctuations [8].

4.4. Baseline Comparison Experiments

Based on the aforementioned experimental settings and baseline models, we conducted comparative experiments to evaluate the effectiveness of the proposed model. All experiments were performed on the SZ dataset and MATR_LA data, and the results for all models are presented in Tables 2 and 3. As shown in the table, our MIFA-ST-MGCN model demonstrates superior performance compared with other models, as evidenced by its leading results in the majority of evaluation metrics. Compared with the traditional HA and SVR models, the RMSE of our model decreases by approximately 4.10% and 2.74%, while the prediction accuracy improves by 1.70% and 1.11%, respectively. Relative to LSTM and GRU, which only capture temporal dependencies, MIFA-ST-MGCN achieves reductions of 17.54% and 41.24% in RMSE, along with improvements of 9.17% and 38.43% in prediction accuracy. Furthermore, when compared with TGCN, STGCN, A3T-GCN, and AST-GCN, which incorporate spatio-temporal modeling, graph convolution, attention mechanisms, and attribute augmentation, respectively, our model achieves RMSE reductions of 13.17%, 54.91%, 14.20%, and 17.43%, as well as accuracy improvements of 6.39%, 86.13%, 7.01%, and 9.09%. These results validate the effectiveness of the proposed model.

Table 2. Results of comparative experiments (SZ-taxi dataset).

Evaluation Indicators	RMSE	MAE	Accuracy	R^2	Var
HA	4.2740	2.7985	0.7021	0.8325	0.8325
SVR	4.1966	2.7777	0.7075	0.8385	0.8400
LSTM	4.0665	2.7257	0.7166	0.8484	0.8486
GRU	4.2540	2.9616	0.7035	0.8341	0.8347
TGCN	4.3243	3.0398	0.6986	0.8287	0.8302
STGCN	8.3224	6.6436	0.4201	0.3652	0.3652
A3T-GCN	4.1070	2.7913	0.7138	0.8454	0.8456
AST-GCN	4.3240	2.9625	0.6987	0.8286	0.8286
Ours	**4.0600**	**2.6951**	**0.7171**	**0.8489**	**0.8491**

Bold is the best indicator.

Table 3. Results of comparative experiments (METR-LA dataset).

Evaluation Indicators	RMSE	MAE	Accuracy	R^2	Var
HA	7.4271	4.4556	0.8506	0.7442	0.7442
SVR	9.6734	6.1117	0.8397	0.6318	0.6326
LSTM	7.8820	5.5922	0.8658	0.6759	0.6877
GRU	8.4899	5.7926	0.8554	0.6240	0.6241
TGCN	6.8520	4.2676	0.8834	0.7551	0.7555
STGCN	10.3812	6.2751	0.7931	0.6536	0.6544
A3T-GCN	7.0729	4.5617	0.8555	0.7312	0.7317
AST-GCN	6.9356	4.8788	0.8663	0.7762	0.7762
Ours	**6.2740**	**4.2419**	**0.8932**	**0.7946**	**0.7952**

Bold is the best indicator.

From Tables 2 and 3, our model outperforms competing methods on both the SZ_taxi and METR_LA datasets. On SZ_taxi, because the margin over the LSTM baseline is relatively small, we conducted repeated trials comparing our model with LSTM. The results show that our model's RMSE is consistently lower by 0.0059–0.0957 across runs, corroborating the reliability of the observed performance gains and validating the effectiveness of our improvements.

4.5. Comparison of Prediction Performance for Different Time Periods

To comprehensively evaluate the performance of the model under different prediction horizons, we conducted experiments on both the proposed model and the baseline models with forecasting intervals of 15, 30, 45, and 60 min. The results are presented in Figure 7. The experiments show that our model consistently outperforms the baselines across all horizons. Although performance declines as the prediction horizon increases, our model still achieves better forecasting results than the baseline methods and exhibits a considerably smaller performance drop compared to the other models.

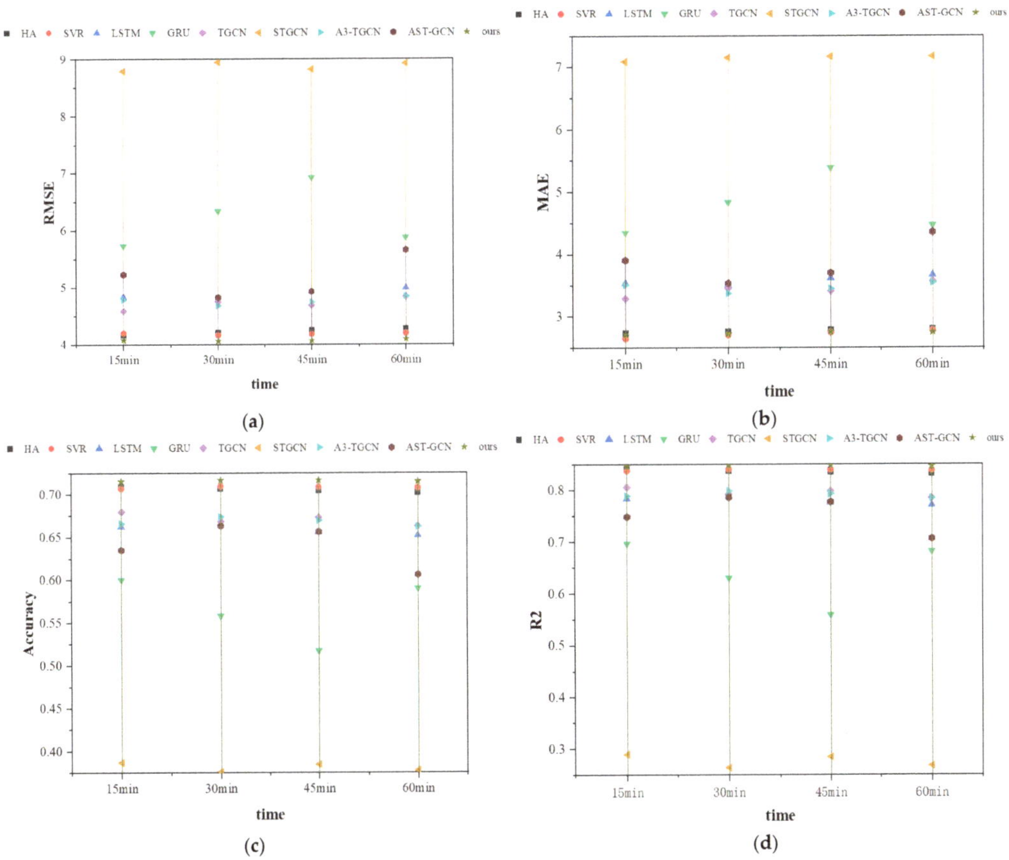

Figure 7. Performance of competing models across different forecasting horizons: (**a**) RMSE (km/h; lower is better), (**b**) MAE (km/h; lower is better), (**c**) Accuracy (higher is better; see Section 4.2 for definition), and (**d**) R^2 (higher is better). The x-axis lists horizons of 13, 30, 45, and 60 min. Results are reported on the SZ_taxi dataset (Jan-2015). Model names match Tables 2 and 3 (HA, SVR, LSTM, GRU, TGCN, STGCN, A3T-GCN, AST-GCN, Ours).

4.6. Ablation Experiments

To evaluate the effectiveness of the key components in the proposed forecasting model, we designed a series of ablation experiments. These experiments can be broadly categorized into two types. The first type is based on TGCN, where POI data and weather data are incorporated to verify the effectiveness of attribute data fusion. The second type is based on our MIFA-ST-MGCN model, where the attribute data fusion module, multi-graph convolution module, and spatio-temporal attention module are individually removed to examine their contributions. The results of the ablation experiments are summarized in Table 4.

Table 4. Results of ablation experiments. Y denotes the added data or module.

Modules					Evaluation Indicators			
Poi	**Weather**	**Poi + Weather**	**ST-Attention**	**MGCN**	**RMSE**	**MAE**	**Acc**	**R^2**
					4.3243	3.0398	0.6986	0.8287
Y					4.2794	2.9656	0.6893	0.8076
	Y				4.2712	2.9702	0.6911	0.8089
		Y			4.2679	2.9467	0.6978	0.8107
		Y		Y	4.2082	2.8621	0.7123	0.8113
		Y	Y		4.0988	2.7776	0.7145	0.8464
			Y	Y	4.1070	2.7913	0.7138	0.8456
		Y	Y	Y	4.0600	2.6951	0.7171	0.8489

4.7. Perturbation Analysis

Given the inherent complexity and uncertainty of real-world scenarios, traffic flow data inevitably contain noise, missing values, and outliers. To evaluate the stability and reliability of the proposed model under such conditions, we conducted a perturbation analysis by injecting Gaussian noise and Poisson noise into the original data. The Gaussian noise follows a normal distribution $N \in (0, \sigma^2)$ with $\sigma \in \{0.2, 0.4, 0.6, 0.8, 1.0\}$, while the Poisson noise follows a distribution $P(\lambda)$ with $\lambda \in \{1, 2, 4, 8, 16\}$. The experimental results, shown in Figure 8, indicate that the evaluation metrics remain relatively stable across different noise types and intensities. These findings demonstrate that our model maintains strong adaptability in complex real-world environments.

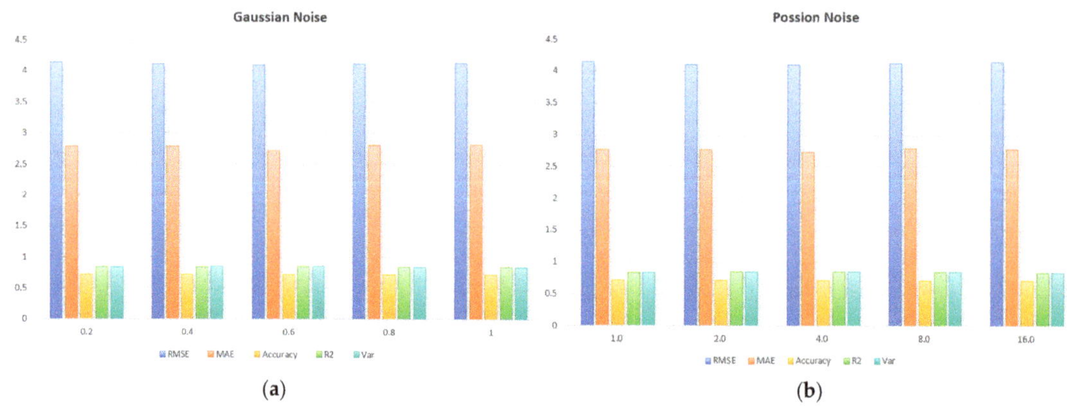

Figure 8. Results of noise experiments. Sub-figures illustrate (a) Gaussian Noise, (b) Poisson Noise

4.8. Visualization of Predictions

To more clearly demonstrate the predictive capability of our model, we conducted a visualization analysis by comparing the predicted results of different models with the ground truth speeds in the test set. Furthermore, we provided a more in-depth interpretation and evaluation of the models from two perspectives:

(1) Predictions of Different Models

Based on 2.5 h of historical traffic flow data, we employed different models to predict traffic speed for the following hour. The prediction results are presented in Figure 9. As illustrated, our model produces predictions that are closer to the ground truth than those of other models, both in terms of overall traffic flow trends and peak prediction accuracy.

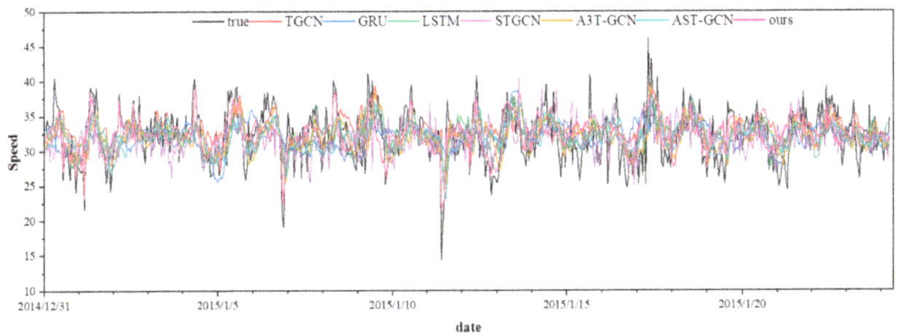

Figure 9. Visualization of predictions. Given 2.5 h of history to forecast the following 1 h, we plot the ground truth and predictions from all models (TGCN, GRU, LSTM, STGCN, A3T-GCN, AST-GCN, Ours). The y-axis reports speed in km/h.

(2) Effectiveness of Different Modules

To further investigate the importance of different modules, we visualized the results of the ablation experiments. Figures 10–12 present a comparison between the baseline TGCN model and the models incorporating external information fusion, spatiotemporal attention, and multi-graph convolution. Figure 10 illustrates the prediction results of the model enhanced with POI and weather data. Compared with TGCN, the inclusion of POI and weather information improves the model's perception capability and prediction accuracy. For instance, on 5 January 2015, when heavy rainfall occurred, the model incorporating weather data produced predictions that were much closer to the ground truth, thereby demonstrating the effectiveness of external information fusion. Figures 11 and 12 show the prediction results of the models with spatiotemporal attention and multi-graph convolution, respectively. Compared with TGCN, the model with spatiotemporal attention exhibits superior performance in capturing overall traffic flow trends, while the model with multi-graph convolution achieves better accuracy in peak prediction.

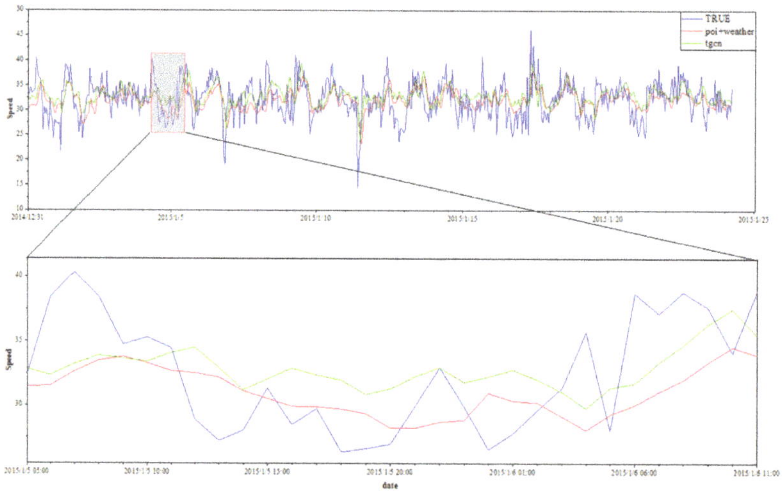

Figure 10. Prediction results of the baseline model versus the model enhanced with external information. We plot the ground truth, TGCN (baseline), and TGCN + POI + Weather in the same time window (Speed in km/h). On 5 January 2015 with heavy rainfall, the externally enhanced model aligns more closely with the ground truth, evidencing the benefit of information fusion.

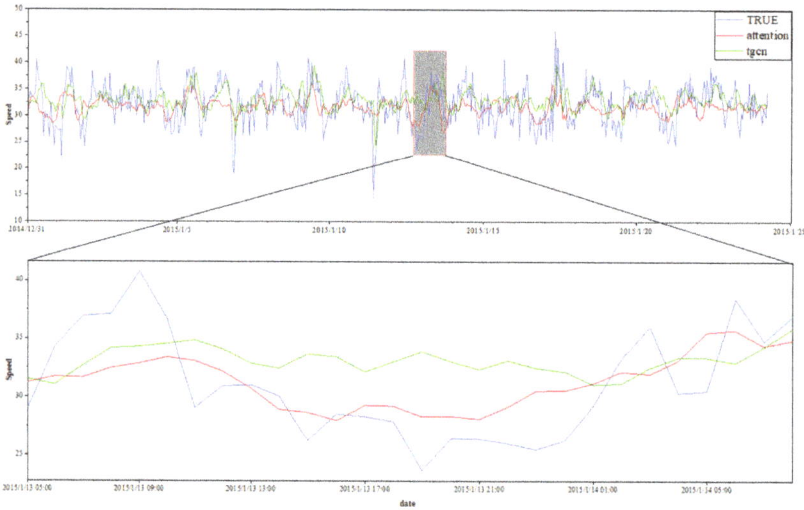

Figure 11. Prediction results of the baseline model versus the spatio-temporal attention-enhanced model. Compared with TGCN (baseline), the attention-enhanced model better captures overall traffic-flow trends in the same time window (Speed in km/h).

Results show that our model adheres more closely to the ground-truth curves during sharp peak periods such as the morning and evening rush hours. This advantage arises from two complementary mechanisms. First, the spatial branch employs multi-graph convolution with Softmax-normalized, learnable fusion weights, enabling the characterization of corridor-like inter-segment couplings and functional similarities—particularly effective in areas with simultaneous demand surges (e.g., commercial districts and transportation hubs). Second, the temporal branch incorporates multi-head self-attention to reweight the most predictive lags during the phases of queue formation and dissipation, thereby improving relative alignment and amplitude depiction around peaks. Moreover, by incorporating weather and POI information, the model can identify traffic-flow variations induced by meteorological conditions or POI types, allowing rapid adaptation to changing exogenous factors and enhancing both the realism and interpretability of the forecasts.

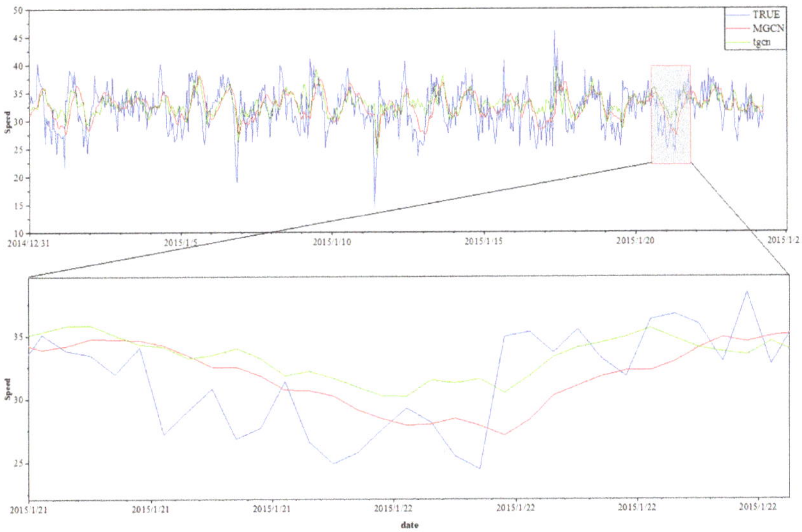

Figure 12. Prediction results of the baseline model versus the multi-graph convolution-enhanced model. Relative to TGCN (baseline), TGCN + MGCN achieves better accuracy on peak depiction. Curves are shown for the same time window (Speed in km/h).

5. Conclusions

We propose a spatio-temporal multi-graph convolution traffic flow prediction model, termed MIFA-ST-MGCN, which integrates multi-source information fusion and attention enhancement to improve prediction accuracy and robustness. By combining Temporal Graph Convolutional Networks (TGCN), multi-source information fusion, and spatio-temporal attention mechanisms, we design a predictive framework capable of effectively capturing spatio-temporal dependencies in traffic flow while accounting for external environmental factors. Comparative experimental results demonstrate that MIFA-ST-MGCN outperforms existing baseline models across multiple evaluation metrics, validating its effectiveness in complex traffic scenarios. Ablation studies further confirm the contributions of feature fusion, multi-graph convolution, and spatio-temporal attention, while perturbation experiments show that the model maintains strong adaptability under noisy and uncertain conditions.

Despite the strong predictive performance demonstrated by the proposed model, several limitations remain. First, the model's scalability is constrained by the computational complexity inherent in its key components. The spatio-temporal multi-head self-attention mechanism, while effective, incurs a computational cost of $O(T \times N^2 \times F)$ for spatial attention and $O(N \times T^2 \times F)$ for temporal attention, where N is the number of nodes, T is the sequence length, and F is the feature dimension. This quadratic dependency on N and T can become a bottleneck when applied to large-scale metropolitan networks with thousands of nodes. Future work will explore more efficient attention mechanisms, such as linearized or sparse attention, to mitigate this issue. Second, the model's performance is sensitive to certain hyperparameters, such as the number of graph convolution layers, the number of attention heads, and the dimensionality of GRU hidden states. Although we conducted extensive parameter tuning for this study, a comprehensive sensitivity analysis was not included. The depth of the multi-graph convolution module, in particular, requires careful design; preliminary ablation experiments (varying layers from 1 to 3) indicated that while a 2-layer structure offered the best trade-off, deeper architectures risk over-smoothing and increased computational overhead without commensurate gains in accuracy. Third, the model's reliance on multi-source data (e.g., POI, weather), while beneficial for accuracy, introduces practical deployment challenges in data-scarce environments. The requirement for high-quality, synchronized external data may limit the model's applicability in regions where such data are incomplete or unavailable. Future iterations could investigate semi-supervised or self-supervised learning strategies to reduce dependency on extensive labeled and auxiliary data. Lastly, the current framework assumes a static graph topology, which restricts its ability to adapt to dynamic network changes caused by traffic incidents or temporary road closures. Integrating dynamic graph construction techniques or temporal graph networks could enhance the model's responsiveness to real-time structural variations.

In summary, the proposed MIFA-ST-MGCN model provides an effective solution to the traffic flow forecasting problem, carrying both theoretical significance and practical value. With the increasing complexity of urban transportation systems and the growing volume of data, the model can be further optimized and extended in future work to better adapt to more dynamic and complex traffic environments.

Author Contributions: Conceptualization, W.L. and Z.S.; methodology, W.L.; software, Y.W.; validation, W.L., Z.S. and Y.W.; formal analysis, Z.S.; investigation, Z.S.; resources, W.L.; data curation, Z.S.; writing—original draft preparation, Z.S.; writing—review and editing, W.L.; visualization, Z.S.; supervision, W.L.; project administration, Z.S.; funding acquisition, W.L. All authors have read and agreed to the published version of the manuscript.

Funding: This work was funded by the Wuhan Key Research and Development Program, project number 2024050702030122.

Institutional Review Board Statement: Not applicable.

Informed Consent Statement: Not applicable.

Data Availability Statement: The original contributions presented in this study are included in the article. Further inquiries can be directed to the corresponding author.

Conflicts of Interest: The authors declare no conflict of interest.

References

1. Dickerson, A.; Peirson, J.; Vickerman, R. Road accidents and traffic flows: An econometric investigation. *Economica* **2000**, *67*, 101–121. [CrossRef]
2. Retallack, A.E.; Ostendorf, B. Relationship between traffic volume and accident frequency at intersections. *Int. J. Environ. Res. Public Health* **2020**, *17*, 1393.
3. Leich, A.; Nippold, R.; Schadschneider, A.; Wagner, P. Physical models of traffic safety at crossing streams. *Phys. A Stat. Mech. Its Appl.* **2024**, *640*, 129669.
4. Wang, X.; Zhuang, X.; Gao, Z.; Dong, Y. A Survey on Highway Traffic Flow Prediction: Methods and Advances. *Comput. Technol. Dev.* **2025**, *35*, 1–9.
5. Wang, Q.; Wu, Y.; Zhu, C.; Wang, Y. Short-term traffic flow prediction studies integrated with external properties. *Appl. Res. Comput.* **2022**, *39*, 2974–2978.
6. Zhang, Y.; Zhou, C.; Chen, Y. Highway Traffic Flow Prediction Based on a Feature Fused Spatio-Temporal Graph Mixed Networks. *J. Dalian Jiaotong Univ.* **2025**, *46*, 15–25.
7. Lei, B.; Li, J.; Zhang, P.; Li, W.; Chen, C. Long Term Prediction on Urban Traffic Flow Based on Multi-source Spatio-temporal Graph Convolutional Neural Network Model. *J. Highw. Transp. Res. Dev.* **2024**, *41*, 204–213.

8. Zhu, J.; Wang, Q.; Tao, C.; Deng, H.; Zhao, L.; Li, H. AST-GCN: Attribute-Augmented Spatiotemporal Graph Convolutional Network for Traffic Forecasting. *IEEE Access* **2021**, *9*, 35973–35983.
9. Zong, X.; Yan, H.; Qi, Y. Recent Advances in Multi-source Data Fusion for Traffic Flow Prediction: A Review. *Arch. Comput. Methods Eng.* 2025, *Prepublish*.
10. Sun, Y.; Zhang, G.; Yin, H. Passenger Flow Prediction of Subway Transfer Stations Based on Nonparametric Regression Model. *Discret. Dyn. Nat. Soc.* **2014**, *2014*, 1–8. [CrossRef]
11. Klepsch, J.; Klüppelberg, C.; Wei, T. Prediction of functional ARMA processes with an application to traffic data. *Econ. Stat.* **2017**, *1*, 128–149. [CrossRef]
12. Kumar, P.B.; Hariharan, K. Time series traffic flow prediction with hyper-parameter optimized ARIMA models for intelligent transportation system. *J. Sci. Ind. Res.* **2022**, *81*, 408–415. [CrossRef]
13. Balawi, M.; Tenekeci, G. Time series traffic collision analysis of London hotspots: Patterns, predictions and prevention strategies. *Heliyon* **2024**, *10*, e25710. [CrossRef] [PubMed]
14. Huang, Z.; Lin, P.; Lin, X.; Zhou, C.; Huang, T. Spatiotemporal attention mechanism-based multistep traffic volume prediction model for highway toll stations. *Arch. Transp.* **2022**, *61*, 21–38. [CrossRef]
15. Feng, X.; Ling, X.; Zheng, H.; Chen, Z.; Xu, Y. Adaptive multi-kernel SVM with spatial–temporal correlation for short-term traffic flow prediction. *IEEE Trans. Intell. Transp. Syst.* **2018**, *20*, 2001–2013.
16. Xia, J.; Wang, S.; Wang, X.; Xia, M.; Xie, K.; Cao, J. Multi-view Bayesian spatio-temporal graph neural networks for reliable traffic flow prediction. *Int. J. Mach. Learn. Cybern.* **2024**, *15*, 65–78.
17. Zhu, H.; Xie, Y.; He, W.; Sun, C.; Zhu, K.; Zhou, G.; Ma, N. A novel traffic flow forecasting method based on RNN-GCN and BRB. *J. Adv. Transp.* **2020**, *2020*, 7586154. [CrossRef]
18. Chaoura, C.; Lazar, H.; Jarir, Z. Traffic Flow Prediction at Intersections: Enhancing with a Hybrid LSTM-PSO Approach. *Int. J. Adv. Comput. Sci. Appl.* **2024**, *15*, 494–501. [CrossRef]
19. Zhang, W.; Li, X.; Li, A.; Huang, X.; Wang, T.; Gao, H. Sgru: A high-performance structured gated recurrent unit for traffic flow prediction. In Proceedings of the 2023 IEEE 29th International Conference on Parallel and Distributed Systems (ICPADS), Ocean Flower Island, China, 17–21 December 2023; IEEE: Dallas, TX, USA, 2023; pp. 467–473.
20. Ma, X.; Dai, Z.; He, Z.; Ma, J.; Wang, Y.; Wang, Y. Learning traffic as images: A deep convolutional neural network for large-scale transportation network speed prediction. *Sensors* **2017**, *17*, 818. [CrossRef]
21. Yu, B.; Yin, H.; Zhu, Z. Spatio-temporal graph convolutional networks: A deep learning framework for traffic forecasting. *arXiv* **2017**, arXiv:1709.04875.
22. Li, W.; Sui, L.; Zhou, M.; Dong, H. Short-term passenger flow forecast for urban rail transit based on multi-source data. *EURASIP J. Wirel. Commun. Netw.* **2021**, *2021*, 9.
23. Xu, Z.; Yuan, J.; Yu, L.; Wang, G.; Zhu, M. Machine learning-based traffic flow prediction and intelligent traffic management. *Int. J. Comput. Sci. Inf. Technol.* **2024**, *2*, 18–27. [CrossRef]
24. Wang, K.; Liu, L.; Liu, Y.; Li, G.; Zhou, F.; Lin, L. Urban regional function guided traffic flow prediction. *Inf. Sci.* **2023**, *634*, 308–320. [CrossRef]
25. Du, M.; Yang, L.; Tu, J. A novel approach to calculate the spatial–temporal correlation for traffic flow based on the structure of urban road networks and traffic dynamic theory. *Sensors* **2021**, *21*, 4725. [CrossRef]
26. Chen, M.; Yuan, H.; Jiang, N.; Bao, Z.; Wang, S. Urban traffic accident risk prediction revisited: Regionality, proximity, similarity and sparsity. In Proceedings of the 33rd ACM International Conference on Information and Knowledge Management, Boise, ID, USA, 21–25 October 2024; pp. 281–290.
27. Chu, L.; Hou, Z.; Jiang, J.; Yang, J.; Zhang, Y. Spatial-temporal feature extraction and evaluation network for citywide traffic condition prediction. *IEEE Trans. Intell. Veh.* **2023**, *9*, 5377–5391. [CrossRef]
28. Lee, E.H.; Lee, E. Congestion boundary approach for phase transitions in traffic flow. *Transp. B Transp. Dyn.* **2024**, *12*, 2379377.
29. Laval, J.A. Traffic Flow as a Simple Fluid: Toward a Scaling Theory of Urban Congestion. *Transp. Res. Rec.* **2024**, *2678*, 376–386. [CrossRef]
30. Lee, E.H.; Kho, S.-Y.; Kim, D.-K.; Cho, S.-H. Travel time prediction using gated recurrent unit and spatio-temporal algorithm. *Proc. Inst. Civ. Eng. Munic. Eng.* **2021**, *174*, 88–96.
31. Zhao, L.; Song, Y.; Zhang, C.; Liu, Y.; Wang, P.; Lin, T.; Deng, M.; Li, H. T-GCN: A temporal graph convolutional network for traffic prediction. *IEEE Trans. Intell. Transp. Syst.* **2019**, *21*, 3848–3858. [CrossRef]
32. Sui, D.Z. Tobler's first law of geography: A big idea for a small world? *Ann. Assoc. Am. Geogr.* **2004**, *94*, 269–277. [CrossRef]
33. Li, Y.; Yu, R.; Shahabi, C.; Liu, Y. Diffusion convolutional recurrent neural network: Data-driven traffic forecasting. *arXiv* **2017**, arXiv:1707.01926.

34. Li, Y.; Xu, W. Short-term traffic flow forecasting based on SVR. In Proceedings of the 2018 3rd International Conference on Automation, Mechanical Control and Computational Engineering (AMCCE 2018), Dalian, China, 12–13 May 2018; Atlantis Press: Amsterdam, The Netherlands, 2018; pp. 57–61.
35. Shao, H.; Soong, B.H. Traffic flow prediction with long short-term memory networks (LSTMs). In Proceedings of the 2016 IEEE Region 10 Conference (TENCON), Singapore, 22–25 November 2016; IEEE: Dallas, TX, USA, 2016; pp. 2986–2989.
36. Zhang, D.; Kabuka, M.R. Combining weather condition data to predict traffic flow: A GRU-based deep learning approach. *IET Intell. Transp. Syst.* **2018**, *12*, 578–585.
37. Bai, J.; Zhu, J.; Song, Y.; Zhao, L.; Hou, Z.; Du, R.; Li, H. A3t-gcn: Attention temporal graph convolutional network for traffic forecasting. *ISPRS Int. J. Geo-Inf.* **2021**, *10*, 485. [CrossRef]

applied sciences

MDPI

Article

Iterative Forecasting of Short Time Series

Evangelos Bakalis

Department of Chemistry "G. Ciamician", University of Bologna, V. P. Gobetti 85, 40129 Bologna, Italy; evangelos.bakalis2@unibo.it

Abstract

We forecast short time series iteratively using a model based on stochastic differential equations. The recorded process is assumed to be consistent with an α-stable Lévy motion. The generalized moments method provides the values of the scaling exponent and the parameter α, which determine the form of the stochastic term at each iteration. Seven weekly recorded economic time series—the DAX, CAC, FTSE100, MIB, AEX, IBEX, and STOXX600—were examined for the period from 2020 to 2025. The parameter α is always 2 for the four of them, FTSE100, AEX, IBEX, and STOXX600, indicating quasi-Gaussian processes. For FTSE100, IBEX, and STOXX600, the processes are anti-persistent ($H < 0.5$).The rest of the examined markets show characteristics of uncorrelated processes whose values are drawn from either a log-normal or a log-Lévy distribution. Further, all processes are multifractal, as the non-zero value of the mean intermittency indicates. The model's forecasts, with the time horizon always one-step-ahead, are compared to the forecasts of a properly chosen ARIMA model combined with Monte Carlo simulations. The low values of the absolute percentage error indicate that both models function well. The model's outcomes are further compared to ARIMA forecasts by using the Diebold–Mariano test, which yields a better forecast ability for the proposed model since it has less average loss. The ability and accuracy of the model to forecast even small time series is further supported by the low value of the absolute percentage error; the value of 4 serves as an upper limit for the majority of the forecasts.

Keywords: α-Lévy stable motion; α-Lévy stable noise; generalized moments method; iterative stochastic differential equations; ARIMA forecasts

1. Introduction

In diverse fields, processes are typically presented as time series; therefore, it would be very beneficial, if it is feasible through the available data, to forecast the evolution of such a process and make well-informed decisions [1]. The experimental evidence can be categorized as stochastic, noisy deterministic, deterministic, or a combination of these. Among deterministic data, those that show chaotic dynamics are important since they display very complex trajectories [2]. A finite-size trajectory can be either stochastic or chaotic since both are very much alike; hence, data pre-processing will be crucial [3]. Stochastic data can be correlated or uncorrelated in time, a property usually revealed, but not limited by, the form of their power spectra over the frequency domain [4].

A flat power spectrum indicates the existence of a memory-less stochastic process—white noise. However, the probability distribution from which such a process draws values is to be determined [5]. The power spectrum obeys a power law of the form $S(f) \sim f^{-\beta}$, with f being the frequency and β the scaling exponent. A link has been established between the scaling exponent of the spectrum and the scaling exponent H of the process, which is called the Hurst exponent in honor of Harold Edwin Hurst. More specifically, for stationary processes, $\beta = 2H - 1$ with $-1 < \beta < 1$, and $\beta = 2H + 1$ with $1 < \beta < 3$ for the non-stationary [6,7]. Hurst, in his seminal work ref. [8], introduced the adjusted range statistics. Later, Mandelbrot and Wallis established the mathematical foundation for the rescaled range analysis (R/S) that is used today through a heuristic graphical method [9]. They demonstrated how a scaling property for stationary processes permits a link between past and future events. Let $x(t)$ be a stationary time series, and let τ be the length of the window connecting the differences between past and future events. If there is a unique scaling, the equation $x(\tau) = c\tau^H$ holds; note that c is a positive constant and H is the scaling exponent, $H \in (0, 1)$. For $0 < H < \frac{1}{2}$, the process is anti-persistent, meaning that every new value is likely to go in the opposite direction of the previous one. For $\frac{1}{2} < H < 1$, the process is called persistent, and every new value is likely to follow the trajectory of the previous one. For $H = \frac{1}{2}$, the process has no memory, and every new value is entirely independent of the values that preceded it.

Various methods can be used to calculate the Hurst exponent, including the generalized Hurst exponent (GHE) [10,11], the generalized moments method (GMM) [12–14], dispersional analysis (DA) [15], power spectral density (PSD) [16], rescaled range analysis (R/S) [9], detrended fluctuation analysis (DFA) [17], multifractal detrended fluctuation analysis (MF-DFA) [18,19], and others; see, for example, [20]. The accuracy with which each technique provides the scaling exponent depends on the length of the data and the type of distribution the data satisfy (existence or nonexistence of heavy tails). It is questionable if PSD, DFA, R/S, and DA can "reveal" the characteristics of multiscaling, or multifractal, processes because they each produce a distinct scaling exponent (monofractality). Moreover, it has been demonstrated that R/S overestimates the true Hurst exponent [21]. In contrast to GHE and GMM, which are the best methods at capturing multifractality, MF-DFA is not appropriate for data with heavy tails and small sample sizes [22]. Furthermore, GMM works well for short time series and is among the most dependable methods for analyzing non-stationary time series [12,22]. It is worth mentioning that GMM works only for weak stationary time series: time series whose increments form a stationary realization. When GMM operates on a stationary process, it returns a scaling exponent close to zero.

Time series data from seismic, meteorological, and financial sources offer a wealth of information for testing any model that seeks to apply forecasting models or explore the underlying stochastic nature of a process. The values of an index comprise a financial time series. They are recorded at equidistant time frames (hour, day, week, month), and are a collection of stocks, businesses, and other economic indicators that characterize a stock market and whose study can provide information about market performance. A different way to look at stock market prices is to assume that they are random variables [23]. This allows us to compute scaling exponents that describe both stationary and non-stationary stochastic processes [24]. Today, the field of studying these exponents in financial markets is actually extremely rich [25–30], and the growing body of empirical data is constantly enhancing our comprehension of their behavior.

Let the price of such an index on time frame i be x_i. The price return is defined as $r_i = log\left(\frac{x_i}{x_{i-1}}\right)$, the logarithmic difference of x_i over two successive recordings. It is widely accepted that the variable r_i is normally distributed, and thus the underlying process is quasi-Gaussian. In this situation, the deviation from the mean, or volatility, is a crucial element for the analysis [23]. If the process is also monofractal, then a unique scaling exponent exists whose value is half of the scaling of volatility. However, research showed that financial time series are described by numerous scalings and thus are best characterized as multifractals [27,30]. In addition, the type of the distribution of the returns has also been challenged, and it has been proposed that an α-Lévy stable distribution is the proper one, where α is the Lévy index, and $\alpha = 2$ yields the normal distribution once more.

For a certain period, price indices on the Milan, Italy, stock exchange exhibit an α-Lévy stable distribution, according to analysis [31]. Furthermore, the same patterns have been noted in European options data, where an α-Lévy stable distribution is responsible for the multifractality [28]. Using either fractional Brownian (fBm) [32] or fractional Lévy motion (fLm) [33], economic data have been analyzed and shown to be better described as multifractals than as monofractals (fBm) [28,33–38]. It is worth noticing that in 1963, Mandelbrot suggested that an α-Lévy stable distribution would be a good model to use for analysing price disparities or logarithmic returns [25].

The present work applies the model we recently presented [39] to short time series, weekly based recordings of stock markets, namely, DAX, CAC, FTSE100, MIB, IBEX, AEX, and STOXX600 over the period 2020 to 2025. On the one hand, the model employs an α-Lévy stable distribution as the underlying stochastic process, and the Lévy index α as well as the Hurst exponent are computed by applying GMM [12–14]. GMM is reliable in delivering the scaling exponent; GMM may be applied even to small data sets [22,40] and is similar to the generalized Hurst exponent approach (GHE), which is frequently utilized for financial data analysis [10]. The structure function [41] is computed by GMM, and its form provides the Hurst exponent, the mean intermittency parameter (C), and the index α. Notice that the scaling exponent of an α-Lévy stable distribution is given by $H - \frac{1}{2} + \frac{1}{\alpha}$ [36]. On the other hand, a stochastic differential equation with a drift and a diffusion term is used for forecasting [23,37,38,42]. The drift term can be calculated using the mean of the process's increments, weighted by its actual value. The diffusion term computation requires the autocorrelation of the underlying noise that arises from the discretization of an α-Lévy stable motion. To forecast the value of the process at the point k + 1, we consider that the previous k points form the "historical data" on which we apply GMM and compute the values of α, H, and C. Next, we generate a number of α-Lévy stable noise sequences (white noises), each with a zero mean and an equal length to the time series that was previously analyzed. Each one of them is transformed to have the correct spectral exponent, and then it follows the inverse Fourier transform for the reconstruction of the sequence in the time domain [5]. The actual value of the index at the forecast point is influenced by the mean of their tail, which is determined by the non-zero portion of their autocorrelation. At every forecast, the values of α and H could fluctuate. Finally, we compare the predictions of the present model with those of a properly chosen ARIMA-based forecast model combined with Monte Carlo (MC) simulations on the basis of the Diebold- Mariano test.

2. Materials and Methods

We provide a brief description of the iterative forecasting model that is utilized in the present study; ref. [39] contains the model's analytical specifics. In order to assist the estimation of the process's scaling exponent and define the type of the general α-Lévy stable distribution, the model first classifies the type of the process by computing the parameters H and α. The latter determines the type of random process that is satisfied by the input data increments, and this result is essential since it defines the diffusion term in the stochastic equation. The final forecasting value is produced by creating sequences of random processes that are of the same length and have the same properties as the time series under analysis, see discussion below.

2.1. Generalized Moments Method (GMM)

Let $\{x_i\}$ represent the measurements of a field, and the index i runs from 1 to N, with N the length of the field at time t. We define the increments of the field $y_n(\Delta) = |x(n+\Delta) - x(n)|$, with $n = 1,2,\dots,N-\Delta$ and Δ being the time lag between two raw data ($\Delta = 1,2,\dots,\frac{N}{10}$), in order to understand how the fluctuations of the field behave. The latter can be achieved by computing the field's scaling properties; "scaling" refers to a process's invariance across a range of scales. Scaling systems can be described using fractal and multifractal theories, the latter of which is an extension of the former. Fractal theory [43] focuses on simple scaling or monofractal processes, and complex events can be described with a small number of parameters. However, multifractal theory deals with multiscaling, which allows one to generalise a process's scaling properties [44,45].

The structure function, or scaling exponent $z(q)$, establishes whether a process is multifractal (having a convex form) or monofractal (a linear function of $z(q)$ versus q). Among all multifractals, universal multifractals, which are essentially log-Lévy multifractals, seem to be ubiquitous [41], with a structure function of the form

$$z(q) = qh - \frac{C}{\alpha - 1}(q^{\alpha} - q) \tag{1}$$

where $\alpha = -\frac{1}{C}\frac{d^2 z(q)}{dq^2}|_{q=1}$ is the Lévy index or index of multifractality. It takes values in the interval $0 < \alpha \leq 2$. For $\alpha = 2$, $z(q) = qh - C(q^2 - q)$, the logarithm of the field is normally distributed [46], for $\alpha = 1$, $z(q) = qh - Cqlog(q)$ the field is distributed according to log-Cauchy, while for all the other values of α in the range (0, 2), the field is distributed according to log-Lévy. The term C is called co-dimension information and measures the mean intermittency, and is defined as, $C = \frac{dz(q)}{dq}|_{q=1} - H$. It takes values in the range $C \in [0, d]$ with d being the dimension of the support, which is 1 for a one-dimensional time series. For $C = 0$, the field is homogeneous; only one scale exists. From lower to higher values of C, the degree of intermittency increases, and some extreme outliers will occur. The parameter $h \in (0,1)$, when the field is multifractal, shows the degree of fractional integration (persistent for $h > 0.5$ and anti-persistent for $h < 0.5$). The value of $z(q)$ for $q = 1$ provides the Hurst exponent for a multifractal field. Since $z(q) = qh$, it follows that $z(q=1) = z(q=2)/2$ when a field is monofractal [7]. Therefore, the analytical form of the scaling exponent or structure function, Equation (1), gives information about the nature of the stochastic mechanisms; see for details refs [7,12–14,19,47]. For further information on how GMM is used in financial time series, see ref. [39].

2.2. Stochastic Differential Equation

A differential equation with a stochastic term given by an α-Lévy stable motion reads [33,48]

$$dx_{\alpha}(t) = \mu(t, x_{\alpha}(t))dt + \sigma(t, x_{\alpha}(t))dL_{\alpha}(t) \tag{2}$$

with $x_{\alpha}(t=0) = x_{\alpha,0}$ being a random variable taking values from an α-Lévy stable motion (Lsm). The terms $\mu(t, x(t))$ and $\sigma(t, x(t))$ describe the drift and the diffusion term, respectively. To a first approximation, these two terms can be expressed as $\mu(t, x(t)) = \mu x(t)$ and $\sigma(t, x(t)) = \sigma x(t)$, and the discrete equivalent of Equation (2) is as follows

$$\Delta x(t) = \mu x(t)\Delta t + \sigma x(t)\Delta\xi(t) \tag{3}$$

with Δt being the minimum time lag. ξ is the generic symbol used for the stochastic term. Actually, it represents a generic process whose values are drawn from a probability distribution, be it fractional Brownian motion (fBm), fractional Lévy stable motion (fLsm), or multifractional Brownian motion (mfBm). In addition, the process $\xi(t)$ is also regarded as weak-stationary, meaning that its increments, $\Delta\xi(t)$, create a stationary process with zero mean and variance that needs to be computed.

Equation (3) can be used iteratively to provide the next value of the process $x(t)$ at each iteration; however, this requires an estimate of the terms μ and σ. Dividing each term of Equation (3) with $x(t)$, we write

$$\frac{\Delta x(t)}{x(t)} = \mu\Delta t + \sigma\Delta\xi(t) \tag{4}$$

Taking the expectation value of Equation (4), and since $< \Delta\xi(t) >= 0$ we find

$$\mu =< \frac{\Delta x(t)}{x(t)} >= \frac{1}{N}\sum_{n=1}^{N}\frac{\Delta x(t_n)}{x(t_n)} \tag{5}$$

We also find that $< \frac{\Delta x(t)}{x(t)}\frac{\Delta x(t')}{x(t')} >$ equals $\mu^2 + \sigma^2 ACF_{\Delta\xi}(|t-t'|)$, with $ACF_{\Delta\xi}(|t-t'|)$ being the autocorrelation of the process $\Delta\xi(t)$. Combining findings we end up with

$$\sigma = (ACF_{\Delta\xi}(0))^{-1/2}\sqrt{< \left(\frac{\Delta x(t)}{x(t)}\right)^2 > - < \left(\frac{\Delta x(t)}{x(t)}\right) >^2} \tag{6}$$

where $ACF_{\Delta\xi}(0)$ is the value of the autocorrelation function at $t = t'$. Equation (3) for a generic stochastic motion $\xi(t)$ in conjunction with Equations (5) and (6) is the main equation utilized in the iterative forecast.

It is worth mentioning that the generic stochastic term can be expressed through the corresponding white noise, the mother process from which $x(t)$ is coming from [5], as $\Delta\xi(t) = w_\xi(t)(\Delta t)^s$ where s is the scaling exponent, that equals H for fBm, $H - \frac{1}{2} + \frac{1}{\alpha}$ for fLsm, and $H(t)$ for mfBm [37]. In addition, the spectral exponent these processes must meet reads $\beta = 2(H - \frac{1}{2} + \frac{1}{\alpha}) + 1$, which returns $\beta = 2H + 1$ for $\alpha = 2$.

3. Results and Discussion

We analyze a number of European stock markets from 2020 to 2025, such as the STOXX600, DAX, CAC, FTSE100, MIB, AEX, and IBEX. The data was retrieved from the website of Statista [49] and accessed on 21 July 2025. The generated time series are shorter than 300 points, with weekly sampling. The values of the examined indices are shown in Figure 1.

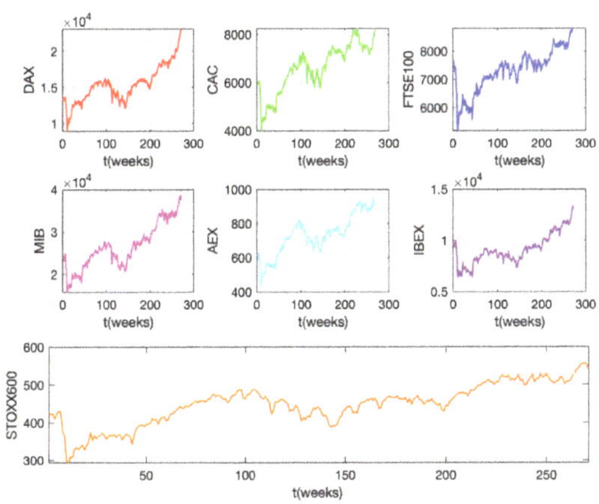

Figure 1. Each one of the panels displays a different index (DAX in red, CAC in green, FTSE100 in blue, MIB in magenta, AEX in cyan, IBEX in purple, and STOXX600 in orange) over a five-year period, from 2020 to 2025. Consecutive events have a distance of one week.

The model assumes that the noise term of the stochastic equation, Equation (3), takes steps from an α- Lévy stable distribution. For financial time series the value of α is frequently 2, and lesser values are discarded. Indeed, Figure 2, which displays the value of α with respect to time, makes it evident that the indices DAX and CAC always have α values below 2. In contrast, the value of α for the FTSE100, AEX, IBEX, and STOXX600 indexes is consistent with the value of 2. These results demonstrate that FTSE100, AEX, IBEX, and STOXX600 are processes that use steps from a Gaussian distribution, specifically a fractional Brownian motion (fBm), if H is constant over time, or a multifractional Brownian motion (mfBm), if H changes, with increments that correspond to a fractional Gaussian noise (fGn). Conversely, the DAX and CAC indexes use a non-Gaussian probability distribution to determine their steps, and an α-Lévy stable noise (α-Lsn) determines their increments. Finally, for the first forecast points, the index MIB's α value is less than two, and after that, it settles at two.

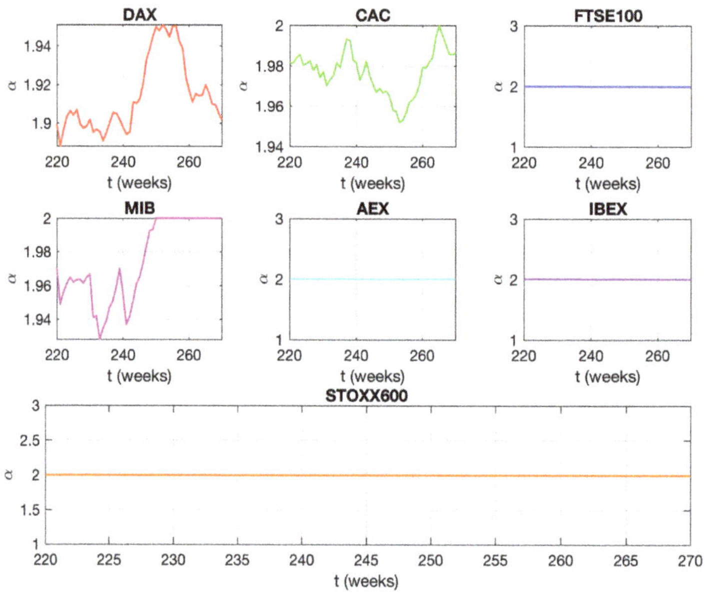

Figure 2. α parameters in time for several indices. The computed values are shown for historical data with lengths that go from $\lfloor 0.8N \rfloor$ to $N-1$, with N being the total length.

The time series displayed in Figure 1 are non-stationary and thus represent processes that are out of equilibrium. To prove that, we consider the increments of each one of them, which are then analyzed by GMM. Each analysis returns a Hurst exponent with a value of zero; results are not shown. Next, we consider the original time series whose thorough investigation based on the generalized moments method approach(GMM) [7,12–14,19,47] delivers the values of H, C, and α. Refer to Figures 3–5, which show the scaling exponent (s), the mean intermittency parameter (C), and the Lévy stable index (α) versus time.

This study uses a minimal value of k equal to $\lfloor 0.8N \rfloor$, although it can be even smaller. A process of length k is analyzed, and its length increases by 1 at each iteration. With N representing the entire length of each trajectory shown in Figure 1, the maximum value the index k can take is $N-1$. Observe that even if smaller trajectories have been treated in the literature, the GMM requires a minimum of 100 points for a successful application [40].

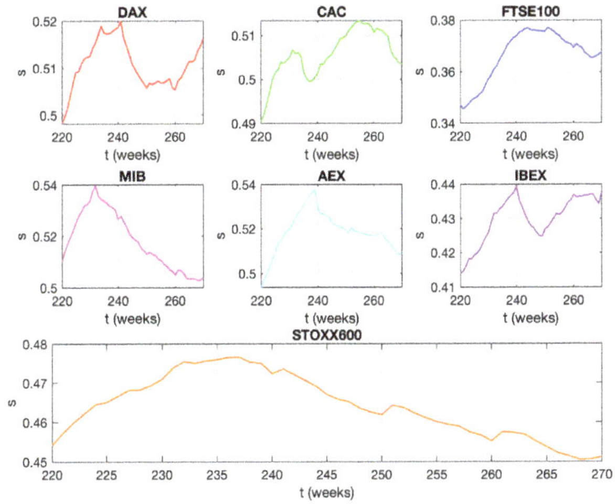

Figure 3. Scaling exponent between a minimum of 220 weeks and a maximum of 270 weeks. With every new trade week, the length of time that data is accumulated increases by one.

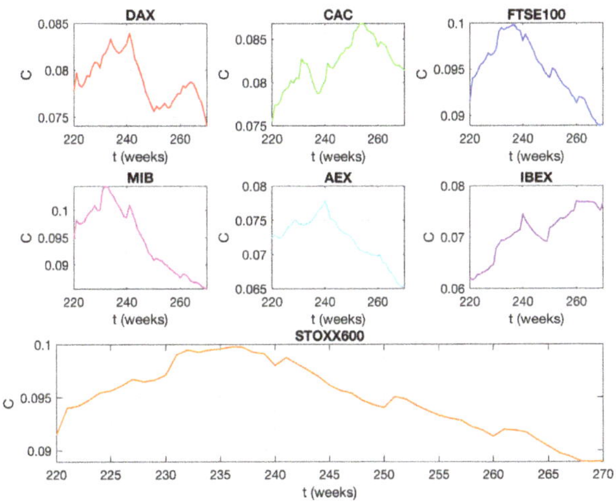

Figure 4. The mean intermittency parameter C for each market. The values are between a minimum of 220 weeks and a maximum of 270 weeks.

From trading week to trading week, the value of the scaling exponent fluctuates; see Figure 3, and its value is directly connected to the type of stochastic process: 0.5 for white noise and coloured otherwise. Notice that there exist various types of white and coloured noises [5]. As it was mentioned, the scaling exponent satisfies the relation $s = H - \frac{1}{2} + \frac{1}{\alpha}$, and thus, when $\alpha = 2$, the scaling exponent coincides with the Hurst exponent ($H = z(q = 1)$). Unlike persistency ($s > 0.5$), which shows the presence of temporal drifts, anti-persistency ($s < 0.5$) shows a propensity for prices to return to their mean. Because their respective scaling exponents are consistently below the value of 0.5 during the course of the analysis window, the FTSE100, IBEX, and STOXX600 return to their average value (anti-persistent behavior). Instead, for DAX, CAC, MIB, and AEX, their scaling fluctuates in a very narrow range around or just above the value of 0.5, which points to an uncorrelated process (white noise). However, since the mean intermittency parameter is non-zero (see below), this uncorrelated process concerns the logarithm of the distribution and not the distribution itself. It is log-normal distributed when $\alpha = 2$, and log-α-Lévy stable when $\alpha < 2$.

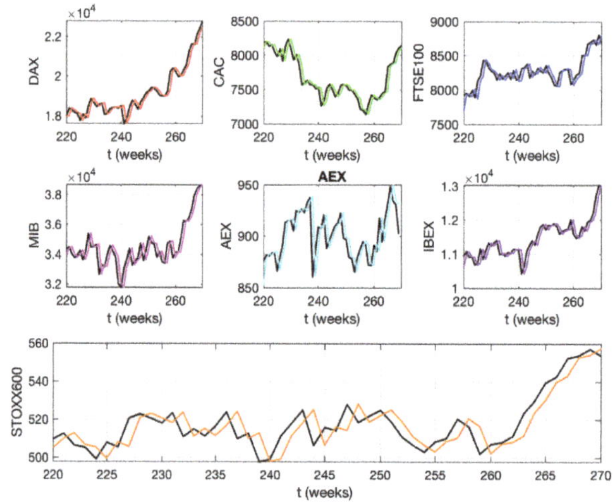

Figure 5. In black displays the evolution of each market index based on a weekly sampling rate over a constrained time frame of five years (2020–2025). The forecasting values at the same time frame are represented by the coloured curves.

The value of the mean of the intermittency parameter (C) approximately falls between 0.05 and 0.11, see Figure 4. Values of C constantly different than 0 point to a multifractal process, which, in conjunction with the values of the parameter α, classify the processes examined here as multifractal fractional Brownian motion (mfBm) [37] for FTSE100, AEX, IBEX, and STOXX600; as fractional Lévy-stable motion (fLsm) for DAX and CAC; and as fLsm that turns to mfBm for MIB. When shuffling input data, GMM analysis yields zero scaling exponents, indicating stationary processes. The latter underlines that multifractality originates from long-range correlations or, in other words, pattern formations that shuffling destroys.

We utilize Equation (3) to forecast the data in Figure 1. At each iterative step, the values of α and H are used to create an α-Lévy stable motion, whose increments describe either a fractional Lévy stable noise or fractional Gaussian noise (fGn). For the creation of fractional Lévy stable noise sequences, we use Matlab's built-in function STBL (Alpha stable distributions for MATLAB R2022b) [50]. This function takes four inputs: the value of α computed by GMM, the skewness parameter that is also available as the third central moment computed by GMM, the scale parameter also available as the second central moment, and the position parameter that is considered zero in all cases. The function returns an α-Lévy stable white noise that must be transformed to have the same properties as the original input data set. We first calculate the power spectrum, which is flat, and then multiply it by $f^{-\beta}$ to produce the correct scale. Remember that $\beta = 2(H - \frac{1}{2} + \frac{1}{\alpha}) + 1$. Next, applying the inverse Fourier transform, we end up with the sequence in the time domain [51,52]. For each iteration of length k, 1500 of such sequences are generated. The last members of them, indicated by the non-vanishing autocorrelation function, contribute to the value of the sequence at the point k+1. The mean of all forecasts for the point k+1 is then used in Equation (3). The forecasts are shown in Figure 5.

The recorded evidence is satisfactorily correlated with the forecasts. For some points, though, the trend between the predicted and actual values differs. The ability of the current model to forecast should be compared to some well-established models. An ARIMA-based forecast combined with Monte Carlo (MC) simulations is used [53]. The first k points of the input data shape the 'historical data', and the time horizon for the forecast is one-step-ahead. Various ARIMA models have been fitted to 'historical data', and the ARIMA(2,1,2) best performs based on both the Akaike information criterion (AIC) and the Bayesian information criterion (BIC), both of which have lower values at the same time. First ARIMA(2,1,2) is used to fit the data, then it makes a forecast for the time horizon, and finally, MC based on this forecast creates a number of paths whose mean at each point is the final forecast. For every forecast, 1500 simulations are performed. Matlab built-in functions have been used for the calculations [50], and the results are illustrated in Figure 6.

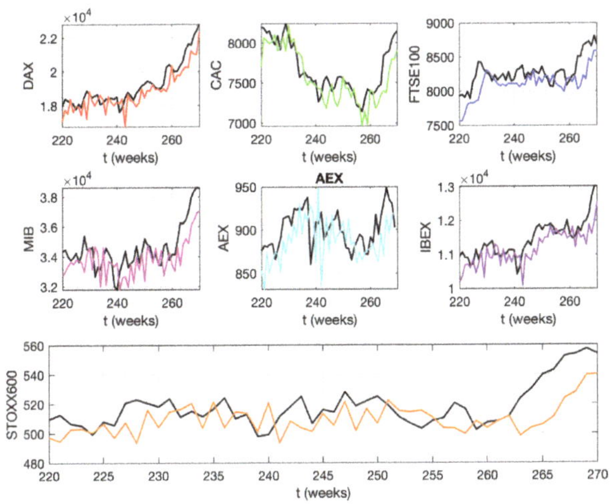

Figure 6. It shows the same information as Figure 5 but for forecasts based on ARIMA(2,1,2). Again, black for the evolution of each market and coloured curves for the forecasts.

The ARIMA(2,1,2) based predictions combined with MC simulations correlate also well with the recorded values; see Figure 6. Comparing the outcomes of the two models, it is important to keep in mind that each forecast is based on historical data of varying lengths; therefore, the idea of the mean cannot be used. Rather, a metric that is not dependent on the actual values of the forecasts is needed. Such a metric is the absolute percentage error (APE), which is defined as $APE(t) = \frac{x(t)-x_f(t)}{x(t)}$ where $x(t)$ and $x_f(t)$ are the actual and the forecast values, respectively, at time t. Figure 7 displays the APE for both models.

Setting the value 5 for the APE to be the upper limit for an accurate forecasting model, we see that both functions well since few values for either model go above this threshold, see Figure 7. Forecasting based on the α-Lévy stable model returns values of APE that have an upper limit that does not exceed the value of 4, while forecasting based on the ARIMA(2,1,2) combined with MC simulation generally yields slightly higher values.

For every market examined in this work, are the predictions made by the two models equally accurate? The Diebold–Mariano (DM) test provides an answer to this question [54,55]. By setting as significant level the value of $p = 0.05$, we run the DM test [56], where the first input corresponds to the forecasts of the present model, the second to the forecasts of ARIMA(2,1,2), and the third one equals one for the forecast horizon. The null hypothesis (both forecasts are equally good) is accepted if $|DM| > 1.96$; if not, it is rejected. For STOXX600/FTSE100/DAX/CAC/MIB/AEX/IBEX the test's computed values are $-3.63/-3.50/-3.21/-2.18/$ $-3.35/-2.15/-3.93$ and for none of them is the null hypothesis satisfied. Moreover, given that all computed values are negative, the forecasts of the present model (first input) perform better than the forecasts of the second one (ARIMA(2,1,2)): negative values indicate less average loss.

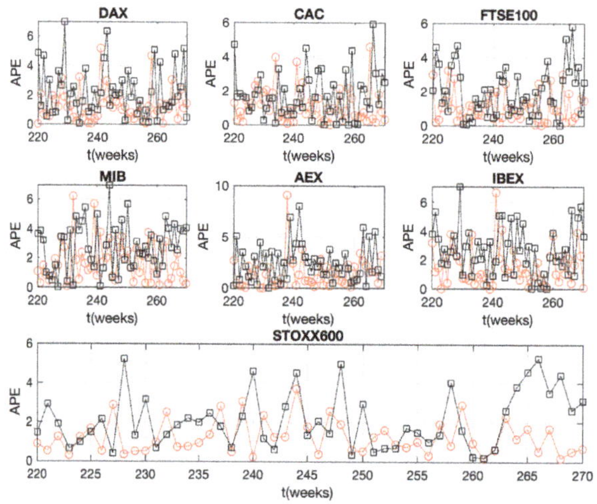

Figure 7. Absolute percentage error (APE) for one-step-ahead forecast. Red for the present model and black for ARIMA(2,1,2).

The forecasts of the present model and actual data patterns are consistent for most of the evidence that has been documented, and there is an acceptable link between the two. Thus, an α-Lévy stable noise, which describes the stochastic term of a forecasting process based on Equation (4), forecasts successfully in a time horizon of one-step-ahead even for small time series. The values of APE show how capable the model is.

4. Conclusions

In this work, we analyze and forecast several European market indices whose experimental evidence forms short time series with lengths shorter than 300 points, a five-year period from 2020 to 2025, and a sampling rate of 1 week. The forecasting is based on an iterative stochastic equation whose diffusive term for each temporal length is defined by a proper α-Lsm, the parameters of which are provided by the method of generalized moments(GMM). STOXX 600, FTSE100, AEX, and IBEX follow a distribution with $\alpha = 2$, as opposed to CAC and DAX, whose values are constantly lower than 2, while MIB partially follows a distribution with $\alpha = 2$. The values of the parameters H, C, and α classify the markets FTSE100, AEX, IBEX, STOXX600 and partially classify MIB as mfBm under the different windows of analysis. DAX, CAC, and partially MIB are classified as multifractal fractional α-Lévy stable motions. The process is anti-persistent for FTSE100, IBEX, and STOXX600, and the other markets reflect processes whose values are uncorrelated and drawn from a log-normal or log-Lévy distribution. The model was tested against the forecasts of a properly chosen ARIMA model combined with MC simulations. The APE for each forecast and for each model has been calculated, showing that both work well with a value of 4 to serve as an upper limit of APE for the majority of the forecasts for the suggested model. In addition, the Diebold–Mariano test showed that forecasting based on the α-Lévy stable model performs better than the corresponding one based on ARIMA. The small values of APE for all forecasts and for all examined markets show that the suggested model works well even if small time series are considered.

Funding: This research received no external funding.

Institutional Review Board Statement: Not applicable.

Informed Consent Statement: Not applicable.

Data Availability Statement: Publicly available datasets were analyzed in this study. This data can be found here: https://www.statista.com/.

Conflicts of Interest: The author declares no conflicts of interest.

Abbreviations

The following abbreviations are used in this manuscript:

fBm	fractional Brownian motion
fGn	fractional Gaussian noise
mfBm	multifractional fractional Brownian motion
fLsm	fractional Lévy stable motion
fLsn	fractional Lévy stable noise
ARIMA	Autoregressive Integrated Moving Average
MC	Monte Carlo
DM	Diebold–Mariano test

References

1. Börner, K.; Rouse, W.B.; Trunfio, P.; Stanley, H.E. Forecasting innovations in science, technology, and education. *Proc. Natl. Acad. Sci. USA* **2018**, *115*, 12573–12581. [CrossRef]
2. Baker, G.L.; Gollub, J.P. *Chaotic Dynamics an Introduction*, 2nd ed.; Oxford University Press: Oxford, UK, 1996.
3. Prado, T.L.; Boaretto, B.R.; Corso, G.; dos Santos, L.G.Z.; Kurths, J.; Lopes, S.R. A direct method to detect deterministic and stochastic properties of data. *New J. Pjys.* **2022**, *24*, 033027. [CrossRef]
4. Hänggi, P.; Thomas, H. Stochastic processes: Time evolution, symmetries and linear response. *Phys. Rep.* **1982**, *88*, 207–319. [CrossRef]
5. Bakalis, E.; Lugli, F.; Zerbetto, F. Daughter Coloured Noises: The Legacy of Their Mother White Noises Drawn from Different Probability Distributions. *Fractal Fract.* **2023**, *7*, 600. [CrossRef]
6. Eke, A.; Herman, P.; Bassingthwaighte, J.; Raymond, G.; Percival, D.; Cannon, M.; Balla, I.; Ikrényi, C. Physiological time series: distinguishing fractal noises from motions. *Pflugers Arch.* **2000**, *439*, 403–415. [CrossRef] [PubMed]
7. Bakalis, E.; Gavriil, V.; Cefalas, A.-C.; Kollia, Z.; Zerbetto, F.; Sarantopoulou, E. Viscoelasticity and Noise Properties Reveal the Formation of Biomemory in Cells. *J. Phys. Chem. B* **2021**, *125*, 10883–10892. [CrossRef]
8. Hurst, H.E. Long term storage capacity of reservoirs. *Trans. Am. Soc. Eng.* **1951**, *116*, 770–799. [CrossRef]
9. Mandelbrot, B.B.; Wallis, J.R. Robustness of the rescaled range R/S in the measurement of noncyclic long run Statistical dependence. *Water Resour. Res.* **1969**, *5*, 967–988. [CrossRef]
10. Di Matteo, T. Multi-scaling in finance. *Quant. Financ.* **2007**, *7*, 21–36. [CrossRef]
11. Barabási, A.-L.; Viscek, T. Multifractality of self-affine fractals. *Phys. Rev. A* **1991**, *44*, 2730. [CrossRef]
12. Bakalis, E.; Höfinger, S.; Venturini, A.; Zerbetto, F. Crossover of two power laws in the anomalous diffusion of a two lipid membrane. *J. Chem. Phys.* **2015**, *142*, 215102. [CrossRef] [PubMed]
13. Sändig, N.; Bakalis, E.; Zerbetto, F. Stochastic analysis of movements on surfaces: The case of C60 on Au(1 1 1). *Chem. Phys. Lett.* **2015**, *633*, 163–168. [CrossRef]

14. Bakalis, E.; Mertzimekis, T.J.; Nomikou, P.; Zerbetto, F. Breathing modes of Kolumbo submarine volcano (Santorini, Greece). *Sci. Rep.* **2017**, *7*, 46515. [CrossRef]
15. Caccia, D.C.; Percival, D.; Cannon, M.J.; Raymond, G.; Bassingthwaighte, J.B. Analyzing exact fractal time series: Evaluating dispersional analysis and rescaled range methods. *Physica A* **1997**, *246*, 609–632. [CrossRef] [PubMed]
16. Eke, A.; Herman, P.; Kocsis, L.; Kozak, L.R. Fractal characterization of complexity in temporal physiological signals. *Physiol. Meas.* **2002**, *23*, R1–R38. [CrossRef]
17. Peng, C.-K.; Buldyrev, S.V.; Havlin, S.; Simons, M.; Stanley, H.E.; Goldberger, A.L. Mosaic organization of DNA nucleotides. *Phys. Rev. E* **1994**, *49*, 685–1689. [CrossRef]
18. Kantelhardt, J.W.; Zschiegner, S.A.; Koscielny-Bunde, E.; Havlin, S.; Bunde, A.; Stanley, H.E. Multifractal detrended fluctuation analysis of nonstationary time series. *Physica A* **2002**, *316*, 87–114. [CrossRef]
19. Bakalis, E.; Ferraro, A.; Gavriil, V.; Pepe, F.; Kollia, Z.; Cefalas, A.-C.; Malapelle, U.; Sarantopoulou, E.; Troncone, G.; Zerbetto, F. Universal Markers Unveil Metastatic Cancerous Cross-Sections at Nanoscale. *Cancers* **2022**, *14*, 3728. [CrossRef]
20. Gómez-Águila, A.; Trinidad-Segovia, J.E.; Sánchez-Granero, M.A. Improvement in Hurst exponent estimation and its application to financial markets. *Financ. Innov.* **2022**, *8*, 86. [CrossRef]
21. Taqqu, M.S.; Teverovsky, V.; Willinger, W. Estimators for long-range dependence: An empirical study. *Fractals* **1995**, *3*, 785–798. [CrossRef]
22. Barunik, J.; Kristoufek, L. On Hurst exponent estimation under heavy-tailed distributions. *Physica A* **2010**, *389*, 3844–3855. [CrossRef]
23. Black, F.; Scholes, M. The pricing of options and corporate liabilities. *J. Political Econ.* **1973**, *81*, 637–654. [CrossRef]
24. Plerou, V.; Gopikrishnan, P.; Rosenow, B.; Amaral, L.A.N.; Stanley, H.E. Econophysics: Financial time series from a statistical physics point of view. *Physica A* **2000**, *279*, 443–456. [CrossRef]
25. Mandelbrot, B. The Variation of Certain Speculative Prices. *J. Bus.* **1963**, *36*, 394–419. [CrossRef]
26. Mantegna, R.; Stanley, H.E. Scaling behaviour in the dynamics of an economic index. *Nature* **1995**, *376*, 46–49. [CrossRef]
27. Bouchaud, J.P.; Potters, M.; Meyer, M. Apparent multifractality in financial time series. *Eur. Phys. J. B* **2000**, *13*, 595–599. [CrossRef]
28. Scalas, E. Scaling in the market of futures. *Physica A* **1998**, *253*, 394–402. [CrossRef]
29. Calvet, L.; Fisher, A. Multifractality in Asset Returns: Theory and Evidence. *Rev. Econ. Stat.* **2002**, *84*, 381–406. [CrossRef]
30. Ruipeng, L.; Di Matteo, T.; Lux, T. Multifractality and long-range dependence of asset returns: The scaling behaviour of the Markov-switching multifractal model with lognormal volatility components. *Adv. Complex Syst.* **2008**, *11*, 669–684. [CrossRef]
31. Mantegna, R.N. Lévy walks and enhanced diffusion in Milan stock exchange. *Physica A* **1991**, *179*, 232–242. [CrossRef]
32. Mandelbrot, B.B.; Van Ness, J.W. Fractional brownian motions, fractional noises and applications. *SIAM Rev.* **1968**, *10*, 422–437. [CrossRef]
33. Samorodnitsky, G.; Taqqu, M. *Stable Non-Gaussian Random Processes: Stochastic Models with Infinte Variance*, 1st ed.; Chapman and Hall: New York, NY, USA, 1994.
34. Fama, E.F.; Roll, R. Some properties of symmetric stable distributions. *Am. Stat. Assoc. J.* **1968**, *63*, 817–836. [CrossRef]
35. Kogon, S.M.; Manolakis, G. Signal modeling with self-similar $\alpha/$-stable processes: The fractional Levy stable motion model. *IEEE Trans. Signal Process.* **1996**, *44*, 1006–1010. [CrossRef]
36. Samorodnitsky, G.; Taqqu, M. Linear models with long-range dependence and with finite and infinite variance. In *New Directions in Time Series Analysis. The IMA Volumes in Mathematics and Its Applications*, 1st ed.; Brillinger, D., Caines, P., Geweke, J., Parzen, E., Rosenblatt, M., Taqqu, M.S., Eds.; Springer: New York, NY, USA, 1993; pp. 325–340.
37. Song, W.; Cattani, C.; Chi, C.-H. Multifractional Brownian motion and quantum-behaved particle swarm optimization for short term power load forecasting: An integrated approach. *Energy* **2020**, *194*, 116847. [CrossRef]
38. Liu, H.; Song, W.; Li, M.; Kudreyko, A.; Zio, E. Fractional Lévy stable motion: Finite difference iterative forecasting model. *Chaos* **2020**, *133*, 109632. [CrossRef]
39. Bakalis, E.; Zerbetto, F. Iterative Forecasting of Financial Time Series: The Greek Stock Market from 2019 to 2024. *Entropy* **2025**, *27*, 497. [CrossRef]
40. Bakalis, E.; Parent, L.R.; Vratsanos, M.; Chiwoo, P.; Gianneschi, N.C.; Zerbetto, F. Complex Nanoparticle Diffusional Motion in Liquid-Cell Transmission Electron Microscopy. *J. Phys. Chem. C* **2020**, *124*, 14881. [CrossRef]
41. Schertzer, D.; Lovejoy, S. Physical modeling and Analysis of Rain and Clouds by Anisotropic Scaling of Multiplicative Processes. *J. Geophys. Res.* **1987**, *92*, 9693–9714. [CrossRef]
42. Laskin, N.; Lambadaris, I.; Harmantzis, F.C.; Devetsikiotis, M. Fractional Lévy motion and its application to network traffic modeling. *Comput. Netw.* **2002**, *40*, 363–375. [CrossRef]
43. Mandelbrot, B.B. *The Fractal Geometry of Nature*, 1st ed.; W. H. Freeman & Company: New York, NY, USA, 2008; p. 480.

44. Lyra, M.L.; Tsallis, C. Nonextensivity and multifractality in low-dimensional dissipative systems. *Phys. Rev. Lett.* **1998**, *80*, 53–56. [CrossRef]
45. Jiang, Z.-Q.; Xie, W.-J.; Zhou, W.-X.; Sornette, D. Multifractal analysis of financial markets: A review. *Rep. Prog. Phys.* **2019**, *82*, 125901. [CrossRef] [PubMed]
46. Kolmogorov, A.N. A refinement of previous hypotheses concerning the local structure of turbulence in a viscous incompressible fluid at high Reynolds number. *J. Fluid Mech.* **1962**, *13*, 82–85. [CrossRef]
47. Bakalis, E.; Fujie, H.; Zerbetto, F.; Tanaka, Y. Multifractal structure of microscopic eye–head coordination. *Physica A* **2018**, *512*, 945–953. [CrossRef]
48. Janicki, A.; Weron, A. Can One See α-Stable Variables and Processes? *Stat. Sci.* **1994**, *9*, 109–126. [CrossRef]
49. Available online: https://www.statista.com/search/?q=weekly+indexes+2020+-+2025+europe&Search=&p=1 (accessed on 21 July 2025).
50. The MathWorks Inc. *MATLAB*, Version: 9.13.0 (R2022b); The MathWorks Inc.: Natick, MA, USA, 2022.
51. Kasdin, N. Discrete simulation of colored noise and stochastic processes and $1/f^{\alpha}$ power law noise generation. *Proc. IEEE* **1995**, *83*, 802–827. [CrossRef]
52. Zhivomirov H. Pink, Red, Blue and Violet Noise Generation with Matlab Implementation, Version 1.6. Available online: https://www.mathworks.com/matlabcentral/fileexchange/42919-pink-red-blue-and-violet-noise-generation-with-matlab-implementation (accessed on 15 November 2017).
53. Chen, Z. Asset Allocation Strategy with Monte-Carlo Simulation for Forecasting Stock Price by ARIMA Model. In Proceedings of the IC4E '22: Proceedings of the 2022 13th International Conference on E-Education, E-Business, E-Management, and E- Learning, Tokyo, Japan, 14–17 January 2022 ; pp. 481–485. [CrossRef]
54. Diebold, F.X.; Mariano, R.S. Comparing Predictive Accuracy. *J. Bus. Econ. Stat.* **1995**, *13*, 253–263. [CrossRef]
55. Harvey, D.; Leybourne, S.; Newbold, P. Testing the equality of prediction mean squared errors. *Int. J. Forecast.* **1997**, *13*, 281–291. [CrossRef]
56. Ibisevic, S. Diebold-Mariano Test Statistic. MATLAB Central File Exchange. Available online: https://ch.mathworks.com/matlabcentral/fileexchange/33979-diebold-mariano-test-statistic (accessed on 15 October 2025).

applied sciences

MDPI

Article

A Novel Hybrid Framework for Stock Price Prediction Integrating Adaptive Signal Decomposition and Multi-Scale Feature Extraction

Junqi Su [1,*], Raymond Y. K. Lau [1], Yuefeng Du [2], Jia Yu [3] and Hui Zhang [4]

1 Department of Information Systems, City University of Hong Kong, Tat Chee Avenue, Kowloon, Hong Kong
2 Department of Computer Science, City University of Hong Kong, Tat Chee Avenue, Kowloon, Hong Kong
3 Department of Logistics and Maritime Studies, Hong Kong Polytechnic University, Hung Hom, Kowloon, Hong Kong
4 Department of Management, City University of Hong Kong, Tat Chee Avenue, Kowloon, Hong Kong
* Correspondence: junqisu2-c@my.cityu.edu.hk

Abstract

To address the issue of low prediction accuracy due to the inherent high noise and non-stationary characteristics of stock price series, this paper proposes a novel stock price prediction framework (CVASD-MDCM-Informer) that integrates adaptive signal decomposition with multi-scale feature extraction. The framework first employs a CVASD module, which is a variational mode decomposition (VMD) method adaptively optimized by a porcupine optimization (CPO) algorithm, to decompose the original stock price series into a series of intrinsic mode functions (IMFs) with different frequency characteristics, effectively separating noise and multi-frequency signals. Subsequently, the decomposed components are input into a prediction network based on Informer. In the feature extraction phase, this paper designs a multi-scale dilated convolution module (MDCM) to replace the standard convolution of the Informer, enhancing the model's ability to capture short-term fluctuations and long-term trends by using convolution kernels with different dilation rates in parallel. Finally, the prediction results of each component are integrated to obtain the final predicted value. Experimental results on three representative industry datasets (Information Technology, Finance, and Consumer Staples) of the US S&P 500 index show that, compared to several advanced baseline models, the proposed framework demonstrates significant advantages in multiple evaluation metrics such as MAE, MSE, and RMSE. Ablation experiments further validate the effectiveness of the two core modules, CVASD and MDCM. The study indicates that the framework can effectively handle complex financial time series, providing a new solution for stock price prediction.

Keywords: stock price prediction; multi-step forecasting; informer; VMD; multi-scale feature

1. Introduction

Stock market data is characterized by high noise, non-stationarity, and non-linearity, which makes predicting future trends from historical prices a significant yet challenging task in financial research [1]. The prediction of stock prices is directly related to risk mitigation, asset allocation, and investment decisions, serving as a critical basis for market analysis by investors and regulators [2,3]. However, stock prices are influenced by numerous factors such as macroeconomic conditions, policy changes, and market sentiment, exhibiting large fluctuations and elusive patterns [4]. In the context of this paper, we define the challenge of 'high noise' from a signal processing perspective, referring to the high-frequency, non-stationary, and irregular fluctuations that co-exist with and obscure the underlying medium-term cycles and long-term trends. Traditional statistical models, which assume linearity and stationarity, are suitable for small-scale, low-noise scenarios but have limited applicability to the large-scale, non-stationary, and noisy stock price series encountered in reality [5]. Consequently, identifying models capable of handling noise and capturing non-linear dynamics has become a key focus of recent research.

Early stock price prediction primarily relied on traditional time series models such as ARIMA/ARMA. These models, based on linear regression and differencing, can handle stationary series but perform inadequately on complex non-linear patterns [6]. Subsequently, machine learning methods like Support Vector Machines (SVM) and Random Forests were introduced. While they offer advantages in non-linear modeling, they require manual feature engineering and struggle to extract long-term dependencies from large-scale sequences [7]. In recent years, deep learning models have demonstrated powerful feature extraction capabilities. Notably, Recurrent Neural Networks (RNNs) and Long Short-Term Memory (LSTM) networks can capture long-term dependencies in time series through gating mechanisms, effectively mitigating the vanishing gradient problem [8].

Following the success of the Transformer architecture in natural language processing, attempts have been made to apply it to long-sequence time-series forecasting. The Transformer captures long-range dependencies via a self-attention mechanism, but its computational complexity grows quadratically with sequence length, making it difficult to apply directly to long sequences [9]. To address this, the Informer model introduced the ProbSparse self-attention mechanism and a self-attention distilling technique, reducing the complexity to $O(L \log L)$ and significantly improving the efficiency of long-sequence forecasting [10]. However, Transformer and its variants still face two primary limitations. First, directly using the original series, which mixes these different components, can impede model learning. Numerous studies have shown that raw financial data contains significant high-frequency noise and components of different frequencies (e.g., short-term volatility, long-term trends), and feeding this mixed signal directly into deep models reduces learning efficiency [11]. Our goal is not to model the noise itself (as in stochastic models like gBm), but to separate it from the predictable signal components. Consequently, many studies have started to combine signal decomposition techniques (e.g., EMD, CEEMDAN, VMD) with deep learning, adopting a "decomposition-prediction-ensemble" paradigm to break down complex series into simpler sub-series before forecasting [12]. Second, the expressive power of the feature extraction module is limited. The convolution units embedded in models like Informer typically use one-dimensional convolutions with fixed receptive fields to generate queries, keys, and values, making it difficult to simultaneously capture both rapid fluctuations and slow trends at different time scales. Although recent research has proposed multi-scale dilated convolution architectures that use varying dilation rates to perceive multi-scale information concurrently [13], such techniques have not yet been widely applied in the domain of financial time series forecasting.

To address the aforementioned limitations, this paper proposes a novel stock price prediction framework, CVASD-MDCM-Informer, which integrates adaptive signal decomposition with multi-scale feature extraction. The core idea of this model lies in its "decomposition-prediction-ensemble" paradigm. First, the model utilizes a Conditional Optimization-based VMD Adaptive Signal Decomposition (CVASD) module to process the original stock price series. We employ Variational Mode Decomposition (VMD) because, compared to Empirical Mode Decomposition (EMD), it has a more solid mathematical foundation and can effectively avoid the mode mixing problem [14]. To overcome the challenge of empirically setting VMD's key parameters—the number of modes K and the penalty factor α—this paper introduces an improved Conditional Optimization (CPO) algorithm for automatic searching. CPO balances global and local search capabilities, promising to achieve superior decomposition results compared to existing algorithms like the Grasshopper Optimization Algorithm or GMPSO [15]. In the feature extraction stage, we design a Multi-scale Dilated Convolution Module (MDCM) to replace the one-dimensional convolutional layers that generate queries, keys, and values in the original Informer. This module leverages different dilation rates within its convolutional kernels to capture short-term, medium-term, and long-term dependencies in parallel while maintaining a low parameter count [13], thereby effectively capturing both local fluctuations and long-term trends in the stock price series and enriching the feature representation. Finally, the model aggregates the prediction results from each decomposed mode to obtain the final forecast.

The main contributions of this work are as follows:

- We propose a novel stock price prediction framework that fuses adaptive signal decomposition with multi-scale feature extraction. By combining VMD, CPO-based parameter optimization, and multi-scale dilated convolutions, the framework significantly enhances long-sequence forecasting capabilities.
- We are the first to apply the Conditional Optimization (CPO) algorithm to optimize VMD parameters. By automatically searching for the number of modes and the penalty factor, CPO resolves the issue of empirical parameter setting in traditional VMD and achieves superior decomposition performance on complex financial series.
- We design a Multi-scale Dilated Convolution Module (MDCM). This module utilizes different dilation rates to construct multi-scale receptive fields, substantially strengthening Informer's ability to capture features at various time scales.
- Extensive experiments on real-world stock market data validate the superiority of the proposed model. Compared to baseline models, CVASD-MDCM-Informer demonstrates significant improvements in prediction accuracy, stability, and computational efficiency.

The remainder of this paper is organized as follows: Section 2 reviews the relevant literature on deep learning for time series forecasting and the 'decomposition-ensemble' paradigm. Section 3 provides a detailed methodological breakdown of our proposed framework, including the Informer backbone, the CPO-VMD optimization (CVASD) process, and the Multi-scale Dilated Convolution Module (MDCM). Section 4 describes the experimental setup, datasets, baseline models, and presents a comprehensive analysis of the comparative and ablation study results. Finally, Section 5 concludes the paper, discusses the study's limitations, and outlines directions for future work.

2. Related Work

2.1. Deep Learning in Time Series Forecasting

2.1.1. RNN/LSTM-Based Models

Recurrent Neural Networks (RNNs) such as Long Short-Term Memory (LSTM) and Gated Recurrent Unit (GRU) excel at processing temporal relationships, and are capable of capturing long-range dependencies and mitigating the vanishing gradient problem through their gating mechanisms [16]. Research has shown that LSTM achieves high accuracy in stock price prediction, outperforming traditional methods, particularly in handling long-term dependencies and non-linearities [1]. However, the primary drawback of RNN architectures is their difficulty in capturing dependencies in extremely long sequences, their slow training speed, and the degradation of learning performance as sequence length increases due to gradient propagation issues [17].

2.1.2. Transformer-Based Models

The Transformer, which uses a self-attention mechanism to capture global dependencies within a sequence, has achieved tremendous success in natural language processing and computer vision. For long-sequence forecasting tasks, Informer proposed the ProbSparse self-attention and self-attention distilling strategies, reducing the time complexity to $O(L \log L)$ and significantly enhancing the efficiency of long-sequence prediction [10]. Nevertheless, the original Transformer architecture suffers from limitations such as high computational complexity, large memory consumption, and slow step-by-step decoding, making it challenging to apply directly to long-sequence forecasting [9]. To address these issues, researchers have proposed several improved models:

- **Autoformer**: Decomposes the input series into trend and seasonal components and captures periodic information using a stacked auto-correlation mechanism for long-term time series forecasting [18].
- **FEDformer**: Combines seasonal-trend decomposition with frequency domain analysis, introducing a frequency-enhanced module that enables the Transformer to capture global properties in the frequency domain while reducing computational complexity to a linear level [19].
- **Other Models**: Models such as LogSparse Transformer, Longformer, and Reformer employ sparse or local attention strategies to reduce time complexity [10].

Although these models have made progress in efficiency or performance, they generally assume high-quality input sequences and have limited capability in handling high-noise, non-stationary financial data. Furthermore, their internal feature extraction modules often use one-dimensional convolutions, making it difficult to capture multi-scale features.

2.2. The Decomposition–Ensemble Paradigm

2.2.1. Signal Decomposition Techniques

To handle the noise and complex frequency components in financial time series, many studies employ signal decomposition methods to break down the original series into several relatively stationary sub-series. Empirical Mode Decomposition (EMD) iteratively identifies local extrema to extract Intrinsic Mode Functions (IMFs). However, this method is sensitive to noise and sampling, often leading to mode mixing. The improved CEEMDAN (Complete Ensemble Empirical Mode Decomposition) mitigates mode mixing by adding symmetric noise and has been widely applied in financial forecasting [20]. In contrast, Variational Mode Decomposition (VMD) proposes the simultaneous decomposition of all modes by solving a variational model to determine the central frequency of each mode. This method has a solid theoretical foundation, can accurately separate harmonic components with close frequencies, and suppresses mode mixing. Therefore, VMD has become the mainstream choice for financial time series decomposition in recent years. Ref. [16] pointed out that VMD can decompose stock price series into multiple smooth IMFs, thereby reducing data complexity and minimizing noise interference.

2.2.2. Hybrid Prediction Models

Given the challenges of noise and complex frequency components, a 'decomposition–ensemble' paradigm has become a highly effective and mainstream approach in financial forecasting. This paradigm operates on a simple principle: first, use a decomposition method (like VMD or EMD) to split the complex, noisy original series into several simpler, more stationary sub-series (e.g., IMFs). Then, use one or more prediction models to forecast each sub-series individually. Finally, aggregate the individual forecasts to reconstruct the final prediction.

Numerous studies have validated this approach. For example, ref. [16] employed VMD to decompose stock data before feeding it into an LSTM (part of a VMD–TMFG–LSTM model), demonstrating improved accuracy by simplifying the learning task. Similarly, ref. [21] found that training an LSTM on VMD-processed series yielded superior performance and stability compared to using the raw series. This principle extends beyond VMD; other works combining EMD/CEEMDAN with models like CNNs and Transformers have also achieved remarkable results in related fields such as energy and traffic forecasting [22]. These works collectively establish a strong precedent: reducing data complexity and separating noise from signal before prediction is a critical step for enhancing the performance of deep learning models on volatile time series.

2.3. Intelligent Optimization of VMD Parameters

The effectiveness of VMD decomposition is highly dependent on the choice of the number of modes K and the penalty factor α. Manual tuning is time-consuming and often fails to find the optimal combination. Therefore, researchers have proposed using metaheuristic algorithms to automatically optimize VMD parameters. The literature [23] indicates that algorithms such as the Grasshopper Optimization Algorithm (GOA), Cuckoo Search Algorithm (CSA), and Genetic Mutation Particle Swarm Optimization (GMPSO) can effectively search for the optimal parameter combination, thereby improving decomposition quality. The application of GMPSO-VMD in bearing fault diagnosis demonstrated that using particle swarm mutation can avoid getting trapped in local optima, with minimum envelope entropy serving as the fitness function to select the best parameters [24].

Other research has explored methods like the Snake Optimizer Algorithm, Beluga Whale Optimization, and Artificial Bee Colony to optimize the number of modes and the penalty factor of VMD, proposing application-specific fitness functions such as maximum kurtosis or minimum instantaneous frequency error. These works show that using intelligent algorithms to optimize VMD parameters can significantly enhance the effectiveness of signal decomposition, thus providing more reliable inputs for subsequent prediction models.

In summary, existing research has revealed the challenges in financial time series forecasting and has proposed various methods combining deep learning with signal decomposition. Building on this foundation, this paper further innovates by integrating CPO-optimized VMD with multi-scale dilated convolutions, aiming to enhance the accuracy and generalization capability of stock price prediction.

3. Methodology

The prediction framework proposed in this paper uses Informer as its backbone, adopting its efficient encoder-decoder architecture and ProbSparse self-attention mechanism. To address the inherent high-noise and multi-scale characteristics of financial time series, we introduce two key modifications to the original Informer model:

(1) **CPO-VMD Adaptive Signal Decomposition (CVASD) Module**: Prior to data input, we first apply a rolling decomposition to the original stock price series using a Variational Mode Decomposition (VMD) method optimized by the Crested Porcupine Optimizer (CPO). This module decomposes the complex original series into a set of Intrinsic Mode Functions (IMFs), each possessing distinct frequency characteristics. This process effectively separates trends and fluctuations occurring at different time scales. These components are then fed individually into the subsequent network for prediction, which effectively reduces noise interference and enhances the model's ability to capture detailed features.

(2) **Multi-scale Dilated Convolution Module (MDCM)**: To more comprehensively extract temporal features from each decomposed component series, we replace the standard one-dimensional convolutional layers in the Informer's encoder with a Multi-scale Dilated Convolution Module (MDCM). This module employs three parallel convolutional kernels with different dilation rates, enabling it to simultaneously capture short- and long-term dependencies across various receptive fields, and enhances the model's feature representation capabilities through feature fusion.

(3) **Differentiation from Built-in Decomposition Transformers**: It is important to distinguish our "decomposition–prediction–ensemble" paradigm from Transformer variants that embed a fixed trend/seasonal split (e.g., Autoformer [18]). Autoformer's in-network decomposition is highly effective for series with pronounced periodicity (such as energy consumption or meteorology). In contrast, financial time series are dominated by high noise, nonstationarity, regime shifts, and irregular volatility rather than stable seasonality. To address this, the proposed CVASD module leverages CPO-optimized VMD to adaptively partition the input into frequency-based Intrinsic Mode Functions (IMFs), effectively separating high-frequency noise from medium-term oscillations and low-frequency trends. The subsequent MDCM then performs multi-scale, dilation-based feature extraction on each IMF, producing component-aware representations that align better with the heterogeneous, nonstationary structure of financial data than a general-purpose trend/seasonal decomposition.

Through these modifications, the model first performs a refined 'decomposition-and-reconstruction' processing of the signal, significantly enhancing its capacity to model the complex dynamics of financial time series. The overall architecture of the proposed model is illustrated in Figure 1.

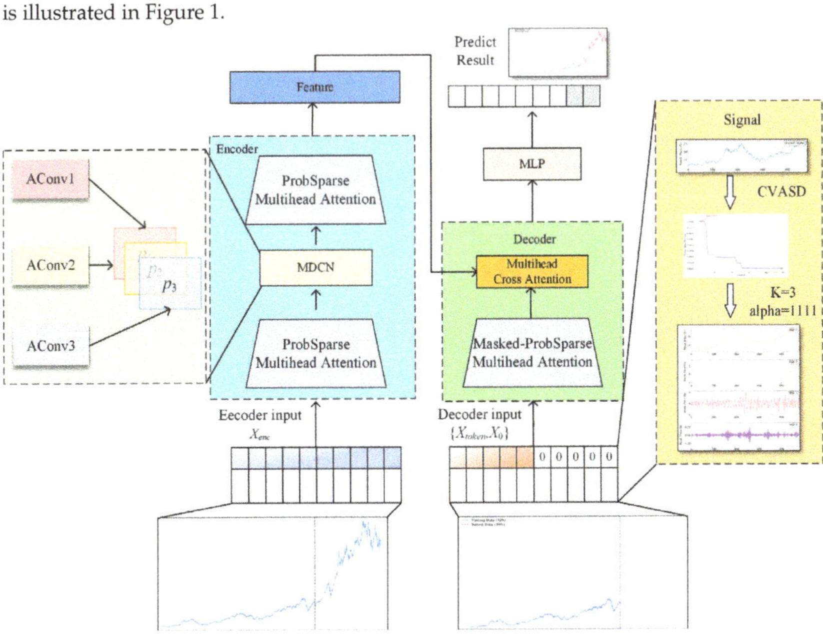

Figure 1. The overall architecture of the proposed CVASD-MDCM-Informer framework. The process follows a 'decomposition-prediction-ensemble' paradigm: (1) The original stock series is decomposed into K IMFs by the CVASD module. (2) Each IMF is fed into a shared-weight Informer encoder, which uses the MDCM module for multi-scale feature extraction. (3) The decoder predicts the future 48 steps for each IMF. (4) The final prediction is obtained by aggregating the results from all IMFs.

3.1. The Informer Model

The Informer model is an enhancement of the Transformer model. It utilizes a Transformer-based encoder architecture but incorporates a ProbSparse Self-Attention mechanism, which reduces the computational time complexity and significantly shortens computation time. It also introduces a self-attention distilling mechanism, which adds one-dimensional convolutional and max-pooling layers to each self-attention block to further reduce the computational overhead. Furthermore, Informer employs a generative-style decoder that performs a single forward pass to generate all predicted outputs at once for long sequences, thereby greatly increasing the prediction speed for long time series. The architecture of the Informer is shown in Figure 2.

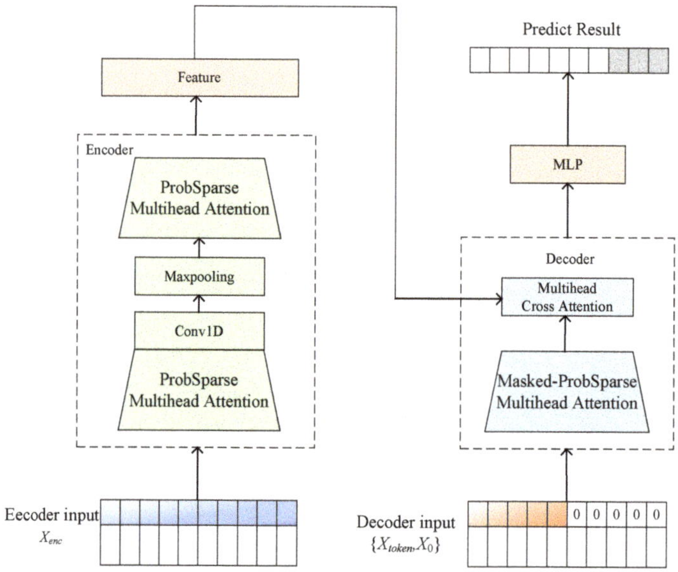

Figure 2. The standard Informer architecture [10], serving as the backbone of our model. It features an encoder with ProbSparse self-attention and distilling, and a generative decoder for efficient long-sequence forecasting.

3.1.1. ProbSparse Self-Attention Mechanism

(1) The formula for the dot-product attention in the traditional self-attention mechanism is:

$$\text{Attention}(Q, K, V) = \text{Softmax}\left(\frac{QK^{\top}}{\sqrt{d}}\right)V \tag{1}$$

where Q is the matrix of query vectors, K is the matrix of key vectors, V is the matrix of value vectors, and d is the dimension of the input features.

(2) The traditional self-attention mechanism has a quadratic complexity, which leads to long computation times for long sequences. To address this issue, the ProbSparse self-attention mechanism defines the sparsity measurement for the i-th query by approximating the KL-divergence, dropping the constant term:

$$M(q_i, K) = \ln \sum_{j=1}^{L_k} e^{\frac{q_i K_j^{\top}}{\sqrt{d}}} - \frac{1}{L_k} \sum_{j=1}^{L_k} \frac{q_i K_j^{\top}}{\sqrt{d}} \tag{2}$$

where the first term is the Log-Sum-Exp (LSE) of the dot products of q_i with all keys K_j, and the second term is their arithmetic mean.

(3) By selecting only the queries with the highest sparsity scores to participate in the subsequent computation, the final attention is calculated as:

$$\text{Attention}(Q, K, V) = \text{Softmax}\left(\frac{\bar{Q}K^{\top}}{\sqrt{d}}\right)V \tag{3}$$

where $\bar{Q}$ is the matrix composed of the selected high-sparsity queries.

3.1.2. Self-Attention Distilling

To further reduce the computational overhead, the Informer encoder adds a one-dimensional convolutional layer and a max-pooling layer after each self-attention layer for a "distilling" operation:

$$X_{j+1} = \text{MaxPool}(\text{ELU}(\text{Conv1d}([X_j]_{\text{AB}}))) \tag{4}$$

where X_{j+1} represents the output of the distilling operation from layer j to layer $j+1$, $[\cdot]_{\text{AB}}$ denotes the output of the attention block, Conv1d is a one-dimensional convolution operation, ELU is the activation function, and MaxPool is the max-pooling operation.

When the convolutional layer distills features from the current layer to the next, the feature map is compressed to half its original length, further reducing computational costs while preserving the dominant features.

3.1.3. Generative-Style Decoder

Using a generative-style decoder, all output data points are generated in a single step, which significantly improves prediction efficiency. The input to the decoder is constructed as follows:

$$X_{\text{de}} = \text{Concat}(X_{\text{token}}, X_0) \in \mathbb{R}^{(L_{\text{token}}+L_y)\times d_{\text{model}}} \tag{5}$$

where X_{de} is the input vector for the decoder, $X_{\text{token}} \in \mathbb{R}^{L_{\text{token}}\times d_{\text{model}}}$ is the start token, and $X_0 \in \mathbb{R}^{L_y\times d_{\text{model}}}$ is a placeholder for the target sequence to be predicted (typically filled with zeros).

3.2. *CPO-Based VMD Parameter Optimization*

To effectively address the non-stationarity and high-noise characteristics of stock price series, this paper adopts the "decomposition-prediction-ensemble" paradigm as the first step of its core strategy. Specifically, we employ Variational Mode Decomposition (VMD) to preprocess the original signal. However, the performance of VMD is highly dependent on its key parameters—the number of decomposed modes K and the quadratic penalty factor α—which are traditionally set based on experience or trial-and-error, making it difficult to guarantee optimality. To this end, this section proposes an adaptive VMD parameter optimization scheme based on the Crested Porcupine Optimizer (CPO) to ensure the effectiveness and robustness of the signal decomposition.

3.2.1. Variational Mode Decomposition (VMD) Principle

VMD is an advanced, adaptive, and non-recursive signal processing technique. Its core idea is to construct and solve a constrained variational problem to decompose a real-valued input signal $x(t)$ into a series of discrete modal components $u_k(t)$ (i.e., Intrinsic Mode Functions, IMFs) with specific sparsity properties. Each mode u_k is assumed to be an AM-FM signal that is compact around a center frequency ω_k.

The optimization objective of VMD is to find K modal components such that the sum of their estimated bandwidths is minimized, subject to the constraint that the sum of these modes accurately reconstructs the original signal. This constrained variational problem is formulated as follows:

$$\begin{cases} \min\limits_{\{u_k\},\{\omega_k\}} \left\{ \sum\limits_{k=1}^{K} \left\| \partial_t \left[\left(\delta(t) + \frac{j}{\pi t} \right) * u_k(t) \right] e^{-j\omega_k t} \right\|_2^2 \right\} \\ \text{s.t.} \ \sum\limits_{k=1}^{K} u_k(t) = x(t) \end{cases} \tag{6}$$

where $\{u_k\}$ and $\{\omega_k\}$ are the sets of K modal components and their respective center frequencies; ∂_t denotes the partial derivative with respect to time t; $\delta(t)$ is the Dirac delta function; j is the imaginary unit; and $*$ represents the convolution operation.

To solve the above problem, the constrained variational problem is transformed into an unconstrained one by introducing a quadratic penalty factor α and a Lagrange multiplier $\lambda(t)$. The augmented Lagrangian function L is defined as:

$$\begin{aligned} L(\{u_k\},\{\omega_k\},\lambda) = \alpha \sum_{k=1}^{K} \left\| \partial_t \left[\left(\delta(t) + \frac{j}{\pi t} \right) * u_k(t) \right] e^{-j\omega_k t} \right\|_2^2 \\ + \left\| x(t) - \sum_{k=1}^{K} u_k(t) \right\|_2^2 + \left\langle \lambda(t), x(t) - \sum_{k=1}^{K} u_k(t) \right\rangle \end{aligned} \tag{7}$$

This optimization problem can be solved iteratively using the Alternating Direction Method of Multipliers (ADMM). The modal components $\hat{u}_k$, center frequencies ω_k, and the Lagrange multiplier $\hat{\lambda}$ are alternately updated in the frequency domain until a convergence criterion is met. The iterative update formulas are as follows:

$$\begin{cases} \hat{u}_k^{n+1}(\omega) = \dfrac{\hat{x}(\omega) - \sum_{i \neq k} \hat{u}_i^n(\omega) + \frac{\hat{\lambda}^n(\omega)}{2}}{1 + 2\alpha(\omega - \omega_k^n)^2} \\ \omega_k^{n+1} = \dfrac{\int_0^\infty \omega |\hat{u}_k^{n+1}(\omega)|^2 d\omega}{\int_0^\infty |\hat{u}_k^{n+1}(\omega)|^2 d\omega} \\ \hat{\lambda}^{n+1}(\omega) = \hat{\lambda}^n(\omega) + \tau \left(\hat{x}(\omega) - \sum\limits_{k=1}^{K} \hat{u}_k^{n+1}(\omega) \right) \end{cases} \tag{8}$$

where $\hat{\cdot}$ denotes the Fourier transform, and τ is an update parameter.

3.2.2. CPO-VMD Parameter Optimization Process

Although VMD has theoretical advantages, its decomposition effectiveness is extremely sensitive to the parameters K and α. To achieve adaptive parameter optimization, this paper introduces the Crested Porcupine Optimizer (CPO) algorithm. CPO is a metaheuristic algorithm that simulates the defensive behavior of crested porcupines against predators. It performs an efficient search in the solution space by simulating two mechanisms: the "scent strategy" (global exploration) and the "physical strategy" (local exploitation). This algorithm is characterized by its strong global search capability, fast convergence speed, and a reduced tendency to become trapped in local optima.

As illustrated in Figure 3, we formulate the VMD parameter search as an optimization task. The specific process is as follows:

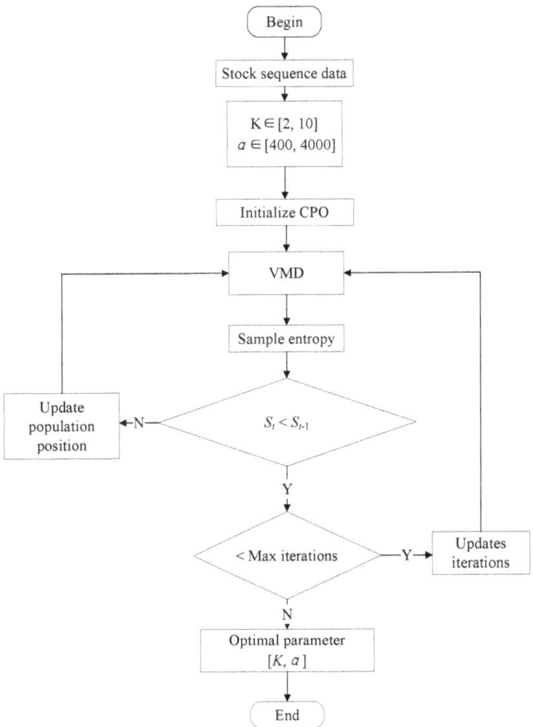

Figure 3. Flowchart of the CPO-VMD signal decomposition process. For each rolling window of data, the CPO algorithm iteratively searches for the optimal VMD parameters (K, α) by minimizing the average sample entropy (the fitness function) of the resulting IMFs.

1. **Define the Fitness Function**: We select Sample Entropy as the fitness function. Sample Entropy is a key metric for measuring the complexity of a time series; a smaller value indicates stronger regularity and lower noise levels. An ideal VMD decomposition should produce a series of stationary and physically meaningful modal components, which typically have low Sample Entropy values. Therefore, we set the average Sample Entropy of all modal components as the optimization objective. That is, the goal of the CPO algorithm is to find a set of parameters (K, α) that minimizes this fitness function.
2. **Execute CPO Optimization**: Within a predefined parameter range (e.g., $K \in [2, 10]$, $\alpha \in [400, 4000]$), the CPO algorithm searches for the optimal parameter combination by iteratively updating the population's positions.
3. **Sliding Decomposition Strategy**: A critical aspect of our methodology is to strictly prevent data leakage. We perform CPO-VMD decomposition using a sliding window approach. While our model's input sequence length (seq_len) is 96, we found that VMD requires a longer sequence for stable decomposition. Therefore, to generate the input for time t (which requires data from $t-95$ to t), we perform the decomposition on a longer preceding window of 960 time steps (i.e., the interval $[t-960+1, t]$). Crucially, from the resulting K decomposed IMFs (each of length 960), we only extract the final 96 time steps (the interval $[t-95, t]$) to serve as the input for the Informer network. This rolling decomposition ensures that at no point does the model use future information. For example, to predict from time 961, we decompose the series from time 1 to 960,

and use the IMF segments from time 865 to 960 as input. This design assumes a 'warm-up' period of data (960 − 96 = 864 steps) is available before the first prediction.

Through this process, we find the most suitable decomposition parameters for each input sequence. To balance computational efficiency and specificity, we optimized these parameters per industry on the training set. We selected a representative company from each sector (e.g., AAPL for Information Technology, AFL for Financials, and ADM for Consumer Staples) to find a unified set of parameters for that sector. The parameters used in this study are as follows:

- **Information Technology: $K = 4$, $\alpha = 1111$**
- **Financials: $K = 4$, $\alpha = 1536$**
- **Consumer Staples: $K = 3$, $\alpha = 1625$**

The convergence curve for the Information Technology sector optimization is shown in Figure 4. Figure 5 visualizes the decomposition result for an AAPL stock series using these parameters.

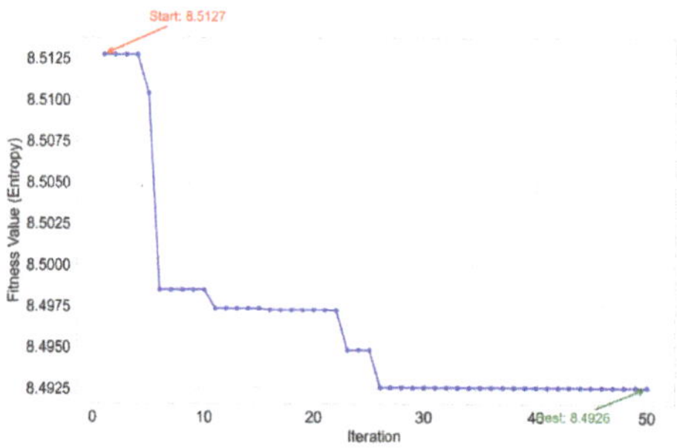

Figure 4. Convergence curve of the fitness function (Average Sample entropy) during the CPO algorithm optimization process for the AAPL dataset (Information Technology sector). The algorithm rapidly converges to a stable minimum value, demonstrating its efficiency in finding optimal VMD parameters.

3.3. Multi-Scale Dilated Convolution

3.3.1. Fundamental Theory

Dilated Convolution, also known as Atrous Convolution (AConv), is a special type of convolution where a standard kernel is expanded by inserting a number of zeros ('holes') between consecutive values. It was first proposed in the *Alogrithme à Trous* for wavelet decomposition of time-varying signals and was later widely applied in semantic segmentation. The operation of AConv can be expressed as:

$$h(i) = \sum_{k=1}^{K} d(i + r \times k) \cdot X(k) \tag{9}$$

where $d(i)$ is a one-dimensional discrete signal, $h(i)$ is the i-th output feature value of $d(i)$ after the dilated convolution operation, r is the dilation rate of the kernel, and K is the size of the kernel.

Global Average Pooling (GAP) refers to the operation of taking the spatial average of the output feature map from the last convolutional or pooling layer. Compared to fully-connected layers, a GAP layer has no parameters to train, thus significantly reducing the number of parameters and the computational load of the network. The GAP operation can be expressed as:

$$u^c = \frac{1}{n}\sum_{i=1}^{n} v_i^c \tag{10}$$

where c is the number of output channels, v_i^c is the i-th feature value in the c-th channel, n is the number of feature values per channel, and u^c is the global average value output for the c-th channel.

Batch Normalization (BN) can reduce the distribution difference between the features of training and testing samples, thereby improving the network's performance and generalization ability. The BN operation, when placed after a convolutional layer, can be expressed as:

$$z_i = \frac{x_i - \mu_B}{\sqrt{\sigma_B^2 + \varepsilon}}, \quad y_i = \gamma z_i + \beta \tag{11}$$

where x_i is the i-th value of the input feature map in layer l; μ_B is the mean of the input samples in each batch; σ_B^2 is the variance of the input samples in each batch; ε is a small constant to prevent division by zero; z_i is the i-th value of the standardized feature map in layer l; γ and β are the scale and shift factors, respectively; and y_i is the i-th output result of batch normalization in layer l.

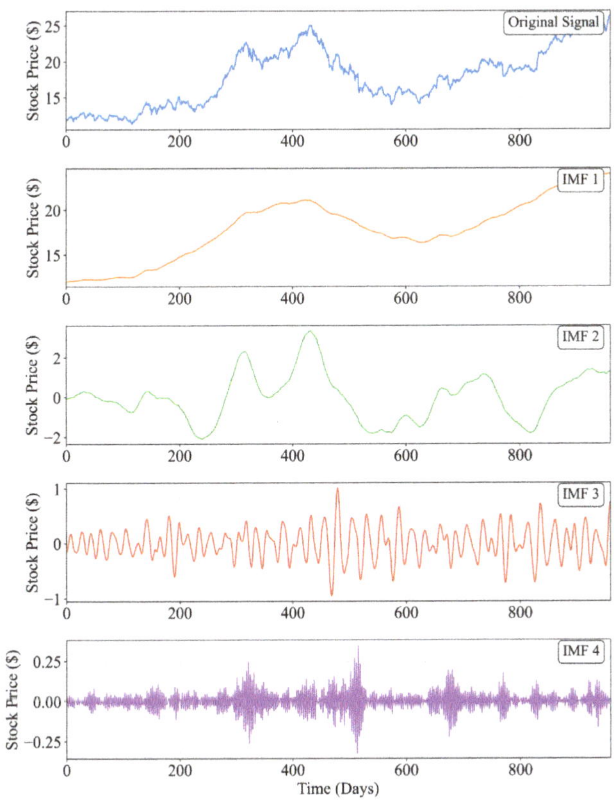

Figure 5. Example of CPO-VMD decomposition results for the AAPL stock price series (Information Technology sector). The original complex series (top) is decomposed into four distinct modal components (IMF1 to IMF4) with frequencies ranging from high (IMF1, capturing noise) to low (IMF4, capturing the main trend).

3.3.2. Multi-Scale Dilated Convolution Module

First, standard convolutional kernels and dilated convolutional kernels of different sizes are applied to the input layer of the model to fully extract features at various scales from different receptive fields. Then, each of the three large-kernel convolutional layers is followed by a small 1×1 convolutional kernel to further integrate the features extracted by the large kernels in each channel, as shown in Figure 6.

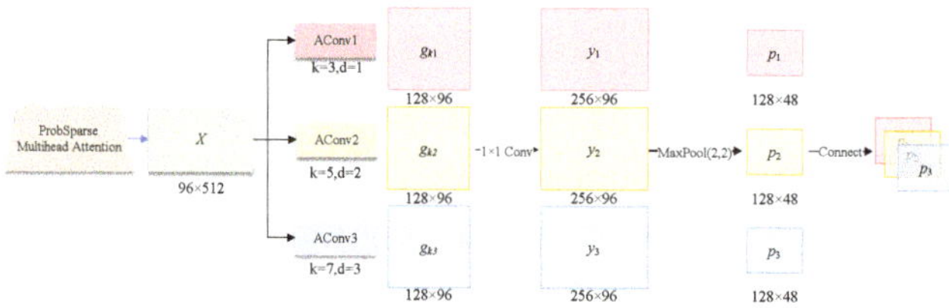

Figure 6. The architecture of the proposed Multi-scale Dilated Convolution Module (MDCM). It replaces the standard 1D convolution in Informer. Input features are processed in parallel by three branches with different dilation rates (e.g., 1, 2, 4) to capture dependencies at multiple time scales. The outputs are concatenated to produce a rich, multi-scale feature representation.

(1) The mathematical form of the multi-scale convolution is:

$$g_k^i = \text{Conv}_k^i(X, W_k^i) + b_k^i \tag{12}$$

where X is the input feature, W_k^i is the i-th kernel of size k, b_k^i is the bias added to the output feature of the i-th kernel of size k, g_k^i is the i-th output feature from the kernel of size k, and Conv is the convolution operation.

(2) The feature maps output by the 1×1 small-kernel convolutions are processed with Batch Normalization according to Equation (11).

(3) A non-linear mapping is applied to the normalized output features to enhance their expressive power and separability. This paper uses the commonly adopted Rectified Linear Unit (ReLU) as the activation function, whose mathematical form is:

$$y_i^l = f(\alpha_i^l) = \max(0, \alpha_i^l) \tag{13}$$

where α_i^l is the i-th value of the input feature map in layer l after being normalized by Equation (11), and y_i^l is the i-th output result of α_i^l in layer l after ReLU activation.

(4) A Max Pooling (MaxPool) operation is performed on the features output after ReLU activation. Its mathematical form is:

$$p_i^{l+1} = \max_{t \in S}(q_i^l(t)) \tag{14}$$

where $q_i^l(t)$ is the output of the t-th neuron in the i-th channel of layer l, S is the size of the pooling kernel, and p_i^{l+1} is the output result of the i-th channel in layer $l+1$.

(5) The feature maps extracted by the convolutions at three different scales are denoted as X_1, X_2, and X_3. They are then stacked along the depth dimension using tensor concatenation to achieve feature fusion. The mathematical form of this feature stacking is:

$$Y = \text{concat}(X_1, X_2, X_3) \tag{15}$$

where concat is the tensor concatenation operation, and Y is the output result of stacking X_1, X_2, and X_3. The final fused and distilled features are then input into the Informer's decoder.

4. Experiments

4.1. Dataset

The experimental data for this study is sourced from the constituent stocks of the S&P 500 index. We selected a time span from 3 January 2011, to 12 December 2022. The data is sampled at a daily frequency (one data point per trading day), and we filtered for 352 companies with complete data during this period. To test the model's generalization ability and its adaptability to different market dynamics, we categorized these companies into three representative sectors based on the Global Industry Classification Standard (GICS) [25]:

- **Information Technology**: This sector is typically characterized by high growth and is sensitive to market risk appetite. Its stock price dynamics are well-described by features such as momentum, earnings forecast revisions, and R&D intensity.
- **Financials:** This sector is highly sensitive to interest rate changes. Its performance is closely correlated with exogenous factors like macroeconomic policies (e.g., policy rates), market interest rates (e.g., long- and short-term government bond yields), and credit spreads.
- **Consumer Staples**: This sector possesses defensive characteristics. Companies in this sector have relatively stable earnings and lower volatility, often exhibiting strong downside resistance during market-wide drawdowns.

The stock price prediction task is formulated as a multivariate time series regression problem, where we use historical data from the past 96 time steps to predict the closing price for the next 48 time steps. To construct a rich feature representation, the model's input includes basic trading data such as opening price (OPEN), closing price (CLOSE), highest price (HIGH), lowest price (LOW), and trading volume (VOLUME), as well as multiple quantitative technical factors from the 'mytt' factor library. These factors are grouped into three categories to comprehensively capture market dynamics: (1) **Trend-following factors**, such as the MACD series (DIF, DEA, MACD) and Bollinger Bands (UPPER, MID, LOWER), to identify the market's primary direction and long-term trends; (2) **Momentum/Oscillator factors**, such as the Stochastic Oscillator (KDJ), Bias Ratio (BIAS), Commodity Channel Index (CCI), and Momentum (MTM), to capture short-term overbought/oversold signals and pullback opportunities; and (3) **Sentiment factors**, such as the Psychological Line (PSY), Popularity Index (AR), and Willingness to Buy/Sell Index (BR), to quantify the collective psychological state of market participants and assist in identifying market turning points. The specific factors are detailed in Table 1. By jointly utilizing these factors from different dimensions, the model can obtain a more comprehensive understanding of the market state. All input data is subjected to Z-score normalization before being fed into the model. To prevent data leakage, the mean and standard deviation required for normalization are computed only on the training dataset. These statistics are then saved and applied consistently to normalize the validation and test datasets.

Table 1. Quantitative Factors.

Factor Type	Factor Name
Basic Factors	OPEN, CLOSE, HIGH, LOW, VOLUME
Technical Factors	DIF, DEA, MACD KDJ, BIAS1, BIAS2, BIAS3, UPPER, MID, LOWER, PSY, PSYMA, CCI, MTM, MTMMA, AR, BR

4.2. Baseline Models

To validate the effectiveness of the framework proposed in this paper, we selected several state-of-the-art time series forecasting models as baselines for comparison:

1. **Transformer** [26]: This architecture employs a self-attention mechanism and an encoder-decoder structure, supporting parallel processing of sequential data to effectively capture long-range dependencies. However, its quadratic time/memory complexity ($O(L^2)$) limits its direct application in long-sequence time series forecasting.
2. **Autoformer** [18]: A decomposition-based Transformer with an auto-correlation mechanism for long-term time series forecasting. The model incorporates a series decomposition block as a built-in operator to progressively separate trend-cyclical and seasonal components. The auto-correlation mechanism replaces self-attention, mining sub-series dependencies based on periodicity via Fast Fourier Transform and aggregating similar sub-series. It has achieved state-of-the-art performance on datasets for energy, traffic, and weather.
3. **Informer** [10]: An efficient Transformer for long-sequence time series forecasting. The model addresses the limitations of the Transformer through three innovations: ProbSparse self-attention, self-attention distilling, and a generative-style decoder. It outperforms traditional models on large-scale datasets such as ETT and Traffic.
4. **Informer-CGRU** [27]: A model based on Informer for non-stationary time series forecasting. It introduces multi-scale causal convolutions to capture causal relationships across time scales and uses a GRU layer to re-learn causal features, enhancing robustness to non-linearity and high noise.
5. **iTransformer** [28]: An inverted Transformer architecture optimized for time series forecasting. This model improves the input processing and attention mechanism of the Transformer to enhance temporal pattern capturing. By reformulating the attention computation paradigm, it improves the efficiency of modeling long-range dependencies in time series data.
6. **FEDformer** [19]: A frequency-enhanced decomposition Transformer for long-term forecasting. The model integrates seasonal-trend decomposition with frequency-domain transforms. Frequency-enhanced blocks and attention mechanisms replace self-attention and cross-attention, ensuring linear complexity by randomly selecting frequency components. This model improves the distributional consistency between predicted and true sequences.
7. **DAT-PN** [29]: A Transformer-based dual-attention framework for financial time series forecasting. The framework consists of two parallel networks: a Price Attention Network (PAN) processes historical price data using multi-head masked self-attention (MMSA) and multi-head cross-attention (MCA) to capture temporal dependencies, while a Non-price Attention Network uses a convolutional LSTM, a bidirectional GRU, and self-attention to extract features from non-price financial data.
8. **VLTCNA** [12]: A stock sequence prediction model that integrates Variational Mode Decomposition (VMD) with dual-channel attention. VMD decomposes the stock sequence into stable frequency sub-windows using a sliding window to reduce volatility. A dual-channel network (LSTMA + TCNA) comprises an LSTMA (2 LSTM layers + self-attention) and a TCNA (4 F-TCN modules + self-attention), where the former captures long-term dependencies and the latter captures local/short-term patterns.

4.3. Experimental Environment and Parameter Settings

4.3.1. Experimental Environment

The specific settings for the experimental environment are shown in Table 2:

Table 2. Experimental Environment Settings.

Environment Component	Specification
Runtime Environment	Python 3.7
Deep Learning Framework	Pytorch 1.9.1 + CUDA 11.1
Operating System	Windows 10
CPU Model	Intel Core i7-12700K @ 4.9 GHz
Memory (RAM)	64 GB (4 × 16 GB) DDR4 @ 3600 MHz
GPU Model	NVIDIA GeForce RTX 3090 (24 GB)

4.3.2. Hyperparameter Settings

The specific hyperparameter settings for the experiment are shown in Table 3:

Table 3. Hyperparameter Settings.

Parameter Name	Description	Value
batch_size	Batch Size	64
learn_rate	Learning Rate	0.0001
seq_len	Input Sequence Length	96
label_len	Label Sequence Length	48
pred_len	Output Sequence Length	48
factor	Sparsity Factor	5
n_heads	Number of Attention Heads	8
e_layers	Number of Encoder Layers	2
d_layers	Number of Decoder Layers	1
d_model	Model Dimension	512

The key hyperparameters listed in Table 3, as well as the specific parameters for the MDCM module, were determined using a grid search on the validation set of the Information Technology dataset. The search space and final choices for the MDCM module and other key network parameters are detailed in Table 4.

Table 4. Grid search space and selected hyperparameters for key network parameters.

Parameter	Search Space	Selected Value
Learning rate	{0.0001, 0.0005, 0.001}	0.0001
Batch size	{32, 64, 128}	64
Encoder layers (*e_layers*)	{2, 3, 4}	2
MDCM kernel sizes	{(1, 3, 5), (3, 5, 7), (1, 5, 9)}	(3, 5, 7)
MDCM dilation rates	{(1, 2, 4), (1, 3, 5), (1, 4, 8)}	(1, 2, 4)
Model dimension (*d*_model)	{256, 512}	512

4.4. Evaluation Metrics

The experiments use Mean Absolute Error (MAE), Mean Squared Error (MSE), and Root Mean Squared Error (RMSE) as performance evaluation metrics. A smaller value for these metrics indicates a smaller prediction error and better model performance. The calculation methods for these metrics are as follows.

$$\text{MAE} = \frac{1}{m}\sum_{i=1}^{m}|y_i - \hat{y}_i| \tag{16}$$

$$\text{MSE} = \frac{1}{m}\sum_{i=1}^{m}(y_i - \hat{y}_i)^2 \tag{17}$$

$$\text{RMSE} = \sqrt{\frac{1}{m}\sum_{i=1}^{m}(y_i - \hat{y}_i)^2} \quad (18)$$

where m is the number of evaluation samples, y_i is the true value, and $\hat{y}_i$ is the predicted value. MAE represents the average of the absolute differences between predicted and true values and is equally sensitive to large and small errors, focusing on the overall model performance. MSE represents the average of the squared differences between predicted and actual values and is more sensitive to large errors. RMSE is the square root of MSE, sharing the same units as the target variable, which allows for a more intuitive interpretation of the error. The smaller the values of these metrics, the smaller the corresponding error and the better the model's performance.

4.5. Comparative Analysis of Experiments

To validate the effectiveness of our proposed model, we selected several recent state-of-the-art methods as baseline models for comparison on three different industry datasets from the S&P 500.

The results of the comparative experiments are shown in Table 5. The best results for each metric are marked in bold, and the second-best results are underlined. As can be seen from the table, our proposed method achieves optimal or near-optimal performance across the three different industry datasets.

Table 5. Comparative experimental results.

Model	Metric	Information Tech	Financials	Consumer Staples
Transformer	MAE	0.3977	0.3978	0.3378
	MSE	0.2228	0.2380	0.2089
	RMSE	0.4728	0.4910	0.4690
Autoformer	MAE	0.3845	0.4240	0.3776
	MSE	0.2395	0.2398	0.2067
	RMSE	0.4890	0.4889	0.4677
Informer	MAE	0.3727	0.4011	0.3578
	MSE	0.2024	0.2386	0.1890
	RMSE	0.4503	0.4876	0.4335
Informer-CGRU	MAE	0.3378	0.3658	0.3349
	MSE	0.1701	0.2210	0.1779
	RMSE	0.4125	0.4613	0.4212
iTransformer	MAE	**0.3089**	0.3678	0.3581
	MSE	<u>0.1489</u>	0.2183	0.1852
	RMSE	<u>0.3852</u>	0.4673	0.4398
DAT-PN	MAE	0.3389	<u>0.3508</u>	0.3238
	MSE	0.1771	<u>0.1910</u>	0.1681
	RMSE	0.4209	<u>0.4384</u>	0.4230
FEDformer	MAE	0.3294	0.3585	<u>0.3224</u>
	MSE	0.1644	0.1994	<u>0.1448</u>
	RMSE	0.4087	0.4470	<u>0.3843</u>
VLTCNA	MAE	0.3268	0.3715	0.3231
	MSE	0.1617	0.2094	0.1767
	RMSE	0.4103	0.4564	0.4109
Ours	MAE	<u>0.3110</u>	**0.3380**	**0.2942**
	MSE	**0.1411**	**0.1809**	**0.1356**
	RMSE	**0.3760**	**0.4279**	**0.3722**

Table 5. *Cont.*

Model	Metric	Information Tech	Financials	Consumer Staples
Improvement	MAE	−0.68%	3.65%	8.74%
	MSE	5.24%	5.29%	6.35%
	RMSE	2.39%	2.40%	3.15%

As the table indicates, our proposed method achieves optimal or near-optimal performance across the three different industry datasets. Specifically: In the highly volatile "Information Technology" sector, where stock prices typically contain more high-frequency noise and abrupt changes, our model achieved the best performance on the MSE and RMSE metrics (0.1411 and 0.3760, respectively). Compared to the second-best model, iTransformer, the MSE was reduced by 5.24%. This suggests that the CVASD module successfully separated high-frequency noise from the main trends, allowing the model to focus more on learning effective information. Simultaneously, the MDCM module, with its multi-scale receptive fields, effectively captured fluctuation characteristics across different periods, leading to outstanding performance on the MSE metric, which penalizes larger errors more heavily. In the "Financials" sector, which is closely linked to the macro-economy and exhibits stronger cyclical patterns, our model achieved the best results across all metrics, with MAE, MSE, and RMSE values of 0.3380, 0.1809, and 0.4279, respectively. Compared to the second-best model, DAT-PN, the MSE was reduced by 5.29%. This demonstrates the effectiveness of the proposed framework in handling sequences with complex cyclical components, where the adaptive decomposition and multi-scale feature extraction prove highly beneficial. For the "Consumer Staples" sector, where stock price movements are relatively stable and more trend-driven, our model also demonstrated comprehensive superiority, ranking first on all metrics. Compared to the second-best model, FEDformer, the MSE was reduced by 6.35%, showing the most significant advantage. This indicates that even in scenarios with relatively low noise and clear trends, our model can uncover subtle patterns that are difficult for a single model to capture, thanks to its fine-grained signal decomposition and multi-scale feature extraction, thereby achieving more accurate predictions. In summary, the proposed CVASD-MDCM-Informer model exhibits excellent performance and good generalization ability across financial time series with different characteristics, validating the effectiveness of combining adaptive signal decomposition with multi-scale feature extraction.

4.6. Ablation Study

To verify the effectiveness of the key modules in our model, we designed an ablation study. The experiment evaluates the contribution of specific components to the overall performance by removing them, as shown in Table 6. "*w/o*" indicates the removal of the corresponding module; bold text indicates the best (lowest) value for that metric within that industry.

From the experimental data in Table 6, it is evident that the complete model (Our Model) significantly outperforms all other variants in the tests across all three industries, achieving the lowest prediction errors for MAE, MSE, and RMSE. In contrast, the original Informer, serving as the baseline model (*w/o* CVASD & MDCM), performed the worst. For instance, on the Information Technology dataset, the full model reduced MAE, MSE, and RMSE by 19.7%, 30.4%, and 19.5%, respectively, compared to the baseline. This strongly demonstrates that the combination of our two proposed improvement modules can effectively enhance the accuracy of stock price prediction, validating the rationality and advancement of the overall model design.

Table 6. Ablation study results across industries (lower is better).

Industry	Variant	MAE	MSE	RMSE
Information Technology	*w/o* CVASD & MDCM	0.3874	0.2028	0.4672
	w/o CVASD	0.3432	0.1745	0.4262
	w/o MDCM	0.3234	0.1670	0.4003
	Our Model	**0.3110**	**0.1411**	**0.3760**
Financials	*w/o* CVASD & MDCM	0.3789	0.2257	0.4766
	w/o CVASD	0.3537	0.1909	0.4460
	w/o MDCM	0.3581	0.2136	0.4578
	Our Model	**0.3380**	**0.1809**	**0.4279**
Consumer Staples	*w/o* CVASD & MDCM	0.3673	0.1830	0.4335
	w/o CVASD	0.3104	0.1689	0.4031
	w/o MDCM	0.3207	0.1589	0.3899
	Our Model	**0.2942**	**0.1356**	**0.3722**

When the CVASD module was removed (*w/o* CVASD), the model's performance showed a significant decline compared to the complete model. This was observed across all datasets, particularly in the Information Technology sector, where removing this module led to a 10.3% increase in MAE. The core function of the CVASD module is to decompose the high-noise, non-stationary original stock price series into a set of more predictable modal components with distinct frequency characteristics. The experimental results show that this "decomposition-prediction-aggregation" strategy can effectively filter market noise and help the model more accurately capture underlying patterns at different time scales, which is crucial for improving prediction stability.

Similarly, removing the MDCM module (*w/o* MDCM) also led to a decrease in the model's predictive capability. The MDCM endows the model with the ability to capture multi-scale temporal dependencies within a single layer by using parallel convolutional kernels with different dilation rates. Compared to the standard convolutions in the original Informer, this design allows for a more comprehensive extraction of both short-term details and long-term trend features from the sequence. The experimental results validate the effectiveness of this design, proving that enhancing the local feature extraction capability at the encoder end is essential for improving overall prediction accuracy.

By comparing the two variants with a single module removed, we observe that the modules exhibit differential importance across different industry datasets. In the Information Technology sector, the performance loss from removing the CVASD module was greater than that from removing the MDCM module, which may imply that noise interference and multi-frequency superposition characteristics in this sector's stock series are the main challenges for prediction. In the Financials sector, however, the impact of removing the MDCM module was more significant, suggesting that accurately capturing complex long- and short-term temporal dependencies is key to forecasting in this domain. In the Consumer Staples sector, removing the CVASD module had a larger impact on MSE and RMSE, while removing the MDCM module had a larger impact on MAE. This reveals that CVASD plays a prominent role in suppressing extreme errors (which are amplified by the squared term in MSE/RMSE), whereas MDCM contributes more to improving the overall average prediction accuracy (as measured by MAE).

In conclusion, the results of the ablation study clearly demonstrate that both the CVASD and MDCM modules are indispensable components of our prediction framework. They optimize the model from the perspectives of signal processing and feature extraction, respectively, and through their synergistic effect, significantly enhance the model's predictive performance and robustness when handling complex financial time series.

5. Conclusions and Future Work

This paper proposed a novel hybrid framework, CVASD-MDCM-Informer, for stock price prediction, designed to address the challenges of high noise and non-stationarity in financial time series. Our 'decomposition-prediction-ensemble' approach first utilizes a CVASD module, where the CPO algorithm adaptively optimizes VMD parameters (K and α) to decompose the raw price series into more stable IMF components. Subsequently, an MDCM module with parallel dilated convolutions replaces the standard convolution in the Informer, enhancing the model's ability to capture multi-scale features from these components.

Extensive experiments on three S&P 500 industry datasets (Information Technology, Financials, and Consumer Staples) demonstrated the superiority of our framework. Compared to several state-of-the-art baseline models, including iTransformer, FEDformer, and DAT-PN, our model achieved significant improvements across all evaluation metrics (MAE, MSE, RMSE). Ablation studies further confirmed the indispensable contributions of both the CVASD and MDCM modules, validating our hypothesis that combining adaptive signal decomposition with multi-scale feature extraction is a highly effective strategy for complex time series forecasting.

5.1. Limitations

Despite the promising results, this study has several limitations that should be acknowledged.

- **Survivorship Bias:** As noted by the reviewer, our dataset construction, which selected 352 companies with complete data from 2011 to 2022, introduces survivorship bias. By excluding delisted or inactive firms, our sample is skewed towards more stable and successful companies, which may artificially inflate the model's predictability. Future work should explore methods to incorporate data from delisted stocks to create a more representative dataset.
- **Computational Overhead and Data Requirement:** The CPO-VMD rolling decomposition strategy, while effective, is computationally intensive. Furthermore, our use of a 960-step window for stable decomposition requires a significant 'warm-up' period (864 data points in our study), making the model less suitable for 'cold-start' scenarios or very short time series.

5.2. Future Work

Building on this work, several avenues for future research are apparent.

- **Broader Asset Classes:** The framework's effectiveness should be tested on other financial instruments, such as exchange-traded funds (ETFs), cryptocurrencies, and macroeconomic indicators, which possess different volatility structures and noise characteristics.
- **Cross-Domain Applications:** The proposed methodology is not limited to finance. It can be extended to other domains characterized by complex, non-stationary time series, such as energy load forecasting, electricity price prediction, and meteorological forecasting.
- **Model Efficiency:** Future research could explore more lightweight adaptive decomposition techniques or investigate knowledge distillation methods to reduce the computational burden of the CVASD module without sacrificing performance.

Author Contributions: Conceptualization, J.S.; Methodology, J.S. and R.Y.K.L.; Software, J.S., Y.D. and J.Y.; Validation, J.S. and J.Y.; Writing—original draft, J.S. and Y.D.; Visualization, J.S., Y.D. and H.Z. All authors have read and agreed to the published version of the manuscript.

Funding: This work was partly supported by a grant from the Research Grants Council of the Hong Kong SAR (Project No.: CityU 11507323).

Data Availability Statement: The relevant data can be obtained by contacting the corresponding author.

Conflicts of Interest: The authors declare that they have no known competing financial interests or personal relationships that could have appeared to influence the work reported in this paper.

References

1. Rezaei, H.; Faaljou, H.; Mansourfar, G. Stock price prediction using deep learning and frequency decomposition. *Expert Syst. Appl.* **2021**, *169*, 114332. [CrossRef]
2. Hossein, A.; Charlotte, C.; Hou, A.J. The effect of uncertainty on stock market volatility and correlation. *J. Bank. Financ.* **2023**, *154*, 106929.
3. Su, J.; Lau, R.Y.K.; Yu, J.; NG, D.C.T.; Jiang, W. A multi-modal data fusion approach for evaluating the impact of extreme public sentiments on corporate credit ratings. *Complex Intell. Syst.* **2025**, *11*, 436. [CrossRef]
4. Barberis, N.; Jin, L.J.; Wang, B. Prospect theory and stock market anomalies. *J. Financ.* **2021**, *76*, 2639–2687. [CrossRef]
5. Peress, J.; Schmidt, D. Noise traders incarnate: Describing a realistic noise trading process. *J. Financ. Mark.* **2021**, *54*, 100618. [CrossRef]
6. Farahani, M.S.; Hajiagha, S.H.R. Forecasting stock price using integrated artificial neural network and metaheuristic algorithms compared to time series models. *Soft Comput.* **2021**, *25*, 8483. [CrossRef] [PubMed]
7. Suthaharan, S. Support vector machine. In *Machine Learning Models and Algorithms for Big Data Classification: Thinking with Examples for Effective Learning*; Springer: Berlin/Heidelberg, Germany, 2016; pp. 207–235.
8. Medsker, L.; Jain, L.C. *Recurrent Neural Networks: Design and Applications*; CRC Press: Boca Raton, FL, USA, 1999.
9. Nagar, P.; Shastry, K.; Chaudhari, J.; Arora, C. SEMA: Semantic Attention for Capturing Long-Range Dependencies in Egocentric Lifelogs. In Proceedings of the IEEE/CVF Winter Conference on Applications of Computer Vision, Waikoloa, HI, USA, 3–8 January 2024; pp. 7025–7035.
10. Zhou, H.; Zhang, S.; Peng, J.; Zhang, S.; Li, J.; Xiong, H.; Zhang, W. Informer: Beyond efficient transformer for long sequence time-series forecasting. In Proceedings of the AAAI Conference on Artificial Intelligence, Virtually, 2–9 February 2021; Volume 35, pp. 11106–11115.
11. Zhang, C.; Sjarif, N.N.A.; Ibrahim, R. Deep learning models for price forecasting of financial time series: A review of recent advancements: 2020–2022. *Wiley Interdiscip. Rev. Data Min. Knowl. Discov.* **2024**, *14*, e1519. [CrossRef]
12. Liu, Y.; Huang, S.; Tian, X.; Zhang, F.; Zhao, F.; Zhang, C. A stock series prediction model based on variational mode decomposition and dual-channel attention network. *Expert Syst. Appl.* **2024**, *238*, 121708. [CrossRef]
13. Ma, R.; Han, T.; Lei, W. Cross-domain meta learning fault diagnosis based on multi-scale dilated convolution and adaptive relation module. *Knowl.-Based Syst.* **2023**, *261*, 110175. [CrossRef]
14. Dragomiretskiy, K.; Zosso, D. Variational mode decomposition. *IEEE Trans. Signal Process.* **2013**, *62*, 531–544. [CrossRef]
15. Abdel-Basset, M.; Mohamed, R.; Abouhawwash, M. Crested Porcupine Optimizer: A new nature-inspired metaheuristic. *Knowl.-Based Syst.* **2024**, *284*, 111257. [CrossRef]
16. Zhang, Z.; Liu, Q.; Hu, Y.; Liu, H. Multi-feature stock price prediction by LSTM networks based on VMD and TMFG. *J. Big Data* **2025**, *12*, 74. [CrossRef]
17. Jain, A.; Zamir, A.R.; Savarese, S.; Saxena, A. Structural-rnn: Deep learning on spatio-temporal graphs. In Proceedings of the IEEE Conference on Computer Vision and Pattern Recognition, Las Vegas, NV, USA, 26 June–1 July 2016; pp. 5308–5317.
18. Wu, H.; Xu, J.; Wang, J.; Long, M. Autoformer: Decomposition transformers with auto-correlation for long-term series forecasting. *Adv. Neural Inf. Process. Syst.* **2021**, *34*, 22419–22430.
19. Zhou, T.; Ma, Z.; Wen, Q.; Wang, X.; Sun, L.; Jin, R. Fedformer: Frequency enhanced decomposed transformer for long-term series forecasting. In Proceedings of the International Conference on Machine Learning, PMLR, Baltimore, MD, USA, 17–23 July 2022; pp. 27268–27286.
20. Lv, P.; Shu, Y.; Xu, J.; Wu, Q. Modal decomposition-based hybrid model for stock index prediction. *Expert Syst. Appl.* **2022**, *202*, 117252. [CrossRef]

21. Gao, Z.; Zhang, J. The fluctuation correlation between investor sentiment and stock index using VMD-LSTM: Evidence from China stock market. *N. Am. J. Econ. Financ.* **2023**, *66*, 101915. [CrossRef]
22. Li, Q.; Wang, G.; Wu, X.; Gao, Z.; Dan, B. Arctic short-term wind speed forecasting based on CNN-LSTM model with CEEMDAN. *Energy* **2024**, *299*, 131448. [CrossRef]
23. Rayi, V.K.; Mishra, S.; Naik, J.; Dash, P.K. Adaptive VMD based optimized deep learning mixed kernel ELM autoencoder for single and multistep wind power forecasting. *Energy* **2022**, *244*, 122585. [CrossRef]
24. Ding, J.; Huang, L.; Xiao, D.; Li, X. GMPSO-VMD algorithm and its application to rolling bearing fault feature extraction. *Sensors* **2020**, *20*, 1946. [CrossRef]
25. S&P Global Market Intelligence. *The Global Industry Classification Standard (GICS®)*; Technical Report; Standard & Poor's Financial Services LLC (S&P), MSCI: New York, NY, USA, 2018.
26. Vaswani, A.; Shazeer, N.; Parmar, N.; Uszkoreit, J.; Jones, L.; Gomez, A.N.; Kaiser, L.; Polosukhin, I. Attention Is All You Need. *arXiv* **2023**, arXiv:1706.03762. [CrossRef]
27. Yang, Y.; Dong, Y. Informer_Casual_LSTM: Causal Characteristics and Non-Stationary Time Series Prediction. In Proceedings of the 2024 7th International Conference on Machine Learning and Natural Language Processing (MLNLP), Chengdu, China, 18–20 October 2024; IEEE: Piscataway, NJ, USA, 2024; pp. 1–6.
28. Liu, Y.; Hu, T.; Zhang, H.; Wu, H.; Wang, S.; Ma, L.; Long, M. itransformer: Inverted transformers are effective for time series forecasting. *arXiv* **2023**, arXiv:2310.06625.
29. Hadizadeh, A.; Tarokh, M.J.; Ghazani, M.M. A novel transformer-based dual attention architecture for the prediction of financial time series. *J. King Saud Univ. Comput. Inf. Sci.* **2025**, *37*, 72. [CrossRef]

applied sciences

MDPI

Article

KOSLM: A Kalman-Optimal Hybrid State-Space Memory Network for Long-Term Time Series Forecasting

Xin Tan, Lei Wang *, Mingwei Wang and Ying Zhang

School of Electronic Information and Artificial Intelligence, Shaanxi University of Science and Technology, Xi'an 710021, China; tanxin@sust.edu.cn (X.T.); wangmingwei@sust.edu.cn (M.W.); 231612117@sust.edu.cn (Y.Z.)
* Correspondence: 231612097@sust.edu.cn

Abstract

Long-term time series forecasting (LTSF) remains challenging, as models must capture long-range dependencies and remain robust to noise accumulation. Traditional recurrent models, such as Recurrent Neural Networks (RNNs) and Long Short-Term Memory (LSTM), often suffer from instability and information degradation over extended horizons. The state-of-the-art method xLSTMTime improves memory retention through exponential gating and enhanced memory-transition rules, but it still lacks principled guidance. To address these issues, we propose the Kalman-Optimal Selective Long-Term Memory (KOSLM) model, which embeds a Kalman-optimal selective mechanism driven by the innovation signal within a structured state-space reformulation of LSTM. KOSLM dynamically regulates information propagation and forgetting to minimize state estimation uncertainty, providing both theoretical interpretability and practical efficiency. Extensive experiments across energy, finance, traffic, healthcare, and meteorology datasets show that KOSLM reduces mean squared error (MSE) by 14.3–38.9% compared with state-of-the-art methods, with larger gains at longer horizons. The model is lightweight, scalable, and achieves up to 2.5× speedup over Mamba-2. Beyond benchmarks, KOSLM is further validated on real-world Secondary Surveillance Radar (SSR) tracking under noisy and irregular sampling, demonstrating robust and generalizable long-term forecasting performance.

Keywords: long-term time series forecasting; LSTM; state-space model; kalman optimality; selective memory; robust prediction; SSR tracking

1. Introduction

Long-term time series forecasting (LTSF) aims to predict future values over extended horizons based on historical observations. Accurate long-range forecasts are critical for applications such as energy scheduling, climate modeling, financial planning, traffic management, and healthcare resource allocation. LTSF presents unique challenges: capturing long-range dependencies, mitigating error accumulation, and adapting to non-stationary temporal dynamics [1].

Traditional recurrent architectures, including Recurrent Neural Networks (RNNs) [2] and Long Short-Term Memory (LSTM) networks [3,4], often struggle with long sequences due to vanishing or exploding gradients [5]. While LSTM gates alleviate some of these issues, their heuristic design lacks explicit structural constraints derived from a principled optimality criterion, potentially leading to suboptimal memory retention over extended horizons. The recently proposed xLSTMTime further enhances long-range modeling by introducing the sLSTM and mLSTM components. The sLSTM employs scalar memory with exponential gating to manage long-term dependencies, while the mLSTM component uses matrix memory and a covariance update rule to enhance storage capacity and relevant information retrieval capabilities. Despite these improvements, xLSTMTime remains constrained by heuristic recurrent design.

The Kalman filter (KF) [6] provides optimal state estimation under Gaussian noise [7]. Recent studies have explored integrating KF with LSTM to improve the accuracy of time-series forecasting. Representative approaches include the following: (i) Deep Kalman Filters [8], which parameterize the state-transition and observation functions of KF using LSTM; (ii) KalmanNet [9], which employs an LSTM to learn residual corrections to the Kalman gain under a known KF model; and (iii) uncertainty-aware LSTM–KF hybrids [10], which estimate the covariance or uncertainty structures of KF through recurrent dynamics. However, these approaches generally maintain a loose coupling between LSTM and KF—they do not embed Kalman-inspired feedback directly into the internal gating dynamics of LSTM.

Recently, selective state space models (SSMs) [11,12] have demonstrated efficient sequence modeling with linear-time complexity. By dynamically modulating SSM's parameters based on the input, they can filter task-irrelevant patterns while retaining critical long-term information. This selective modulation property motivates revisiting LSTM gates from a state-space perspective, into which Kalman-inspired feedback can be injected, thereby endowing the gating mechanism with Kalman-optimal structural constraints.

In this work, we propose the Kalman-Optimal Selective Long-Term Memory (KOSLM) model, which establishes a context-aware feedback pathway that optimally balances memory retention and information updating, providing both theoretical interpretability and practical efficiency.

Our main contributions are as follows:

- **State-space reformulation of LSTM:** We formalize LSTM networks as input- and state-dependent SSMs, where each gate dynamically parameterizes the state-transition and input matrices. This framework provides a principled explanation of LSTM's long-term memory behavior.
- **Kalman-optimal selective gating:** Inspired by the Kalman filter and selective SSMs, we introduce a Kalman-optimal selective mechanism in which the state-transition and input matrices are linearly modulated by a Kalman gain learned from the *innovation term*, establishing a feedback pathway that minimizes state estimation uncertainty.
- **Applications to real-world forecasting:** KOSLM consistently outperforms state-of-the-art baselines across LTSF benchmarks in energy, finance, traffic, healthcare, and meteorology, achieving 14.3–38.9% lower mean squared error (MSE) and a maximum reduction in mean absolute error (MAE) of 25.2%; it also delivers up to 2.5× faster inference than Mamba-2. In real-world Secondary Surveillance Radar (SSR) tracking under noisy and irregular sampling, KOSLM demonstrates strong robustness and generalization.

By bridging heuristic LSTM gating with principled Kalman-optimal estimation, KOSLM provides a robust, interpretable, and scalable framework for long-term sequence modeling, offering both methodological novelty and practical forecasting utility.

2. Background and Theory

2.1. LSTM Networks

RNNs model sequential data through recursive temporal computation. However, traditional RNNs often suffer from vanishing and exploding gradients when capturing long-term dependencies. The LSTM network [3] addresses this issue by introducing gating mechanisms that regulate the flow of information over time. The network structure of the LSTM neuron is shown in Figure 1.

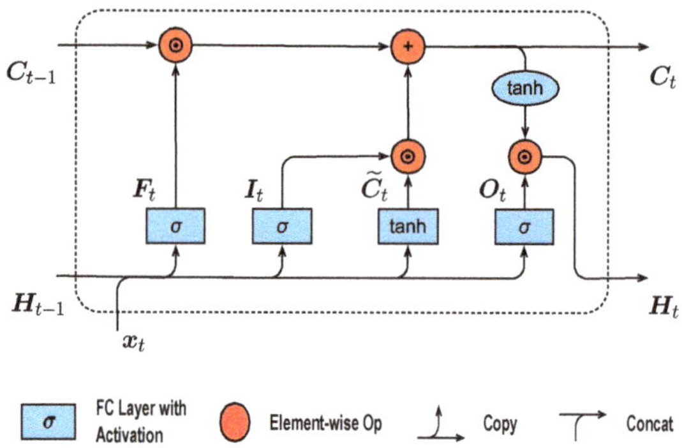

Figure 1. Structure of the LSTM network.

Each LSTM unit maintains a cell state C_t that carries long-term information and a hidden state H_t that provides short-term representations. The cell state is updated according to

$$C_t = F_t \odot C_{t-1} + I_t \odot \tilde{C}_t, \tag{1}$$

where F_t, I_t, and O_t denote the forget, input, and output gates, respectively. These gates are nonlinear functions of the current input x_t and the previous hidden state H_{t-1}. The final output is obtained as follows:

$$H_t = O_t \odot \tanh(C_t). \tag{2}$$

This gating mechanism allows the model to selectively retain or discard information, mitigating gradient degradation. However, the gates are learned heuristically through data-driven optimization rather than derived from an explicit structural optimality constraint, making the LSTM sensitive to noise and unstable for long-term dependencies.

2.2. State Space Models

2.2.1. Selective State Space Models

Selective State Space Models (S6) are a recent class of sequence models for deep learning that are broadly related to RNNs and classical SSMs. They are inspired by a particular system (Equation (3)) that maps a one-dimensional function or sequence $x_t \in \mathbb{R} \to y_t \in \mathbb{R}$ through an implicit latent state $h_t \in \mathbb{R}^N$:

$$h_t = Ah_{t-1} + Bx_t, \tag{3a}$$

$$y_t = Mh_t. \tag{3b}$$

where A is the state transition matrix, defining how the past latent state h_{t-1} influences the current state; B is the input matrix, determining how x_t modulates h_t; and M maps the implicit latent state h_t to the output.

These models integrate the SSM described above into deep learning frameworks and introduce **input-dependent selective mechanisms** (see Appendix A for a detailed discussion), achieving Transformer-level modeling capability with linear computational complexity.

The theoretical connections among LSTM, KF, and SSM provide the foundation for constructing a unified **Kalman-optimal selective memory** framework.

2.2.2. Kalman Filter

The KF [6] is a classical instance of the state-space model, providing the minimum mean-square-error (MMSE) estimate of hidden system states under noisy observations. The system dynamics are expressed as

$$h_t = Ah_{t-1} + w_t, \quad w_t \sim \mathcal{N}(0, Q_t), \tag{4a}$$
$$z_t = Mh_t + v_t, \quad v_t \sim \mathcal{N}(0, R_t), \tag{4b}$$

where M is the same as the M in the SSM introduced above, but in the context of the KF, it is commonly referred to as the observation matrix; w_t and v_t denote process and observation noise with covariances Q_t and R_t, respectively. **Note that the observation z_t can be regarded as the input x_t in the SSMs.**

At each time step, the KF performs two operations: *prediction* and *update*. The prediction step estimates the prior state $\hat{h}_t^-$ and prior error covariance $\hat{P}_t^-$ based on the posterior state $\hat{h}_{t-1}$ and posterior error covariance $\hat{P}_{t-1}$ from the previous time step:

$$\hat{h}_t^- = A\hat{h}_{t-1}, \tag{5}$$

$$\hat{P}_t^- = A\hat{P}_{t-1}A^T + Q_t \tag{6}$$

The update step then refines this prediction using the observation z_t:

$$K_t = \hat{P}_t^- M^T (M\hat{P}_t^- M^T + R_t)^{-1} \tag{7}$$

$$\hat{h}_t = \hat{h}_t^- + K_t(z_t - M\hat{h}_t^-) \tag{8}$$

$$\hat{P}_t = (I - K_t M)\hat{P}_t^- \tag{9}$$

Here, K_t is the Kalman gain, which minimizes the posterior error covariance $\hat{P}_t$. It determines how much new information (the *innovation term* $z_t - M\hat{h}_t^-$) should be incorporated into the updated state ($\hat{h}_t$ is the posterior state). This principle of **optimal selective information integration** forms the theoretical foundation for the *innovation*-driven gating design proposed later. A detailed justification for interpreting the Kalman gain as a prototype of dynamic selectivity is provided in Appendix B.1.

3. Proposed Method

3.1. Reformulating LSTM as a State-Space Model

Following the LSTM formulation in Equation (1), the LSTM cell can be equivalently reformulated as a time-varying SSM, where the cell state C_t evolves under nonlinear, input- and state-dependent dynamics:

$$C_t = A_t C_{t-1} + B_t z_t, \tag{10}$$

Here, z_t serves as the observation in the KF (corresponding to the input x_t in the SSM and LSTM), and the matrices A_t and B_t are determined by the forget gate F_t and input gate I_t, respectively. A detailed derivation of this LSTM-to-SSM reconstruction, including the mapping of gating mechanisms to state-space parameters, is provided in Appendix B.2.

3.2. Kalman-Optimal Selectivity via Innovation-Driven Gain

We introduce the *innovation term* from the KF:

$$Innov_t = z_t - M_t A_{t-1} C_{t-1}, \tag{11}$$

which measures the discrepancy between the observation input z_t and the predicted state based on the previous cell state C_{t-1}, serving as a real-time correction signal between model prediction and actual measurement.

In classical KF, this *innovation* drives the computation of the **Kalman gain** K_t (Equation (7)), which regulates the incorporation of new information and the retention of prior state during the update step. In KOSLM, rather than explicitly solving the Riccati recursion [13], we **learn a functional mapping**:

$$K_t = \phi(Innov_t; \theta_\phi), \tag{12}$$

where $\phi(\cdot)$ is a lightweight multi-layer perceptron (MLP) with sigmoid activation, with its width and depth controlled by two hyperparameters, MLP_d and MLP_layers (e.g., MLP_d = 64, MLP_layers = 2), and parameterized by θ_ϕ. The sigmoid activation ensures that the estimated gain $K_t \in (0,1)$, maintaining physical interpretability as an adaptive weighting coefficient and preventing numerical instability.

The learned gain dynamically regulates the trade-off between prior memory and new information, yielding a learnable yet principled mechanism for **Kalman-optimal selectivity**. The bounded output range of K_t further stabilizes the state-space update and constrains divergence during long-horizon inference. Appendix B.3 provides the theoretical derivation demonstrating that the *innovation term* serves as a sufficient statistic for learning the Kalman gain. Appendix B.4 further presents controlled experiments confirming that the learned gain accurately approximates the oracle Kalman gain across various (Q, R) regimes (Table A1). Controlled ablation studies (see Section 4.2.2 and Table 1) show that increasing the hidden layer size or depth yields only marginal improvements in predictive performance, indicating that KOSLM's performance is largely insensitive to the specific configuration of ϕ. This confirms that the performance gains primarily stem from the Kalman-optimal structural design rather than the network capacity. These results jointly establish the theoretical and empirical foundation for embedding Kalman-optimal selectivity into deep learning models.

Table 1. Capacity ablation (corrected). Performance of KOSLM with varying ϕ network capacities on four long-term forecasting datasets ($L = 720$). Averaged over five runs with standard deviations. Lower values (indicated by the down arrow "↓") correspond to better performance.

Dataset	Variant	#Params	MSE ↓	MAE ↓	Δ MSE vs. SmallMLP
ETTm1	Linear	0.183M	0.480 ± 0.012	0.355 ± 0.008	+46.3%
	SmallMLP	0.255M	**0.328 ± 0.010**	0.282 ± 0.007	—
	MediumMLP	0.389M	0.335 ± 0.011	0.284 ± 0.001	+2.13%
	HighCap	0.898M	0.342 ± 0.015	0.288 ± 0.010	+4.27%
ETTh1	Linear	0.183M	0.440 ± 0.010	0.330 ± 0.004	+30.95%
	SmallMLP	0.255M	**0.336 ± 0.008**	0.291 ± 0.006	—
	MediumMLP	0.389M	0.341 ± 0.010	0.294 ± 0.002	+1.49%
	HighCap	0.898M	0.355 ± 0.012	0.302 ± 0.008	+5.65%
Traffic	Linear	0.183M	0.290 ± 0.011	0.232 ± 0.006	+9.02%
	SmallMLP	0.255M	**0.266 ± 0.010**	0.212 ± 0.007	—
	MediumMLP	0.389M	0.270 ± 0.012	0.216 ± 0.002	+1.50%
	HighCap	0.898M	0.279 ± 0.014	0.222 ± 0.005	+4.89%
Exchange	Linear	0.183M	0.272 ± 0.010	0.305 ± 0.005	+4.62%
	SmallMLP	0.255M	**0.260 ± 0.009**	0.291 ± 0.009	—
	MediumMLP	0.389M	0.264 ± 0.010	0.294 ± 0.007	+1.54%
	HighCap	0.898M	0.285 ± 0.012	0.310 ± 0.009	+9.62%

We then define the state-space evolution as follows:

$$A_t = (I - K_t M_t)A, \quad B_t = K_t, \tag{13}$$

where A is the base state transition matrix, serving as a learnable parameter of the model; A_t and B_t are dynamically modulated by the Kalman gain K_t.

3.3. Structural Overview of KOSLM

The KOSLM cell preserves the computational efficiency of a standard LSTM while embedding a **Kalman-inspired feedback loop**. The KOSLM network architecture is illustrated in Figure 2; at each timestep, the operations are as follows:

1. Initialization: The learnable matrix A is initialized following S4D-Lin and for the real case is S4D-Real [14], which is based on the HIPPO theory [15]. These define the n-th element of A as $-\frac{1}{2} + ni$ and $-(n+1)$ respectively;
2. Compute z_t and M_t via the candidate gate $\tilde{C}_t$ and output gate O_t;
3. Compute the *innovation*: $Innov_t = z_t - M_t A C_{t-1}$;
4. Estimate the Kalman gain: $K_t = \phi(Innov_t)$;
5. Update state transition matrices: $A_t = (I - K_t M_t)A$, $B_t = K_t$;
6. Propagate the hidden state: $C_t = A_t C_{t-1} + B_t z_t$;
7. Compute the output hidden representation: $H_t = M_t C_t$.

To further clarify the conceptual and structural differences between KOSLM and existing Kalman-based neural architectures, Table 2 summarizes a direct comparison with the KalmanNet [9].

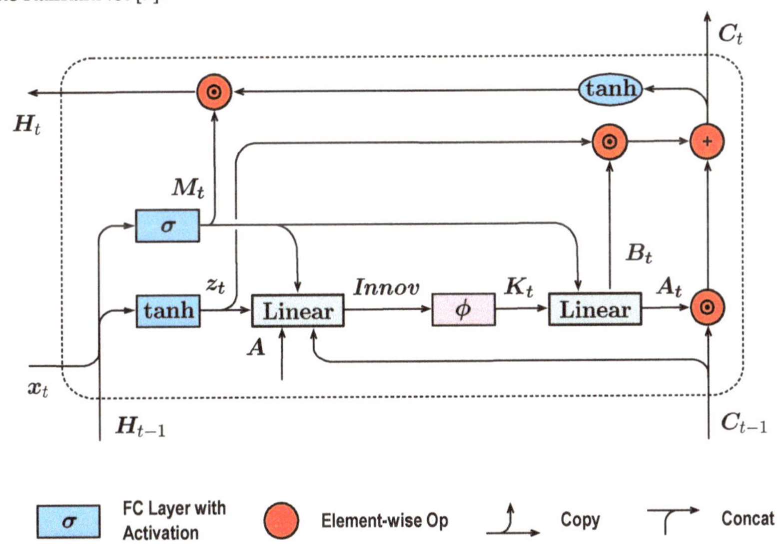

Figure 2. Unfolded KOSLM network layer. The D-dimensional input x_t is dynamically coupled with the N-dimensional hidden state H_{t-1} and mapped to the output H_t through the higher-dimensional cell state $C_t \in \mathbb{R}^T$. Compared to the heuristic forget and output gates of the classical LSTM, our mechanism introduces Kalman-optimal constraints by parameterizing SSM parameters (A_t, B_t) based on the *innovation term*.

Table 2. Core structural comparison between KalmanNet and KOSLM.

Aspect	KalmanNet	KOSLM (Ours)
Core idea	Strictly follows the classical KF architecture (Section 2.2.2); the neural network learns to correct the Kalman gain K_t^{KF} under partially known linear dynamics.	Reinterprets LSTM gating as a Kalman-optimal state estimation problem; state estimation does not strictly adhere to the KF equations, but directly learns K_t from the *innovation term*.
System dynamics	Assumes fixed parameters (A, M, Q, R); suitable for systems with known or partially known dynamics.	Learns (A_t, B_t, M_t) from data; A is a learnable parameter matrix; fully adaptive to nonlinear and nonstationary environments.
Form of gain	$K_t = K_t^{\mathrm{KF}} + f(s_t; \theta_f)$ — learns a residual correction to the classical Kalman gain.	$K_t = \phi(Innov_t; \theta_\phi)$ — directly learns the gain function from the *innovation*.
Gain network input	$s_t = [\hat{h}_{t\|t-1}, z_t - M\hat{h}_{t\|t-1}]$ — uses both predicted state and *innovation* as inputs.	$\mathrm{Innov}_t = z_t - M_t A_{t-1} C_{t-1}$ — relies solely on the *innovation* signal for gain computation (see Appendices B.3 and B.4, which prove its sufficiency).
Output role	Outputs a residual correction ΔK_t added to K_t^{KF}.	Outputs K_t and integrates gain estimation into the state transition: $A_t = (I - K_t M_t)A$, $B_t = K_t$, forming a unified recurrent–estimation pathway.
System dependency	Requires partial knowledge of (A, H, Q, R) for baseline Kalman computation.	Fully data-driven; no analytical Kalman gain or explicit system parameters required.
Theoretical interpretation	Neural residual learning to compensate model mismatch.	Innovation-driven dynamic selectivity that enforces Kalman-optimal information update behavior.

3.4. Theoretical Interpretation

Under linear–Gaussian assumptions and with a sufficiently expressive mapping ϕ, the learned gain K_t converges to the oracle Kalman solution. Consequently, KOSLM inherits the nonlinear expressive power of LSTM while achieving the **minimum-variance estimation property** of the Kalman filter in its linear regime. This leads to improved stability and robustness, particularly in long-horizon or noisy sequence modeling.

3.5. Practical Advantages

KOSLM offers several practical benefits:

- **Robustness**: The feedback structure mitigates error accumulation and improves performance under noise or distributional shifts.
- **Efficiency**: With only 0.24 M parameters, KOSLM achieves up to 2.5× faster inference than Mamba-2, while maintaining competitive accuracy.
- **Versatility**: The model generalizes across diverse domains, from energy demand forecasting to radar-based trajectory tracking.

4. Results

This section provides a comprehensive evaluation of the proposed **KOSLM** model through large-scale long-term forecasting benchmarks (Section 4.1), component ablation studies (Section 4.2), efficiency assessments (Section 4.3), and a real-world radar trajectory tracking case study (Section 4.4). The experimental analyses collectively aim to validate both the *predictive accuracy* and *robustness* of KOSLM across diverse and noisy temporal conditions.

4.1. Main Experiments on Benchmark Datasets

To examine the capability of KOSLM in modeling long-range dependencies and maintaining stability over extended forecasting horizons, we conduct systematic experiments on nine widely used real-world datasets covering domains such as traffic flow, electrical load, exchange rate, meteorology, and epidemiology. The datasets vary in frequency, dimensionality, and temporal regularity, providing a comprehensive benchmark for assessing model generalization.

4.1.1. Dataset Details

We summarize the datasets used in this study as follows. **Weather** [16] contains 21 meteorological variables (e.g., temperature and humidity) recorded every 10 min throughout 2020. **ETT** (Electricity Transformer Temperature) [17] includes four subsets: two hourly-level datasets (ETTh1, ETTh2) and two 15-minute-level datasets (ETTm1, ETTm2). **Electricity** [18], derived from the UCI Machine Learning Repository, records hourly power consumption (kWh) of 321 clients from 2012 to 2014. **Exchange** [19] comprises daily exchange rates among eight countries. **Traffic** [20] consists of hourly road occupancy rates measured by 862 sensors on San Francisco Bay Area freeways from January 2015 to December 2016. **Illness** (ILI) dataset [21] tracks the weekly number of influenza-like illness patients in the United States. Table 3 summarizes the statistical properties of all nine benchmark datasets. All datasets are divided into training, validation, and test subsets with a ratio of 7:1:2.

Table 3. Details of benchmark datasets used in our experiments.

Dataset	Frequency	# Features	Time Steps	Time Span
ETTh1	1 h	7	17,420	2016–2017
ETTh2	1 h	7	17,420	2017–2018
ETTm1	15 min	7	69,680	2016–2017
ETTm2	15 min	7	69,680	2017–2018
Exchange	1 day	8	7588	1990–2010
Weather	10 min	21	52,696	2020
Electricity	1 h	321	26,304	2012–2014
ILI	7 days	7	966	2002–2020
Traffic	1 h	862	17,544	2015–2016

4.1.2. Implementation Details

All models are trained using the Adam optimizer without weight decay. The learning rate is selected from $[1 \times 10^{-3}, 1 \times 10^{-2}]$ via grid search. Batch size is set to 32 by default, adjustable up to 256 depending on GPU memory. Training is performed for 15 epochs, and the checkpoint with the lowest validation loss is used for testing. Experiments are repeated five times, and average results are reported (RMS values are provided in the Supplementary Materials). All models adopt a 2-layer architecture with hidden dimension 64. We use PyTorch's default weight initialization, and no additional regularization (dropout or gradient clipping) is applied. All experiments are implemented in `PyTorch 2.1.0` with `Python 3.11` on NVIDIA RTX 4090 GPUs. Random seeds for Python, NumPy, and PyTorch are fixed to ensure reproducibility.

The gain function ϕ is implemented as a one-layer MLP with hidden dimension 64 and sigmoid activation, followed by a linear projection to the gain matrix K_t. The same ϕ network is shared across all timesteps to ensure parameter consistency.

4.1.3. Experimental Setup of Main Experiments

All experiments follow the evaluation protocol established in xLSTMTime [22], adopting prediction horizons of $T \in \{96, 192, 336, 720\}$ for standard datasets and $T \in \{24, 36, 48, 60\}$ for the weekly-sampled ILI dataset. We compare the proposed KOSLM with nine recent state-of-the-art baselines that represent diverse architectural paradigms, spanning state-space, recurrent, attention-based, and linear modeling frameworks:

- **SSM-based:** FiLM [23], S-Mamba [24];
- **LSTM-based:** xLSTMTime [22];
- **Transformer-based:** FEDformer [25], iTransformer [26], Crossformer [27];
- **MLP/TCN-based:** DLinear [28], PatchTST [29], TimeMixer [30].

This comprehensive selection enables a fair and systematic comparison across diverse sequence modeling paradigms.

4.1.4. Overall Performance

Tables 4 and 5 report the long-term multivariate forecasting results on nine real-world datasets, evaluated by MSE and MAE. Across nearly all datasets and prediction horizons, the proposed **KOSLM** achieves the best or near-best performance, highlighting its superior generalization and robustness under diverse temporal dynamics.

Table 4. Long-term forecasting: Multivariate long-term forecasting results on the Traffic, Electricity, Exchange, Weather, and ILI datasets. The prediction length is set to $T \in \{96, 192, 336, 720\}$ for all datasets except ILI, which uses $O \in \{24, 36, 48, 60\}$ due to its weekly resolution. The best results are shown in bold red, and the second-best are underlined purple. All results are averaged over 5 runs. Lower is better.

Models		KOSLM		xLSTMTime		FiLM		iTransformer		FEDFormer		S-Mamba		Crossformer		DLinear		PatchTST		TimeMixer	
Metric		MSE	MAE	MSE	MAE	MSE	MAE	MSE	MAE	MSE	MAE	MSE	MAE	MSE	MAE	MSE	MAE	MSE	MAE	MSE	MAE
Traffic	96	**0.290**	**0.230**	0.358	0.242	0.416	0.294	0.395	0.268	0.587	0.366	0.382	0.261	0.522	0.290	0.650	0.396	0.462	0.295	0.462	0.285
	192	**0.260**	**0.210**	0.378	0.253	0.408	0.288	0.417	0.276	0.604	0.373	0.396	0.267	0.530	0.293	0.598	0.370	0.466	0.296	0.473	0.296
	336	**0.227**	**0.179**	0.392	0.261	0.425	0.298	0.433	0.283	0.621	0.383	0.417	0.276	0.558	0.305	0.605	0.373	0.482	0.304	0.498	0.297
	720	**0.290**	**0.232**	0.434	0.287	0.520	0.353	0.467	0.302	0.626	0.382	0.460	0.300	0.589	0.328	0.645	0.394	0.514	0.322	0.506	0.313
+31.96%	Avg	**0.266**	**0.212**	0.391	0.261	0.442	0.308	0.428	0.282	0.610	0.376	0.434	0.287	0.550	0.304	0.625	0.383	0.481	0.304	0.485	0.297
Electricity	96	**0.136**	**0.184**	0.128	0.221	0.154	0.267	0.148	0.240	0.193	0.308	0.139	0.235	0.219	0.314	0.197	0.282	0.181	0.270	0.153	0.247
	192	**0.131**	**0.184**	0.150	0.243	0.164	0.258	0.162	0.253	0.201	0.315	0.159	0.255	0.231	0.322	0.196	0.285	0.188	0.274	0.166	0.256
	336	**0.120**	**0.177**	0.166	0.259	0.188	0.283	0.178	0.269	0.214	0.329	0.176	0.272	0.246	0.337	0.209	0.301	0.204	0.293	0.185	0.277
	720	**0.158**	**0.205**	0.185	0.276	0.236	0.332	0.225	0.317	0.246	0.355	0.204	0.298	0.280	0.363	0.245	0.333	0.246	0.324	0.225	0.310
+13.37%	Avg	**0.136**	**0.187**	0.157	0.250	0.186	0.285	0.178	0.270	0.214	0.327	0.170	0.265	0.244	0.334	0.212	0.300	0.205	0.290	0.182	0.272
Exchange	96	0.135	0.212	–	–	**0.086**	**0.204**	**0.086**	0.206	0.148	0.278	**0.086**	0.207	0.256	0.367	0.088	0.218	0.088	0.205	0.095	0.207
	192	0.279	**0.292**	–	–	0.188	**0.292**	0.177	0.299	0.271	0.315	0.182	0.304	0.470	0.509	0.176	0.315	0.176	0.299	**0.151**	0.293
	336	0.341	**0.339**	–	–	0.356	0.433	0.331	0.417	0.460	0.427	0.332	0.418	1.268	0.883	0.313	0.427	0.301	0.397	**0.264**	0.361
	720	**0.282**	**0.322**	–	–	0.727	0.669	0.847	0.691	1.195	0.695	0.867	0.703	1.767	1.068	0.839	0.695	0.901	0.714	0.586	0.602
+5.47%	Avg	**0.259**	**0.291**	–	–	0.339	0.400	0.360	0.403	0.519	0.429	0.367	0.408	0.940	0.707	0.354	0.414	0.367	0.404	0.274	0.365
Weather	96	**0.144**	**0.171**	**0.144**	0.187	0.199	0.262	0.174	0.214	0.217	0.296	0.165	0.210	0.158	0.230	0.196	0.255	0.177	0.218	0.163	0.209
	192	0.224	**0.236**	**0.192**	**0.236**	0.228	0.288	0.221	0.254	0.276	0.336	0.214	0.252	0.206	0.277	0.237	0.296	0.225	0.259	0.208	0.250
	336	0.249	**0.248**	**0.237**	0.272	0.267	0.323	0.278	0.296	0.339	0.380	0.274	0.297	0.272	0.335	0.283	0.335	0.278	0.297	0.251	0.287
	720	**0.169**	**0.175**	0.313	0.326	0.319	0.361	0.358	0.347	0.403	0.428	0.350	0.345	0.398	0.418	0.345	0.381	0.354	0.348	0.339	0.341
+11.26%	Avg	**0.197**	**0.208**	0.222	0.255	0.253	0.309	0.258	0.278	0.309	0.360	0.251	0.276	0.259	0.315	0.265	0.317	0.259	0.281	0.240	0.271
ILI	24	**1.160**	**0.507**	1.514	0.694	1.970	0.875	3.154	1.235	3.228	1.260	–	–	3.041	1.186	2.215	1.081	1.319	0.754	1.453	0.827
	36	**1.262**	**0.561**	1.519	0.722	1.982	0.859	2.544	1.083	2.679	1.150	–	–	3.406	1.232	1.963	0.963	1.579	0.870	1.627	0.903
	48	**1.098**	**0.545**	1.500	0.725	1.868	0.896	2.489	1.112	2.622	1.080	–	–	3.459	1.221	2.130	1.024	1.553	0.815	1.644	0.914
	60	**1.118**	**0.583**	1.418	0.715	2.057	0.929	2.675	1.034	2.857	1.078	–	–	3.640	1.305	2.368	1.096	1.470	0.788	1.633	0.908
+22.04%	Avg	**1.160**	**0.549**	1.488	0.714	1.969	0.890	2.715	1.116	2.847	1.170	–	–	3.387	1.236	2.169	1.041	1.480	0.807	1.589	0.888

Table 5. **Long-term forecasting:** Forecasting results on the ETT datasets with prediction lengths $T \in \{96, 192, 336, 720\}$. The best results are shown in bold red, and the second-best are underlined purple. All results are averaged over 5 runs. Lower is better.

Models		KOSLM		xLSTMTime		FiLM		iTransformer		FEDFormer		S-Mamba		Crossformer		DLinear		PatchTST		TimeMixer	
Metric		MSE	MAE	MSE	MAE	MSE	MAE	MSE	MAE	MSE	MAE	MSE	MAE	MSE	MAE	MSE	MAE	MSE	MAE	MSE	MAE
ETTm1	96	**0.204**	**0.221**	<u>0.286</u>	<u>0.335</u>	–	–	0.334	0.368	0.379	0.419	0.333	0.368	0.404	0.426	0.345	0.372	0.329	0.367	0.320	0.357
	192	**0.233**	**0.237**	<u>0.329</u>	<u>0.361</u>	–	–	0.377	0.391	0.426	0.441	0.376	0.390	0.450	0.451	0.380	0.389	0.367	0.385	0.361	0.381
	336	0.414	**0.339**	**0.358**	**0.379**	–	–	0.426	0.420	0.445	0.459	0.408	0.413	0.532	0.515	0.413	0.413	0.399	0.410	**0.390**	0.404
	720	0.481	**0.368**	**0.416**	<u>0.411</u>	–	–	0.491	0.459	0.543	0.490	0.475	0.448	0.666	0.589	0.474	0.453	0.454	0.439	<u>0.458</u>	0.441
+4.03%	Avg	**0.333**	**0.291**	<u>0.347</u>	<u>0.372</u>	–	–	0.407	0.410	0.448	0.452	0.398	0.405	0.513	0.496	0.403	0.407	0.387	0.400	0.382	0.395
ETTm2	96	**0.100**	**0.232**	<u>0.164</u>	<u>0.250</u>	0.165	0.256	0.180	0.264	0.203	0.287	0.179	0.263	0.287	0.366	0.193	0.292	0.175	0.259	0.175	0.258
	192	**0.132**	**0.274**	<u>0.218</u>	<u>0.288</u>	0.222	0.296	0.250	0.309	0.269	0.328	0.250	0.309	0.414	0.492	0.284	0.362	0.241	0.302	0.237	0.299
	336	**0.244**	0.382	<u>0.271</u>	**0.322**	0.277	0.333	0.311	0.348	0.325	0.366	0.312	0.349	0.597	0.542	0.369	0.427	0.305	0.343	0.298	<u>0.340</u>
	720	**0.268**	0.408	0.361	<u>0.380</u>	0.371	0.389	0.412	0.407	0.421	0.415	0.411	0.406	1.730	1.042	0.554	0.522	0.402	0.400	<u>0.275</u>	**0.323**
+24.39%	Avg	**0.186**	0.324	0.254	<u>0.310</u>	0.259	0.319	0.288	0.332	0.305	0.349	0.288	0.332	0.757	0.610	0.350	0.401	0.281	0.326	<u>0.246</u>	**0.306**
ETTh1	96	**0.298**	**0.277**	<u>0.368</u>	<u>0.395</u>	–	–	0.386	0.405	0.376	0.419	0.386	0.405	0.423	0.448	0.386	0.400	0.414	0.419	0.375	0.400
	192	**0.337**	**0.306**	<u>0.401</u>	<u>0.416</u>	–	–	0.441	0.436	0.420	0.448	0.443	0.437	0.471	0.474	0.437	0.432	0.460	0.445	0.479	0.421
	336	**0.395**	**0.334**	<u>0.422</u>	<u>0.437</u>	–	–	0.487	0.458	0.459	0.465	0.489	0.468	0.570	0.546	0.481	0.459	0.501	0.466	0.484	0.458
	720	**0.471**	**0.364**	<u>0.441</u>	<u>0.465</u>	–	–	0.503	0.491	0.506	0.507	0.502	0.489	0.653	0.621	0.519	0.516	0.500	0.488	0.498	0.482
+8.09%	Avg	**0.375**	**0.320**	<u>0.408</u>	<u>0.428</u>	–	–	0.454	0.447	0.440	0.460	0.455	0.450	0.529	0.522	0.456	0.452	0.469	0.454	0.459	0.440
ETTh2	96	**0.194**	**0.338**	<u>0.273</u>	<u>0.333</u>	–	–	0.297	0.349	0.358	0.397	0.296	0.348	0.745	0.584	0.333	0.387	0.302	0.348	0.289	0.341
	192	**0.238**	**0.384**	<u>0.340</u>	<u>0.378</u>	–	–	0.380	0.400	0.429	0.439	0.376	0.396	0.877	0.656	0.477	0.476	0.388	0.400	0.372	0.392
	336	**0.258**	**0.394**	<u>0.373</u>	<u>0.403</u>	–	–	0.428	0.432	0.496	0.487	0.424	0.431	1.043	0.731	0.594	0.541	0.426	0.433	0.386	0.414
	720	**0.314**	**0.432**	<u>0.398</u>	<u>0.430</u>	–	–	0.427	0.445	0.463	0.474	0.426	0.444	1.104	0.763	0.831	0.657	0.431	0.446	0.412	0.434
+27.46%	Avg	**0.251**	**0.387**	<u>0.346</u>	<u>0.386</u>	–	–	0.383	0.407	0.437	0.449	0.381	0.405	0.942	0.684	0.559	0.515	0.387	0.407	0.364	0.395

- **Consistent superiority across domains:** KOSLM outperforms all competing baselines, particularly under complex and noisy datasets. In terms of average MSE and MAE reductions, KOSLM achieves relative improvements of **31.96% and 18.77%** on Traffic, **13.37% and 25.20%** on Electricity, **5.47% and 20.27%** on Exchange, **11.26% and 18.43%** on Weather, and **22.04% and 23.09%** on ILI. Moreover, on the four ETT benchmarks, KOSLM achieves MSE improvements of up to **27.46%** and MAE improvements of up to **25.2%**, demonstrating strong adaptability to varying periodic and nonstationary patterns. These consistent gains verify that the Kalman-inspired selective updating mechanism effectively filters noise and dynamically adjusts to regime shifts, ensuring stable forecasting accuracy over long horizons.
- **Stable error distribution and reduced variance:** The MSE–MAE gap of KOSLM remains narrower than that of other baselines, implying reduced large deviations and more concentrated prediction errors. This indicates more stable error behavior, which is crucial for long-horizon forecasting where cumulative drift often occurs. The *innovation*-driven Kalman gain estimation provides adaptive correction at each timestep, ensuring smooth and consistent prediction trajectories under uncertain dynamics.
- **Strong scalability and generalization:** KOSLM achieves leading performance not only on large-scale datasets (*Traffic*, *Electricity*) but also on small, noisy datasets (*ILI*), confirming robust generalization across different temporal resolutions and noise levels. Its consistent advantage over Transformer-based (e.g., iTransformer, FEDFormer, PatchTST), recurrent (e.g., xLSTMTime), and state-space models (e.g., S-Mamba, FiLM) demonstrates that the proposed Kalman-optimal selective mechanism provides an effective inductive bias for modeling long-term dependencies.
- **Advantage over LSTM-based architectures:** Compared with advanced LSTM-based models such as xLSTMTime, KOSLM achieves consistently better results across nearly all datasets and horizons. This verifies that replacing heuristic gates with Kalman-optimal selective gating enhances memory retention and update stability. While xLSTMTime alleviates gradient decay via hierarchical memory, KOSLM further refines state updates through *innovation*-driven gain estimation, thereby achieving a more principled and stable information flow.

To further demonstrate the long-horizon stability of the proposed model, we compare **KOSLM** with the recent **S-Mamba** [24], a state-of-the-art state space model that represents the latest advancement in efficient sequence modeling. Figure 3 presents the forecasting trajectories at the long horizon ($T = 720$) across five representative datasets. KOSLM maintains accurate trend alignment and amplitude consistency with the ground truth, showing particularly superior convergence behavior on smoother and more stationary datasets such as *Weather* and *Exchange*, where the Kalman gain adaptation stabilizes long-term predictions. In contrast, S-Mamba exhibits slight temporal lag and amplitude attenuation under extended forecasting conditions. These results visually confirm the advantage of the proposed Kalman-based feedback selectivity in preserving long-term temporal fidelity.

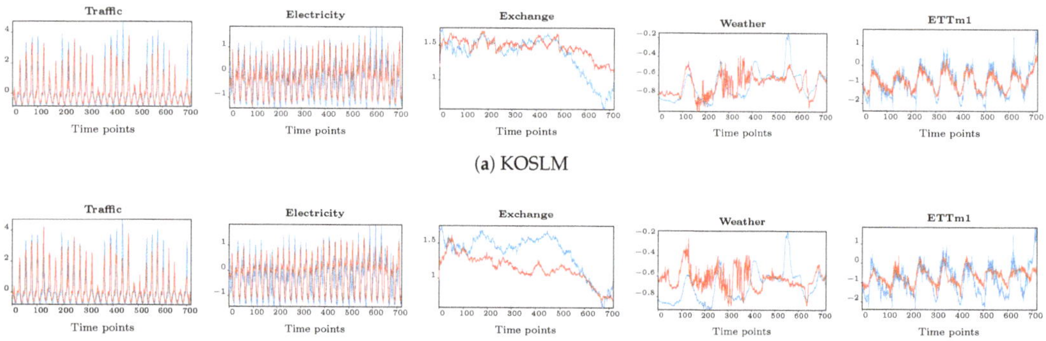

(**a**) KOSLM

(**b**) S-Mamba

Figure 3. KOSLM vs. S-Mamba: Forecasting comparison on five representative datasets with both input and prediction horizons set to 720. The blue line denotes the ground truth, and the red line indicates model predictions. KOSLM demonstrates superior long-horizon stability and trend consistency compared to S-Mamba.

4.2. Ablation Study

To further analyze the effectiveness of each proposed component, we conduct ablation experiments on four widely used long-term time-series forecasting datasets (ETTm1, ETTh1, Traffic, and Exchange) with a prediction length of $L = 720$, following the data preprocessing described in Section 4.1.1. All model variants are trained and tested following the default implementation settings described in Section 4.1.2, and the results are averaged over five runs to ensure statistical reliability.

4.2.1. Structural Ablation.

We first evaluate the contribution of the Kalman-based structure by comparing three variants: **(i) Full (KOSLM)**: the complete model where $K_t = \phi(\text{Innov})$ and $A_t = (I - K_t M_t)A$, $B_t = K_t$; **(ii) No-Gain**: ϕ receives the *innovation* but directly outputs (A_t, B_t), removing the explicit computation of the Kalman gain K_t, i.e., the pathway Innov $\rightarrow K_t \rightarrow (A_t, B_t)$ is replaced by Innov $\rightarrow (A_t, B_t)$; **(iii) No-Innov**: the *innovation* statistic is removed. ϕ receives the standard network input (e.g., $[x_t; H_{t-1}]$), while K_t is still computed and mapped to (A_t, B_t) via the Kalman form, i.e., $[x_t; H_{t-1}] \rightarrow K_t \rightarrow (A_t, B_t)$.

As shown in Table 6, removing either the gain computation or the *innovation* input consistently leads to higher errors across all datasets. For instance, on ETTm1, the MSE increases by 32.9% when the gain path is removed, confirming that both the *innovation* statistic and Kalman gain are essential for stable long-horizon prediction. This finding aligns with the Kalman filtering principle that the *innovation* serves as a sufficient statistic for state correction.

Table 6. Structural ablation. Comparison of KOSLM and its structural variants on four long-term forecasting datasets ($L = 720$). Removing either the Kalman gain path or the *innovation* input leads to consistent performance degradation. All results are averaged over five runs with mean ± standard deviation. Lower values (indicated by the down arrow "↓") correspond to better performance.

Dataset	Model	MSE ↓	MAE ↓	Δ MSE vs. Full
ETTm1	Full	**0.498 ± 0.011**	**0.416 ± 0.007**	—
	No-Gain	0.662 ± 0.017	0.553 ± 0.014	+32.9%
	No-Innov	0.583 ± 0.012	0.487 ± 0.008	+17.1%
ETTh1	Full	**0.510 ± 0.012**	**0.378 ± 0.010**	—
	No-Gain	0.605 ± 0.022	0.448 ± 0.015	+18.6%
	No-Innov	0.554 ± 0.016	0.411 ± 0.010	+8.6%
Traffic	Full	**0.321 ± 0.008**	**0.195 ± 0.016**	—
	No-Gain	0.428 ± 0.011	0.260 ± 0.008	+33.4%
	No-Innov	0.375 ± 0.009	0.228 ± 0.007	+16.9%
Exchange	Full	**0.266 ± 0.014**	**0.341 ± 0.023**	—
	No-Gain	0.299 ± 0.018	0.383 ± 0.023	+12.3%
	No-Innov	0.294 ± 0.015	0.377 ± 0.019	+10.6%

4.2.2. Capacity Ablation.

To verify that the performance gains of KOSLM mainly stem from the Kalman-optimal structural design rather than the network capacity, we conduct a capacity ablation study. We evaluate the ϕ network under four progressively larger configurations: **(i) Linear**: A single linear layer with input dimension equal to the number of input features and output dimension 64, without any activation; **(ii) SmallMLP**: A two-layer MLP with 128 hidden units and sigmoid activation; **(iii) MediumMLP**: A three-layer MLP with 256 hidden units per layer and sigmoid activation; **(iv) HighCap**: A four-layer MLP with 512 hidden units per layer and sigmoid activation.

All models are trained under identical settings with five independent random seeds, and the results are reported as mean ± standard deviation. This allows us to evaluate both performance trends and statistical reliability. From Table 1, it is evident that increasing the ϕ network capacity beyond SmallMLP does not consistently improve MSE or MAE; in some cases, the error even increases slightly. The small MediumMLP improvements (ETTm1) or slight deterioration (ETTh1, Traffic, Exchange) indicate that performance gains are dominated by the Kalman-optimal structural design rather than network depth or over-parameterization. This confirms the effectiveness of the *innovation*-driven gain estimation in KOSLM for stable long-horizon prediction.

4.3. *Efficiency Benchmark*

To evaluate the computational efficiency of KOSLM, we benchmark it against representative sequence modeling baselines, including Transformer, Mamba-2, and LSTM, across four key metrics: runtime scalability, end-to-end throughput, memory footprint, and parameter count. All experiments are conducted using `PyTorch 2.1.0` with `torch.compile` optimization on a single NVIDIA RTX 4090 GPU.

4.3.1. Runtime Scalability.

We assess runtime performance on both a controlled synthetic setup and the ETTm1 dataset. As illustrated in Figure 4, KOSLM exhibits near-linear growth in inference time with increasing sequence length L. For sequences longer than $L = 2\text{k}$, KOSLM surpasses optimized Transformer implementations (FlashAttention [31]) and achieves up to

2.5× faster execution than the fused-kernel version of Mamba-2 [24] at $L > 4$k. Compared with a standard PyTorch LSTM, KOSLM achieves approximately 1.3–1.9× speedup across the tested sequence lengths, demonstrating robust scalability for long sequences.

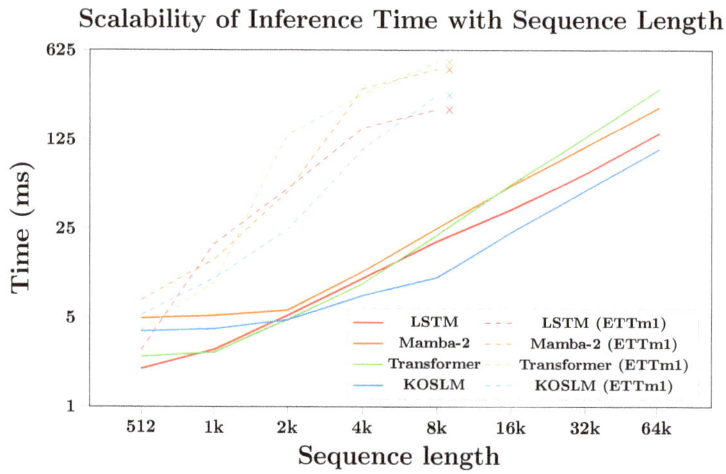

Figure 4. Runtime benchmarks. Inference time as a function of sequence length L. KOSLM demonstrates near-linear scaling and maintains faster inference than Transformer and Mamba-2 for long sequences.

4.3.2. Throughput Analysis.

Figure 5 reports the end-to-end inference throughput measured in million tokens/s (M tokens/s). KOSLM maintains consistently high throughput across all tested sequence lengths, achieving up to 1.9× higher throughput than LSTM and 2.7× higher than Mamba-2 in the 4K–8K regime. The performance remains stable for even longer sequences, highlighting the model's suitability for extended temporal dependencies.

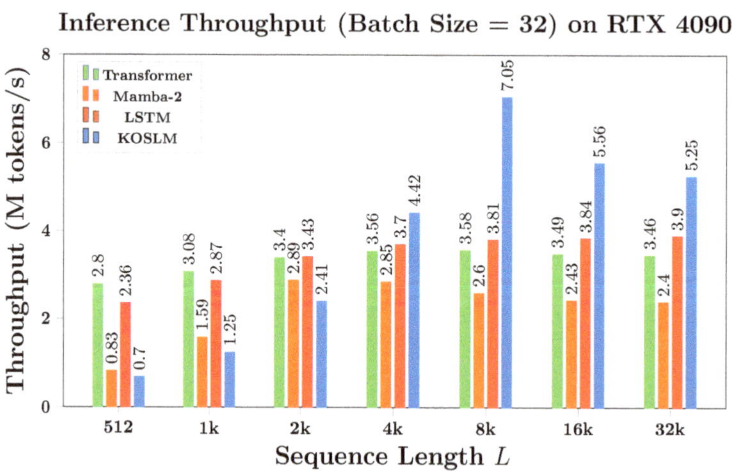

Figure 5. Throughput benchmarks. End-to-end inference throughput on RTX 4090. KOSLM maintains stable throughput as sequence length increases, outperforming LSTM and Mamba-2 for long sequences.

4.3.3. Memory Footprint

Table 7 reports GPU memory usage during training for varying batch sizes with input length 2048 and models containing approximately 0.24 M parameters. KOSLM demonstrates favorable memory efficiency relative to similarly sized Mamba-2 implementations and remains comparable to a standard LSTM with roughly ten times more parameters. This efficiency results from KOSLM's compact Kalman-optimal selective mechanism, which dynamically modulates state transitions with minimal parameter overhead.

Table 7. Memory footprint under different batch sizes, input length = 2048. KOSLM exhibits lower memory usage than Transformer and Mamba-2, while remaining comparable to LSTM despite its smaller size.

Batch Size	Transformer (GB)	Mamba-2 (GB)	LSTM (GB)	KOSLM (GB)
1	0.223	0.085	0.081	0.061
2	0.363	0.158	0.109	0.104
4	0.640	0.290	0.166	0.188
8	1.180	0.561	0.283	0.344
16	2.256	1.103	0.518	0.668
32	4.408	2.188	0.987	1.317
64	8.712	4.357	1.926	2.615
128	17.000	8.696	3.803	5.211

4.3.4. Model Size

Table 8 summarizes the parameter counts. KOSLM contains only 0.24 M parameters, roughly 4.5% of the Transformer and 11% of the LSTM baselines, while achieving comparable or superior forecasting performance. The compact design emphasizes KOSLM's suitability for deployment in resource-constrained environments.

Table 8. Model parameter sizes. KOSLM remains lightweight while providing competitive long-horizon forecasting performance.

Model	Parameters (M)
LSTM	2.17
Transformer	5.33
Mamba-2	0.21
FiLM	1.50
KOSLM	0.24

This efficiency stems directly from the Kalman-optimal selective mechanism, which adaptively regulates the state update through *innovation*-driven gain modulation. Such a formulation not only reduces redundant computation but also provides a principled pathway for scaling to long sequences.

4.4. Real-World Application: Secondary Surveillance Radar (SSR) Target Trajectory Tracking

To further evaluate the robustness and practical generalizability of KOSLM under real-world noisy conditions, we conducted a real-world experiment on SSR target trajectory tracking. SSR is a ground-based air traffic surveillance system where aircraft respond to radar interrogations with encoded transponder signals, generating sparse range–azimuth sequences known as raw SSR plots. These observation sequences present three major challenges for sequential modeling:

- **High stochastic noise:** Measurement noise leads to random fluctuations in the estimated positions;
- **Irregular sampling:** Aircraft maneuvers and radar scan intervals result in uneven temporal spacing;
- **Correlated anomalies:** Spurious echoes or missing detections introduce discontinuities in the trajectories.

These characteristics make SSR a natural yet challenging testbed for assessing model stability, noise resilience, and temporal consistency. KOSLM addresses these challenges by integrating context-aware selective state updates with learnable dynamics under the Kalman optimality principle, enabling robust filtering, smoothing, and extrapolation in partially observed environments.

4.4.1. Experimental Setup

To achieve a balance between realism and experimental controllability, we adopted a semi-physical simulation approach. Training data were derived from Automatic Dependent Surveillance–Broadcast (ADS-B) flight tracks obtained from the OpenSky Network [32]. Controlled Gaussian noise with an SNR of 33 dB was added to emulate SSR observation uncertainty. For evaluation, we deployed KOSLM on real SSR radar data collected from an operational ground-based system (peak transmit power 6 kW), capturing eight live air-traffic targets under operational conditions. Detailed data acquisition and preprocessing procedures are provided in Appendix C.

Unlike the main benchmarking experiments on standardized long-term time series forecasting (LTSF) datasets, this case study aims to demonstrate practical applicability and operational robustness. Due to irregular sampling, sparse observations, and high stochastic noise, conventional quantitative metrics such as MSE or MAE are not meaningful. Therefore, we focus on qualitative trajectory visualizations to illustrate model performance in realistic conditions.

4.4.2. Results and Analysis

Figure 6 presents representative trajectories of eight air-traffic targets, comparing classical KF algorithm, Transformer, Mamba, and KOSLM. The classical KF algorithm exhibits locally inaccurate and jagged trajectories due to its fixed linear dynamics, which cannot adapt to irregular sampling or abrupt maneuvers. The Transformer captures general trends but produces fragmented and temporally inconsistent tracks under sparse and noisy observations. Mamba improves noise robustness but still demonstrates local instability during complex maneuvers.

In contrast, KOSLM generates smooth, coherent, and temporally consistent trajectories that closely follow the true flight paths, highlighting its ability to handle high stochastic noise, irregular sampling, and correlated anomalies. This robustness stems from the *innovation*-driven Kalman gain and context-aware selective state updates, which allow KOSLM to adapt dynamically to non-stationary motion patterns.

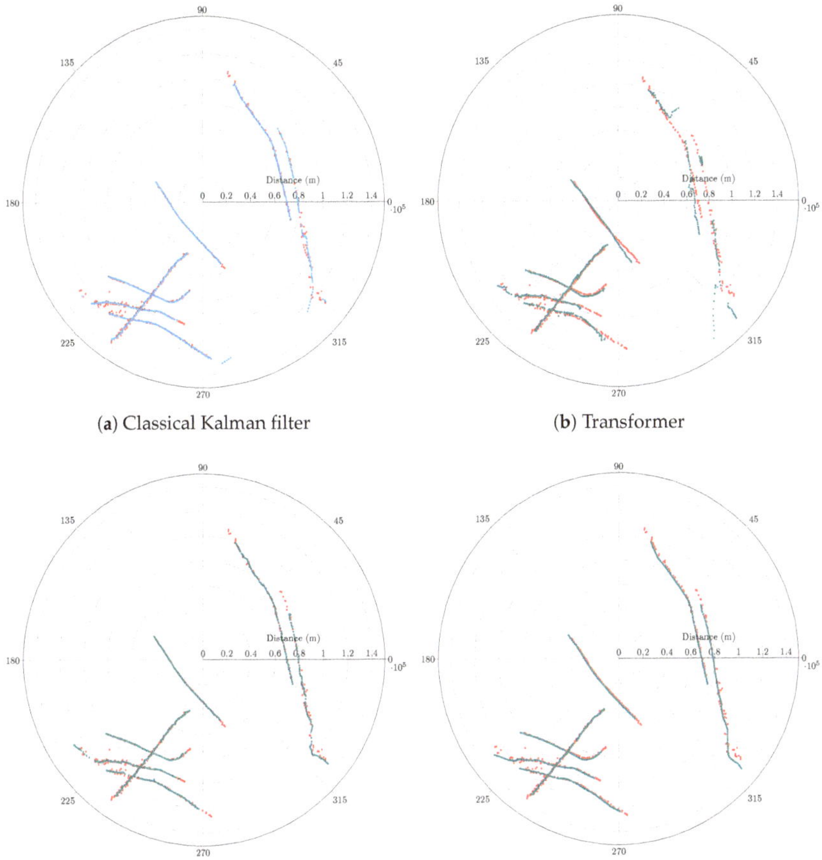

(**a**) Classical Kalman filter
(**b**) Transformer
(**c**) Mamba
(**d**) KOSLM (Ours)

Figure 6. SSR target tracking under real-world operational conditions. Trajectories of eight air-traffic targets tracked from SSR measurements. Red curves denote raw SSR observations, blue curves denote tracking by the conventional KF algorithm (on-site SSR interrogator output), and teal curves indicate deep model predictions (Transformer, Mamba, or KOSLM), followed by KF smoothing. KOSLM achieves smoother, more accurate, and stable tracking under strong noise and irregular sampling conditions.

5. Conclusions

This study addressed a fundamental limitation of recurrent architectures such as LSTM, whose heuristically designed gates lack structural constraints for optimality, leading to instability and information decay over long sequences. To overcome this issue, we proposed the *Kalman-Optimal Selective Long-Term Memory Network (KOSLM)*, which reconstrues LSTM as a nonlinear, input- and state-dependent state-space model, and integrates an *innovation*-driven Kalman-optimal gain path for principled information selection. This formulation unifies LSTM gating, selective state-space modeling, and Kalman filtering into a single theoretically grounded recurrent framework. We note that the proposed KOSLM is lightweight, containing only 0.24 M parameters—significantly smaller than typical LSTM or Transformer baselines—while still achieving competitive or superior performance. This compact design ensures efficient implementation on standard GPU hardware without imposing constraints on memory or computation.

Extensive experiments demonstrate that KOSLM achieves state-of-the-art performance on long-term forecasting benchmarks, while ablation studies confirm the essential role of the *innovation* statistic and Kalman-form gain in stabilizing long-horizon modeling. Moreover, validation on real-world SSR trajectory tracking highlights its robustness under noisy and non-stationary conditions.

In summary, embedding Kalman-optimal principles into deep recurrent networks provides both theoretical insights and practical benefits for robust long-term sequence modeling. Future work will focus on extending KOSLM beyond Gaussian assumptions and applying it to multimodal and cross-domain time-series scenarios.

Supplementary Materials: The following supporting information can be downloaded at: https://www.mdpi.com/article/10.3390/app152312684/s1, Table S1: KOSLM run-to-run variability (RMS) over 5 runs.

Author Contributions: All authors contributed to the conception and design of the study. Material preparation, code implementation, data collection, and analysis were performed by X.T. and L.W. The first draft of the manuscript was written by L.W. and all authors commented on previous versions of the manuscript. X.T. and M.W. supervised the research and provided critical revisions and theoretical guidance. Y.Z. assisted in data verification and manuscript editing. All authors have read and agreed to the published version of the manuscript.

Funding: This research received no external funding.

Data Availability Statement: The datasets and source code used in this study are publicly available at https://github.com/wl822513/KOSLM (accessed on 23 October 2025).

Conflicts of Interest: The authors declare no conflicts of interest.

Appendix A. Selective State Space Models

This section focuses on S6, which provide a unified framework connecting the gating with time-varying state-space representations. Their associated selection mechanisms have become a central design principle in modern SSMs, such as Mamba [11]. These mechanisms improve modeling efficiency by dynamically identifying and retaining task-relevant information over long sequences, effectively compressing the latent state without compromising representational capacity.

Relation to Existing Concepts.

Selection mechanisms are conceptually related to several prior ideas, including **gating**, **hypernetworks**, and **data-dependent parameterization**:

- **Gating** was originally introduced as a signal-control mechanism in recurrent neural networks (RNNs) such as LSTM [3] and GRU [33], where multiplicative gates regulate information updates across time steps. The concept has been generalized to other architectures, including gated convolutions and transformers [34,35], though often without an explicit interpretation in terms of temporal signal control.
- **Hypernetworks** [36] generate network parameters dynamically through auxiliary networks, enabling the model's internal dynamics to adapt based on input signals.
- **Data-dependent parameterization** [37] represents a broader paradigm in which model parameters are directly conditioned on input data. Both gating and hypernetworks can be viewed as specific instances within this larger class.

While selection mechanisms share certain similarities with these paradigms, they form a distinct modeling category. Unlike classical gating or hypernetwork mechanisms that often operate locally or step-wise, selection mechanisms are explicitly designed to route, filter, or suppress sequence-level information in an input- or state-dependent manner, enabling stable long-horizon modeling. In selective SSMs, this is typically realized by parameterizing system matrices (e.g., Δ, A, B, C) as functions of the input at each time step.

From Implicit to Explicit Selection.

Early structured SSMs (S4) [38] encoded fixed inductive biases through learned structured dynamics, providing an implicit, input-independent form of selection via controlled signal propagation. Later models, including Mamba [11], introduced **explicit selection**, where the parameters of the state-space system (e.g., Δ, A, B, C) are conditioned directly on the current input, allowing the model to dynamically emphasize or suppress features at each step.

Semantic Clarification.

As discussed in S6 [11], while selection mechanisms can be loosely related to gating, hypernetworks, or general data dependence, these broader descriptions do not capture the defining characteristic of selection. The term **selection** is reserved for mechanisms that explicitly operate along the sequence dimension, enabling adaptive memory compression and long-range information control. From a dynamical systems perspective, such mechanisms can be interpreted through state-space discretization and signal propagation theory [39,40]. Importantly, classical RNN gating can be viewed as a local, step-wise precursor to these more general selection mechanisms.

Scope and Relevance.

The principle of selection underlies recent progress in linear-time sequence modeling. Building upon this paradigm, our proposed KOSLM extends selective SSMs by introducing an **innovation-driven Kalman-optimal feedback pathway**, transforming heuristic selection into a principled, uncertainty-aware mechanism for long-term sequence modeling.

Appendix B. Detailed Derivation

Appendix B.1. Kalman Gain as a Prototype for Dynamic Selectivity

In classical filtering theory, the Kalman gain $K(t)$ acts as a dynamic weighting factor that determines how prior state estimates and new observations are combined to produce a new state [6]. It is formally derived as the solution to the Riccati differential equation, and its value depends on the evolving uncertainty in the internal system state (captured by the prior covariance $P(t)$) and the noise characteristics of the observation process (captured by the measurement covariance R) [41]. This time-varying gain governs the extent to which incoming measurements correct the state estimate, ensuring minimum mean squared error under Gaussian noise assumptions.

This correction process is essentially a trade-off between the historical state information of the system and the incoming input information. In the context of deep learning, we interpret the observation as the current input sequence, and the prior state estimate as the latent representation of the system's history. From this perspective, the Kalman gain plays a role analogous to a dynamic selection factor that balances the contribution of contextual knowledge (from the model state) and content information (from the input) in generating the updated representation. Therefore, the Kalman gain can be viewed as a principled prototype for content-aware and context-sensitive information selection. This perspective justifies the broader use of dynamic selection strategies in sequence models, especially when aiming to balance stability and adaptability in long-range modeling.

Appendix B.2. LSTM-to-SSM Reconstruction

To endow LSTM networks with structured modeling semantics, in this section we reconstruct them as a form of *nonlinear time-varying state-space model* ((see Equation (3a)). Specifically, we identify that the interactions among the forget gate, input gate, output gate, input signal, and cell state in LSTMs essentially constitute an input-dependent and state-aware state transition–observation process.

As illustrated in Figure 1, the gating mechanisms and the cell state C_t in LSTM establish a pathway for memory propagation across time. The cell state C_t is the core component that preserves long-term memory across the temporal dimension and models long-range dependencies. In terms of representational role, it is both semantically and structurally equivalent to the latent state h_t in SSMs (see Equation (3a)). Therefore, we take C_t as the central axis and analyze its information propagation process.

The forget gate F_t regulates the degree to which short-term memory from cell state C_{t-1} is discarded. It is determined jointly by the current input x_t and the previous hidden state H_{t-1}, thereby embodying an input- and state-dependent transition process:

$$A_t \triangleq F_t = \sigma(W_f X_t + b_f), \tag{A1}$$

where $X_t = [x_t; H_{t-1}]$ denotes the concatenated input. The corresponding retained term is $A_t C_{t-1}$.

The pathway for new information input in LSTM consists of two steps:

1. The candidate cell state $\tilde{C}_t$ can be interpreted as a differentiable nonlinear mapping from the joint input X_t:

$$z_t \triangleq \tilde{C}_t = \tanh(W_c X_t + b_c), \tag{A2}$$

   where the linear transformation $W_c X_t + b_c$ fuses the external input x_t and the hidden state H_{t-1} into the observation space, with $\tanh(\cdot)$ introducing nonlinearity to produce an intermediate representation z_t. We treat this representation as the observation input at time t under the SSM (Here, the observation input denotes the externally observable signal that directly drives the state update process. It is written as z_t in KF and as x_t in SSMs, which are semantically equivalent).

2. The interaction between the input gate I_t and the candidate state $\tilde{C}_t$ establishes a structured pathway for injecting new information into the state dynamics, analogous to the excitation of state evolution by external inputs in SSMs.

$$B_t z_t \triangleq I_t \odot \tilde{C}_t = \sigma(W_i X_t + b_i) \odot \tanh(W_c X_t + b_c), \tag{A3}$$

   where B_t serves as the input matrix, structurally equivalent to the input gate I_t.

Consequently, the update process of the LSTM cell state C_t at time t (Equation (1)) can be rewritten in the form of a state-space transition:

$$C_t = A_t C_{t-1} + B_t z_t, \tag{A4}$$

where A_t and B_t are realized through the forget and input gates driven by the joint input X_t, thereby endowing the model with the ability to selectively forget or memorize long-term information in an input-dependent and context-aware manner.

Finally, the output pathway of LSTM is given by

$$H_t = O_t \odot \tanh(C_t), \tag{A5}$$

which is formally equivalent to the observation equation of a classical SSM:

$$y_t = M_t C_t, \tag{A6}$$

where M_t is formed by the output gate O_t together with the nonlinear transformation $\tanh(\cdot)$.

The above reconstruction shows that the gating mechanism in LSTM can be interpreted as a class of *nonlinear, input- and state-dependent* SSMs. This perspective establishes a valid unification of LSTM gating mechanisms and state-space models, and reveals that LSTM's long-term memory capability originates from the structured realization of both state-space modeling and efficient information selection. Building on this understanding, we further introduce the Kalman gain into this state-space structure, imposing a structural constraint that minimizes the uncertainty of state estimation errors.

Appendix B.3. Kalman-Optimal Selective Mechanism

Recent studies on time series modeling have proposed a class of selective mechanisms based on SSMs, among which the most representative work is S6 [11]. Its core idea points out the following:

> "One method of incorporating a selection mechanism into models is by letting their parameters that affect interactions along the sequence (e.g., the recurrent dynamics of an RNN or the convolution kernel of a CNN) be input-dependent."

Inspired by this, we design a selective mechanism based on Kalman-optimal state estimation within the SSM (Equation (3)). Unlike the input-dependent selection mechanism represented by S6, we make the key parameters (A_t, B_t), which control how the model selectively propagates or forgets information along the sequence dimension, depend on the *innovation term*: the deviation between the observation input and the prior state prediction in the observation space, denoted as *Innov* in this paper. This method constructs a learnable selection path that integrates observational inputs and latent state feedback, with the optimization objective of minimizing the uncertainty in state estimation errors.

Learnable Gain from Innovation.

In the KF algorithm, the closed-form solution of the Kalman gain K_t (Equation (7)) is obtained by minimizing the covariance of the state estimation error [6]. Here, the observation matrix M_t and the observation noise covariance R_t are considered known system priors, usually derived from physical modeling or domain knowledge. Therefore, the dynamics of K_t entirely stem from the statistical properties of the prior estimation error $e_t^- = h_t - \hat{h}_t^-$. Under the linear observation model (4b), e_t^- is mapped into the observation space, yielding the following representation of the *innovation term*:

$$Innov = z_t - M_t\hat{h}_t^- = M_t(h_t - \hat{h}_t^-) + v_t, \tag{A7}$$

This equation shows that *Innov* is a linear mapping of e_t^-, which, although perturbed by additive observation noise v_t, still retains the full uncertainty information of e_t^-. Thus, *Innov* can serve as a sufficient statistic for estimating K_t, providing a theoretical basis for directly estimating K_t from the *innovation term*. The estimation process is defined as

$$K_t = \phi(Innov_t; \theta_\phi), \tag{A8}$$

where $\phi(\cdot)$ is a differentiable nonlinear function, parameterized by a neural network and trained via gradient descent. Theoretical soundness of this formulation is confirmed by a controlled experiment (Appendix B.4), which demonstrates that $\phi(\cdot)$ can reliably recover the oracle Kalman gain from the *innovation* statistics with negligible estimation error.

Optimal Selectivity via Gain.

By substituting the KF prediction process (Equation (5)) into the state update equation (Equation (8)), the two processes can be expressed as a nonlinear time-varying SSM form (Equation (3a)):

$$\hat{h}_t = A_t\hat{h}_{t-1} + B_t z_t, \quad A_t = (I - K_t M_t)A, \quad B_t = K_t, \tag{A9}$$

Within this framework, parameters A_t and B_t are linearly modulated by K_t driven by the *innovation term*, yielding a context-aware Kalman-optimal selective path that minimizes state estimation uncertainty and imposes structural optimality constraints on LSTM.

Appendix B.4. Empirical Validation of Innovation-Driven Kalman Gain Learning

To empirically validate the theoretical claim that the Kalman gain can be dynamically inferred from the *innovation term*, we design a controlled linear–Gaussian synthetic experiment where the optimal gain has a closed-form solution. This allows us to directly compare the learned gain against the ground-truth Kalman gain and quantify their alignment over time.

Experimental Design.

We consider a one-dimensional linear dynamical system as in Equation (4). The analytical Kalman gain K_t is computed from the standard Riccati recursion and serves as the oracle reference. We design three model variants for comparison:

(i) **Oracle-KF:** Standard Kalman filter using the known (A, Q, R, M) parameters to compute K_t.

(ii) **Supervised-ϕ:** A small MLP $\phi(\cdot)$ takes the *innovation* $innov_t = z_t - M\hat{h}_t^-$ as input and predicts $\hat{K}_t = \phi(innov_t)$, trained by minimizing $\mathcal{L} = |\hat{K}_t - K_t|^2$ over all timesteps. During training, the predicted prior state $\hat{h}_t^-$ is provided by the Oracle-KF to isolate the learning of the *innovation*-to-gain mapping. This tests whether the *innovation term* contains sufficient information to recover the oracle gain.

(iii) **End-to-End:** The same MLP is trained to directly map from $(z_t, \hat{h}_t^-)$ to K_t without explicitly constructing the *innovation term*, serving as an ablation to assess the importance of *innovation*-driven modeling.

Setup and Metrics.

We simulate 10,000 trajectories of length $T = 100$ with fixed system parameters $(A = 0.9, M = 1)$. To test robustness under different noise regimes, we consider four parameter settings by varying process noise Q and observation noise R: (i) $Q = 5, R = 0.1$; (ii) $Q = 5, R = 0.5$; (iii) $Q = 10, R = 0.1$; (iv) $Q = 100, R = 0.5$. The supervised models are trained for 200 epochs using Adam optimizer with a learning rate 10^{-3}. Performance is evaluated by

- **MSE of Gain:** $\text{MSE} = \frac{1}{T}\sum_t |K_t - \hat{K}_t|^2$, measuring how well the learned gain matches the oracle trajectory.
- **State Estimation RMSE:** Root-mean-square error between estimated and true state trajectories, where the state estimates are produced by running a Kalman update step with the learned $\hat{K}_t$, verifying that accurate gain learning improves filtering quality.

Results and Analysis.

Figure A1 presents the learned Kalman gain trajectories across all considered noise regimes. Across every (Q, R) configuration, the Supervised-ϕ model closely matches the oracle Kalman gain K_t, producing stable and accurate trajectories throughout both transient and steady phases. In contrast, the End-to-End model produces a noisier and slightly biased gain curve, indicating that omitting the explicit *innovation term* fails to accurately capture the gain dynamics. This qualitative observation is quantitatively confirmed in Table A1: Supervised-ϕ achieves up to **four orders-of-magnitude lower gain MSE** and reduces state-estimation RMSE by 4.5–32.8% under all (Q, R) noise settings. These results demonstrate that the *innovation term* is indeed sufficient for inferring the optimal Kalman gain and that explicitly modeling *innovation*-driven selectivity yields more stable and accurate state estimation. Together, the figure and table provide strong empirical support for our theoretical claim that *innovation*-guided gain learning constitutes a principled and robust mechanism for selective state updates.

Table A1. Quantitative Results of Kalman Gain Learning. Supervised-ϕ achieves **orders-of-magnitude lower gain MSE** and consistently better state estimation RMSE across all noise regimes compared to end-to-end learning, highlighting the critical role of explicitly modeling the *innovation term* in recovering accurate Kalman gain dynamics.

(Q, R)	Supervised-ϕ		End-to-End	
	MSE	RMSE	MSE	RMSE
(5, 0.1)	1.3×10^{-6}	0.317	3.5×10^{-3}	0.368
(5, 0.5)	2.9×10^{-5}	0.684	1.8×10^{-3}	0.694
(10, 0.1)	1.2×10^{-6}	0.216	1.3×10^{-3}	0.341
(100, 0.5)	2.2×10^{-7}	0.310	6.8×10^{-4}	0.747

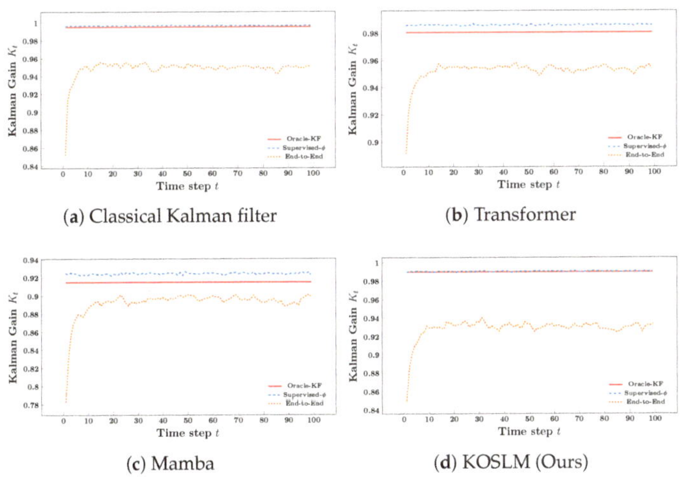

(**a**) Classical Kalman filter (**b**) Transformer

(**c**) Mamba (**d**) KOSLM (Ours)

Figure A1. Kalman gain learning across different noise regimes. Top-left: $(Q = 100, R = 0.5)$; top-right: $(Q = 5, R = 0.1)$; bottom-left: $(Q = 5, R = 0.5)$; bottom-right: $(Q = 10, R = 0.1)$. Supervised-ϕ closely matches the oracle K_t in all cases, while end-to-end trajectories remain noisy and biased. These results confirm that *innovation*-driven modeling provides stable and accurate gain learning across both low- and high-noise conditions.

Appendix C. SSR Case Study: Data and Preprocessing Details

Appendix C.1. Data Acquisition

The semi-physical training dataset is constructed using ADS-B logs from the OpenSky Network, recorded on 15 June 2025, between 10:00 and 10:15 AM. The data cover a 300 km region surrounding Tokyo Haneda, Narita, and Incheon Airports. Each log provides updates every 5 s for 503 aircraft, including longitude, latitude, altitude, and timestamp information. A detailed statistical summary of the dataset is provided in Table A2.

To emulate realistic Secondary Surveillance Radar (SSR) observation noise, zero-mean Gaussian perturbations are applied to each measurement with a signal-to-noise ratio (SNR) of 33 dB, formulated as

$$X_i = X_i + \mathcal{N}\left(0, \frac{\sigma_i^2}{10^{\frac{33}{10}}}\right),$$

where σ_i^2 represents the empirical variance of the i-th feature. This noise configuration reflects typical radar tracking uncertainty under moderate atmospheric interference.

Table A2. Summary of the Semi-physical ADS-B Dataset Used for Training. The dataset consists of 5 s updates for 503 aircraft collected within a 15 min window around major East Asian airports. Each sample includes four state variables: longitude, latitude, altitude, and timestamp.

Targets	Update Frequency	# Features	Total Samples	Time Span
503	5 s	4	166,110	15 June 2025, 10:00–10:15

Appendix C.2. Normalization and Reverse Transformation

All input data are standardized to zero mean and unit variance:

$$y = \frac{x - \bar{x}}{\sigma}, \quad \hat{x} = \sigma\hat{y} + \bar{x}.$$

Appendix C.3. Field Data Collection

For real-world evaluation, we use raw SSR plots collected by a field-deployed radar with a peak transmit power of 6 kW. These sequences contain irregular sampling, strong noise, and missing returns, providing a rigorous test of model robustness.

Appendix C.4. Processing Workflow

The overall pipeline—from semi-physical data generation to field testing—is summarized in Figure A2, where the upper branch denotes the training stage on noisy ADS-B data, and the lower branch represents inference on real SSR plots. This workflow ensures consistency between semi-physical training data and real-world SSR testing scenarios.

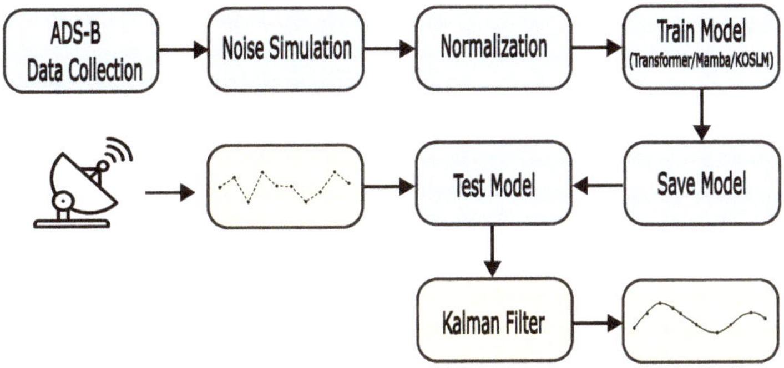

Figure A2. SSR trajectory modeling pipeline. The upper branch shows training using ADS-B data with noise simulation and normalization; the lower branch shows testing on real raw SSR plots using trained models, followed by trajectory prediction and comparison.

References

1. Cheng, M.; Liu, Z.; Tao, X.; Liu, Q.; Zhang, J.; Pan, T.; Zhang, S.; He, P.; Zhang, X.; Wang, D.; et al. A comprehensive survey of time series forecasting: Concepts, challenges, and future directions. *TechRxiv* **2025**. [CrossRef] [PubMed]
2. Salazar, C.; Banerjee, A.G. A distance correlation-based approach to characterize the effectiveness of recurrent neural networks for time series forecasting. *Neurocomputing* **2025**, *629*, 129641. [CrossRef]
3. Hochreiter, S.; Schmidhuber, J. Long short-term memory. *Neural Comput.* **1997**, *9*, 1735–1780. [CrossRef] [PubMed]
4. Benjamin, J.; Mathew, J. Enhancing continuous integration predictions: A hybrid LSTM-GRU deep learning framework with evolved DBSO algorithm. *Computing* **2025**, *107*, 9. [CrossRef]
5. Pascanu, R.; Mikolov, T.; Bengio, Y. On the Difficulty of Training Recurrent Neural Networks. In Proceedings of the 30th International Conference on Machine Learning (ICML'13), Atlanta, GA, USA, 16–21 June 2013; pp. 1310–1318.
6. Kalman, R.E. A New Approach to Linear Filtering and Prediction Problems. *J. Basic Eng.* **1960**, *82*, 35–45. [CrossRef]
7. Yuan, X.; Li, J.; Kuruoglu, E.E. Robustness enhancement in neural networks with alpha-stable training noise. *Digit. Signal Process.* **2025**, *156*, 104778. [CrossRef]
8. Krishnan, R.G.; Shalit, U.; Sontag, D. Deep kalman filters. *arXiv* **2015**, arXiv:1511.05121. [CrossRef]
9. Revach, G.; Shlezinger, N.; Ni, X.; Escoriza, A.L.; Van Sloun, R.J.; Eldar, Y.C. KalmanNet: Neural network aided Kalman filtering for partially known dynamics. *IEEE Trans. Signal Process.* **2022**, *70*, 1532–1547. [CrossRef]
10. Dahan, Y.; Revach, G.; Dunik, J.; Shlezinger, N. Uncertainty quantification in deep learning based Kalman filters. In Proceedings of the ICASSP 2024—2024 IEEE International Conference on Acoustics, Speech and Signal Processing (ICASSP), Seoul, Republic of Korea, 14–19 April 2024; pp. 13121–13125.
11. Gu, A.; Dao, T. Mamba: Linear-time sequence modeling with selective state spaces. *arXiv* **2023**, arXiv:2312.00752. [CrossRef]
12. Dao, T.; Gu, A. Transformers are ssms: Generalized models and efficient algorithms through structured state space duality. *arXiv* **2024**, arXiv:2405.21060. [CrossRef]
13. Aliyu, B.K.; Osheku, C.A.; Adetoro, L.M.; Funmilayo, A.A. Optimal Solution to Matrix Riccati Equation—For Kalman Filter Implementation. In *MATLAB*; Katsikis, V.N., Ed.; IntechOpen: Rijeka, Croatia, 2012; Chapter 4. [CrossRef]
14. Gu, A.; Goel, K.; Gupta, A.; Ré, C. On the parameterization and initialization of diagonal state space models. *Adv. Neural Inf. Process. Syst.* **2022**, *35*, 35971–35983.
15. Gu, A.; Gulcehre, C.; Paine, T.; Hoffman, M.; Pascanu, R. Improving the gating mechanism of recurrent neural networks. In Proceedings of the International Conference on Machine Learning, PMLR, Virtual, 13–18 July 2020; pp. 3800–3809.
16. Weather Dataset. 2024. Weather Data from the BGC Jena Climate Data Archive. Available online: https://www.hypermeteo.com/en/?gad_source=1&gad_campaignid=14712996685&gclid=EAIaIQobChMIvJ_HqOORkQMVSpCDBx3MFTosEAAYASAAEgLx7fD_BwE (accessed on 12 May 2024).
17. ETT Datasets. 2024. Energy Transformer Temperature Dataset (ETT) for Long-Term Forecasting. Available online: https://ieee-dataport.org/documents/merged-ett-dataset-time-series-forecasting (accessed on 12 May 2024).
18. Electricity Dataset. 2024. Electricity Load Diagrams Dataset from UCI Machine Learning Repository. Available online: https://ieee-dataport.org/documents/uci-dataset (accessed on 12 May 2024).
19. Exchange Rate Dataset. 2024. Daily Exchange Rate Dataset for Long-Term Time Series Forecasting, Provided with the ETT Benchmark Collection. Available online: https://github.com/juyongjiang/TimeSeriesDatasets (accessed on 12 May 2024).

20. Traffic Dataset. 2024. Traffic Flow Data from California Performance Measurement System (PeMS). Available online: https://www.smartmicro.com/traffic-sensor?utm_medium=advert&utm_source=google&utm_campaign=search-traffic&utm_content=ad&utm_term=traffic_sensor&gad_source=1&gad_campaignid=22435495111&gclid=EAIaIQobChMIlvLc8OORkQMVQJKDBx3FOSSQEAAYASAAEgITNPD_BwE (accessed on 12 May 2024).
21. Illness (ILI) Dataset. 2024. Weekly Influenza-Like Illness (ILI) Dataset for Forecasting Tasks, Available Within the ETT Benchmark Suite. Available online: https://www.researchgate.net/figure/Influenza-like-illness-ILI-data-from-the-FluID-database-Weekly-scaled-rates-of-ILI_fig1_337718309 (accessed on 12 May 2024).
22. Alharthi, M.; Mahmood, A. xlstmtime: Long-term time series forecasting with xlstm. *AI* **2024**, *5*, 1482–1495. [CrossRef]
23. Zhou, T.; Ma, Z.; Wen, Q.; Sun, L.; Yao, T.; Yin, W.; Jin, R. Film: Frequency improved legendre memory model for long-term time series forecasting. *Adv. Neural Inf. Process. Syst.* **2022**, *35*, 12677–12690.
24. Wang, Z.; Kong, F.; Feng, S.; Wang, M.; Yang, X.; Zhao, H.; Wang, D.; Zhang, Y. Is mamba effective for time series forecasting? *Neurocomputing* **2025**, *619*, 129178. [CrossRef]
25. Zhou, T.; Ma, Z.; Wen, Q.; Wang, X.; Sun, L.; Jin, R. Fedformer: Frequency enhanced decomposed transformer for long-term series forecasting. In Proceedings of the International Conference on Machine Learning, PMLR, Baltimore, MD, USA, 17–23 July 2022; pp. 27268–27286.
26. Liu, Y.; Hu, T.; Zhang, H.; Wu, H.; Wang, S.; Ma, L.; Long, M. itransformer: Inverted transformers are effective for time series forecasting. *arXiv* **2023**, arXiv:2310.06625.
27. Zhang, Y.; Yan, J. Crossformer: Transformer Utilizing Cross-Dimension Dependency for Multivariate Time Series Forecasting. In Proceedings of the Eleventh International Conference on Learning Representations (ICLR), Kigali, Rwanda, 1–5 May 2023.
28. Su, J.; Shen, Y.; Wu, Z. DLinear: A Linear Complexity Approach for Long-Term Time Series Forecasting. *arXiv* **2022**, arXiv:2205.13504.
29. Huang, X.; Tang, J.; Shen, Y. Long time series of ocean wave prediction based on PatchTST model. *Ocean. Eng.* **2024**, *301*, 117572. [CrossRef]
30. Wang, S.; Wu, H.; Shi, X.; Hu, T.; Luo, H.; Ma, L.; Zhang, J.Y.; Zhou, J. Timemixer: Decomposable multiscale mixing for time series forecasting. *arXiv* **2024**, arXiv:2405.14616. [CrossRef]
31. Dao, T. Flashattention-2: Faster attention with better parallelism and work partitioning. *arXiv* **2024**, arXiv:2307.08691.
32. Zhang, J.; Wei, L.; Yanbo, Z. Study of ADS-B data evaluation. *Chin. J. Aeronaut.* **2011**, *24*, 461–466. [CrossRef]
33. Cho, K.; Van Merriënboer, B.; Gulcehre, C.; Bahdanau, D.; Bougares, F.; Schwenk, H.; Bengio, Y. Learning phrase representations using RNN encoder-decoder for statistical machine translation. *arXiv* **2014**, arXiv:1406.1078. [CrossRef]
34. Mehta, H.; Gupta, A.; Cutkosky, A.; Neyshabur, B. Long range language modeling via gated state spaces. *arXiv* **2022**, arXiv:2206.13947. [CrossRef]
35. Hua, W.; Dai, Z.; Liu, H.; Le, Q. Transformer quality in linear time. In Proceedings of the International Conference on Machine Learning, PMLR, Honolulu, HI, USA, 23–29 July 2022; pp. 9099–9117.
36. Ha, D.; Dai, A.; Le, Q.V. Hypernetworks. *arXiv* **2016**, arXiv:1609.09106. [PubMed]
37. Poli, M. Data-dependent neural networks. *arXiv* **2023**, arXiv:2305.13272.
38. Gu, A.; Goel, K.; Ré, C. Combining recurrent, convolutional, and continuous-time models with structured state space. *Adv. Neural Inf. Process. Syst.* **2021**, *34*, 21984–21998.
39. Funahashi, K.I.; Nakamura, Y. Approximation of dynamical systems by continuous time recurrent neural networks. *Neural Netw.* **1993**, *6*, 801–806. [CrossRef]
40. Tallec, C.; Ollivier, Y. Can recurrent neural networks warp time? *arXiv* **2018**, arXiv:1804.11188. [CrossRef]
41. Tahirovic, A.; Redzovic, A. Optimal Estimation for Continuous-Time Nonlinear Systems Using State-Dependent Riccati Equation (SDRE). *arXiv* **2025**, arXiv:2503.10442.

applied sciences

MDPI

Article

MMHFormer: Multi-Source and Multi-View Hierarchical Transformer for Traffic Flow Prediction

Han Wu [1], Guoqing Teng [1,*], Hao Wu [2], Zicheng Qiu [1] and Meng Zhao [1]

1 School of Computer Science and Engineering, Chongqing University of Science and Technology, Chongqing 400054, China; wuhan@cqust.edu.cn (H.W.)
2 School of Electrical and Electronic Engineering, North China Electric Power University, Beijing 102206, China; 120241180417@ncepu.edu.cn
* Correspondence: tengguoqing@cqust.edu.cn

Abstract

Traffic flow prediction is a vital component of Intelligent Transportation Systems (ITSs), playing a key role in proactive traffic management and the optimization of urban mobility. However, the complex spatial–temporal dependencies, dynamic variations, and external factors in traffic networks present significant challenges for accurate predictions. In this paper, we propose MMHFormer, a novel multi-source, multi-view hierarchical Transformer model specifically designed for traffic flow prediction. MMHFormer incorporates three key innovations: (1) a multi-source gated embedding layer that integrates diverse multidimensional inputs, including spatial Laplacian embeddings, temporal periodic embeddings, and traffic occupancy embeddings, to better capture the complex dynamics of traffic conditions; (2) a hierarchical multi-view spatial attention module that models global, local, and dynamic similarity-based spatial dependencies, effectively addressing the spatial heterogeneity of traffic flows; (3) a hierarchical two-stage temporal attention mechanism that captures global temporal dependencies while adapting to node-specific temporal variations. Extensive experiments conducted on four benchmark traffic datasets demonstrate that MMHFormer outperforms state-of-the-art methods, achieving significant improvements in prediction accuracy.

Keywords: traffic prediction; multi-source; multi-view; transformer

1. Introduction

Traffic flow prediction is one of the core technologies of Intelligent Transportation Systems (ITSs). By analyzing historical data and utilizing prediction models, it helps traffic managers forecast future traffic volume and road conditions [1]. With accurate traffic predictions, traffic management departments can develop strategies for road condition optimization, plan and navigate in advance, thereby improving traffic operational efficiency, ensuring travel safety, and optimizing urban transportation networks [2]. However, due to the complex temporal and spatial dependencies of traffic flow, achieving accurate predictions remains challenging.

Traffic flow data, typically collected from road sensors as time-series data, has traditionally been analyzed using linear methods such as Autoregressive Integrated Moving Average (ARIMA) [3], Vector Autoregression (VAR) [4], and Support Vector Regression

(SVR) [5]. However, these methods struggle to effectively capture the nonlinear characteristics of traffic patterns and the spatial attributes of the transportation network, limiting their application in modern traffic flow prediction. With advancements in deep learning techniques and improvements in hardware capabilities, an increasing number of neural network-based models have been employed to more effectively capture the dynamic temporal and spatial dependencies in traffic flow prediction. In the temporal dimension, Recurrent Neural Networks (RNNs) and their variants, such as Long Short-Term Memory (LSTM) [6] and Gated Recurrent Units (GRU) [7], are employed to capture the dynamic temporal features of traffic flow. In the spatial dimension, some researchers utilize Convolutional Neural Networks (CNNs) to partition the road network into distinct grids to extract spatial correlations. Researchers have combined these models to extract spatial–temporal features [8]. However, this approach is only applicable to grid-like Euclidean spatial graphs, while traffic road networks are typically non-Euclidean. Subsequently, Graph Neural Networks (GNNs) have been shown to be more suitable for modeling non-Euclidean traffic road networks [9]. For instance, Zhao et al. proposed the T-GCN model [10], which combines Graph Convolutional Networks (GCNs) and GRU to simultaneously capture both spatial and temporal dependencies in traffic data, achieving significant results.

Despite the significant achievements in exploring the spatial–temporal features of traffic networks, several challenges still remain to be addressed. Firstly, the spatial–temporal features of traffic flow are highly complex and intertwined. In the input embedding stage, previous studies typically input spatial and temporal data into separate spatial and temporal modules for feature extraction, and then fuse the extracted results in subsequent stages. However, this approach fails to effectively integrate spatial–temporal features, leading to insufficient capture and utilization of deeper information correlations. Additionally, existing multidimensional embedding methods have notable limitations, particularly the lack of effective fusion strategies. These methods typically treat data from different dimensions equally, without considering the relative importance and interdependencies of each dimension, thereby limiting the model's performance to some extent. Moreover, most early models primarily rely on a single data source (such as historical traffic flow data) for predictions, neglecting other factors that significantly influence traffic flow. For example, an increase in traffic occupancy during peak hours may lead to congestion, significantly affecting the distribution of traffic flow within the region. Ignoring these critical influencing factors may prevent the model from fully addressing the complexity of real-world scenarios, thereby reducing its reliability and practical applicability.

Secondly, traffic networks exhibit significant spatial heterogeneity. Although some nodes are spatially adjacent (e.g., different locations on the same road segment), their traffic flow and variation trends may be similar. However, as shown in Figure 1b, even when two nodes are far apart and not on the same road segment, they may still exhibit similar traffic patterns, often driven by shared traffic characteristics or demand patterns. This indicates that spatial correlations in traffic networks are not solely dependent on physical distance. However, numerous studies have demonstrated that GNNs tend to suffer from over-smoothing during node feature aggregation [11], which undermines the model's ability to capture long-range dependencies and limits its performance in modeling spatial heterogeneity.

Finally, traffic flow exhibits significant temporal heterogeneity. For example, Figure 1c shows that the traffic flow correlation between Location A and Location B is relatively low during the morning peak period, reflecting a disparity in their traffic patterns [12]. However, during the evening peak period, their traffic patterns tend to align. Conversely, as shown in Figure 1d, the traffic flow correlation between Node B and Node C is stronger during the morning peak, but weaker at other times. Furthermore, Figure 1a shows that even at the same location, traffic patterns on weekdays may differ significantly from those on non-working days. Weekday traffic flows exhibit clear periodicity, whereas non-working days lack such regularity, reflecting differences in travel behaviors on different types of days. Notably, given the significant temporal heterogeneity in traffic flow across different periods, a single temporal processing method may be insufficient to comprehensively capture its complex dynamic variations.

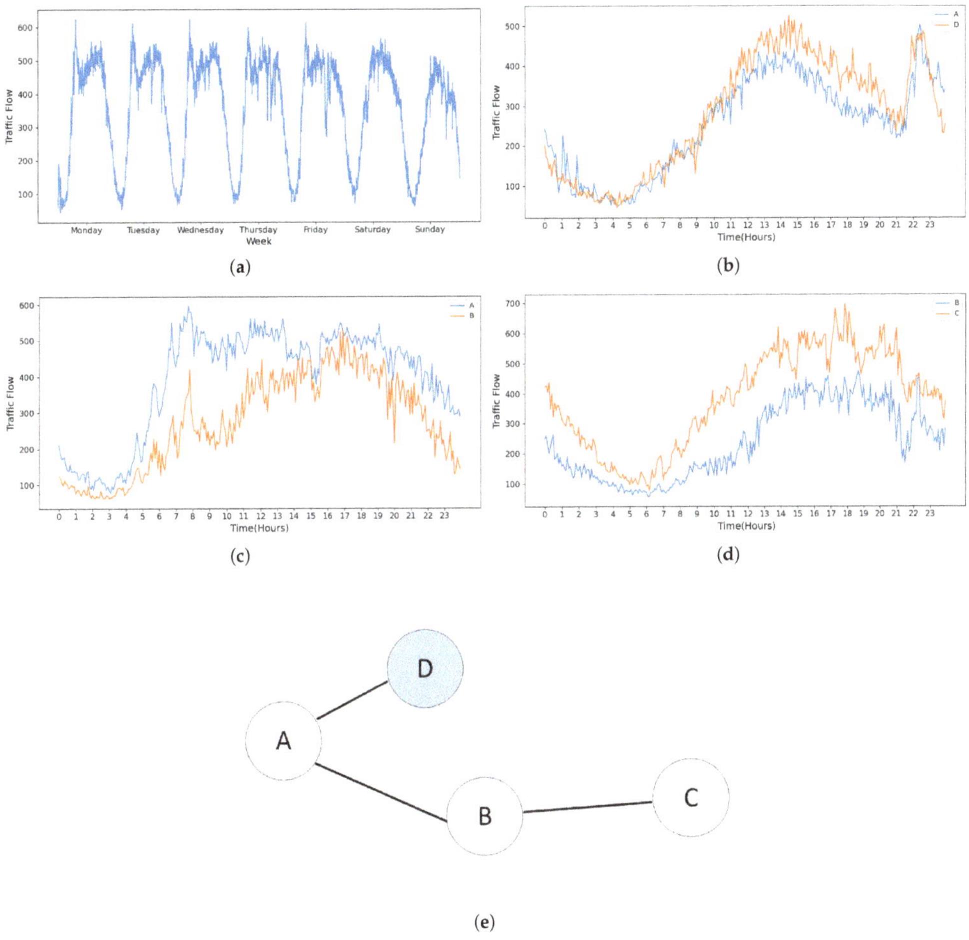

Figure 1. The discoveries about traffic data. (**a**) Periodicity of data, (**b**) Long-range spatial dependencies between A and D, (**c**) Dynamic spatial dependencies between A and B, (**d**) Dynamic spatial dependencies between B and C, (**e**) The connectivity between nodes, where D and A exhibit similar traffic flow patterns.

To address the aforementioned complex traffic flow scenarios, this paper proposes a model based on the Transformer encoder architecture—MMHFormer. To provide enriched input representations for subsequent spatial–temporal encoding, the input embedding layer utilizes gated convolution to integrate traffic features, temporal characteristics, spatial attributes, and external environmental factors. To capture spatial dependencies among traffic nodes from multiple perspectives, we introduce a hierarchical multi-view spatial attention module that explicitly models both global and local spatial dependencies. Additionally, to address cross-node temporal pattern variations, we propose a hierarchical two-stage temporal attention module. In the first stage, a temporal attention mechanism captures global temporal dependencies. In the second stage, convolution is performed at the node level as a secondary query, followed by an additional round of temporal attention to emphasize traffic pattern differences across nodes. The main contributions of MMHFormer are summarized as follows:

1. This paper proposes a novel MMHFormer model, which effectively addresses challenges in multi-source information fusion, multi-view interaction, and dynamic temporal dependency modeling.
2. MMHFormer is designed as a unified framework that integrates multi-source feature fusion, multi-view spatial reasoning, and hierarchical temporal modeling. A multi-source gated embedding layer fuses spatial Laplace, time-period, and traffic occupancy embeddings to represent complex traffic conditions. On this basis, a hierarchical multi-view spatial attention mechanism captures global, geospatial, and dynamic-similarity dependencies, while a two-stage temporal attention module learns both global temporal correlations and node-specific dynamics, enabling adaptive modeling of evolving traffic patterns across time and space.
3. Experimental results on four real-world traffic datasets show that MMHFormer outperforms current state-of-the-art methods, validating the model's effectiveness and generalization capability. Furthermore, we conducted a single-step evaluation of the hourly forecast, demonstrating the model's effectiveness in long-term prediction.

2. Related Works

2.1. Traffic Flow Prediction

Traffic flow prediction is a critical component of intelligent transportation systems, enabling the forecasting of future traffic conditions based on historical data, which significantly aids traffic management and control. Researchers typically use statistical methods, machine learning approaches, and deep learning techniques to predict traffic flow. Traditional statistical models, such as exponential smoothing, ARIMA, VAR, and historical average (HA) [13] approach traffic forecasting by assuming linear dependencies. However, they are less effective at capturing the complex nonlinear variations inherent in traffic flow.

Recognizing the limitations of linear models, researchers shifted their focus to treating traffic prediction as a nonlinear problem. Machine learning techniques, such as support vector machines (SVM) [14], Bayesian networks, autoencoders, and k-nearest neighbors (KNN) [15], were introduced to extract the correlation between data.However, these methods relied on manually crafted features, making feature engineering extremely time-consuming for large datasets, which further limited their effectiveness for large-scale traffic flow prediction.

In recent years, deep learning has gradually become the mainstream method for traffic flow prediction due to its powerful modeling capabilities. For example, Zhang et al. proposed a hybrid deep learning framework combining CNN and LSTM for short-term traffic flow forecasting [15]. The 1D CNN captures spatial features, while LSTM learns short-term variations and periodic features. Liu et al. introduced a module combining convolution and LSTM, referred to as the Conv-LSTM module, to extract spatial–temporal features from traffic flow data [16]. Furthermore, the model includes a Bidirectional LSTM (Bi-LSTM) module to capture the periodic characteristics of traffic flow. Lv et al. proposed the LC-RNN model, which employs a CNN to extract dynamic features from surrounding areas, while an RNN learns long-term temporal patterns [17]. Additionally, LC-RNN integrates periodic information and contextual factors (such as weather and holidays) and adaptively fuses this information to improve prediction accuracy. These models, primarily based on CNNs and RNNs designed for Euclidean data such as grids or sequences, struggle to model traffic networks effectively, as these networks are inherently non-Euclidean and exhibit complex, irregular topologies. Consequently, these methods fail to fully capture the spatial dependencies of road networks, leading to suboptimal performance.

2.2. Graph Convolution Network

Graph Convolutional Networks are designed to analyze non-Euclidean data. Unlike CNNs, which are limited to regular grid-based data, GCNs excel at aggregating feature information from neighboring nodes to compute node representations, effectively leveraging graph topology. Consequently, researchers have applied GCNs to traffic flow prediction to extract spatial features from non-Euclidean traffic networks. For example, Yu et al. proposed the STGCN, which leverages graph convolutional layers to capture spatial topology and temporal convolutional layers to model temporal dynamics, thereby overcoming the limitations of recurrent networks [18]. Guo et al. proposed the ASTGCN [19], which consists of three independent components that model recent, daily, and weekly dependencies in traffic flow, respectively. Each component includes a spatial–temporal attention mechanism to effectively capture dynamic spatial–temporal correlations and a spatial–temporal convolution module that combines graph convolution and standard convolution to capture spatial patterns and temporal features. Although previous studies acknowledge the importance of capturing both spatial and temporal features for traffic forecasting, most methods still rely on predefined adjacency matrices to model spatial correlations, neglecting dynamic features that evolve over time. Therefore, in recent years, some researchers have focused more on the dynamic construction of adjacency matrices to better capture spatial–temporal dependencies between nodes. Wu et al. proposed Graph WaveNet [20], which uses an adaptive dependency matrix to capture hidden spatial dependencies in the data, thereby avoiding reliance on predefined adjacency matrices. In addition, Graph WaveNet employs stacked dilated causal convolutions to capture temporal dependencies, enabling it to handle long time series. For example, Bai et al. proposed the NAPL module and the DAGG module to enhance the capabilities of GCNs [21]. The NAPL module is used to learn node-specific patterns, while the DAGG module automatically generates dependencies between nodes based on the data.

With the advancement of research, the limitations of a single adjacency matrix in capturing complex spatial dependencies have become increasingly apparent. Researchers have begun to explore the use of multiple graph adjacency matrices to more comprehensively capture spatial dependencies. Guo et al. proposed the HGCN [22], which combines micro-level and macro-level graph layers. The micro-level layer represents nodes in a road network and their connections, while the macro-level layer is constructed by clustering the micro-level layer to form larger traffic regions. HGCN uses pooling methods to achieve a hierarchical structure and introduces a dynamic transfer module that facilitates interaction between micro and macro features, thereby enhancing the model's accuracy and generalization ability in traffic forecasting. Ye et al. proposed the DMGNN [23], which constructs multiple types of spatial–temporal graphs to represent different relationships between road nodes, providing richer prior knowledge and better capturing complex relationships within the traffic system. To dynamically reflect changes between nodes, DMGNN includes a dynamic graph adjustment module that updates the adjacency matrix during each training iteration, allowing the model to adapt to temporal variations in dependency relationships. Yin et al. introduced the M-SDCGCN [24], which incorporates multiple types of graphs (including adaptive, dynamic, and static graphs) to model complex dependencies between road nodes, thus capturing interactions at different levels. The model leverages meta-learning to enhance connections between spatial–temporal features and jointly models both static and dynamic factors. Huang et al. presented the MVDGCN [25], which integrates multi-view encoder-decoder modules, dynamic relationship matrix generation, and coupled graph convolutions to capture spatial–temporal dependencies at different time scales. The model extracts traffic flow patterns from hourly, daily, and weekly perspectives, and uses a dynamic fusion module to integrate these features for more accurate traffic flow forecasting.

Beyond the aforementioned graph structures, several studies have increasingly focused on spatial–temporal modeling of heterogeneous traffic flow with heterogeneous spatial–temporal graphs. Zhong et al. [26] proposed a heterogeneous spatial–temporal graph convolution framework that constructs multiple graphs to jointly encode geographical and dynamic correlations, while simultaneously handling missing traffic data via spatial–temporal completion. Xu et al. [27] built a global heterogeneous traffic spatial–temporal graph and introduced HTSTGC to capture interactions among different traffic elements within a unified framework. More recently, Wu et al. [28] presented MHGNet, a multi-heterogeneous graph neural network that jointly represents multiple heterogeneous spatial–temporal graphs and performs prediction on clustered heterogeneous subgraphs. Collectively, these works underscore the importance of explicitly modeling heterogeneous nodes, edges, and relations in traffic networks. However, they mainly focus on structural heterogeneity at the graph level and rarely integrate rich multi-source traffic attributes or explicitly address temporal heterogeneity.

2.3. Attention Mechanism

Transformers [29], known for their powerful sequence modeling capabilities and multi-head attention mechanisms, have achieved significant success in natural language processing and have recently been introduced into the domain of traffic flow forecasting. Compared to traditional models such as GCNs and RNNs, Transformers have a greater advantage in capturing long-term spatial–temporal dependencies, especially when handling dynamic changes and complex patterns in traffic flow. For instance, Jiang et al. proposed PDFormer [30], which employs a spatial self-attention module to capture dynamic spatial dependencies and combines short-range and long-range graph masks to capture local and remote dependencies. Furthermore, PDFormer introduces a delay-aware feature transformation module to explicitly model the delay characteristics of spatial information propagation. Cai et al. introduced LCDFormer [31], which incorporates a temporal aggregation approach that aggregates and compresses historical data over long time windows, effectively retaining long-term historical information while reducing the impact of redundant data. Additionally, LCDFormer proposes a novel spatial–temporal attention module that combines topology-based local attention and node similarity-based global attention, enabling the model to capture local short-term spatial features as well as to discover long-distance dynamic spatial correlations. Geng et al. presented STGAFormer [32], which adopts a Transformer encoder architecture combined with a gated temporal self-attention module and a distance-based spatial self-attention module to extract complex spatial–temporal features. The gated temporal self-attention module enhances the extraction of both local and global temporal features, while the distance-based spatial self-attention module employs thresholding to selectively extract crucial spatial features. Li et al. proposed DDGFormer [33], which introduces direction and distance aware self-attention modules to capture relative position and directionality in traffic flow sequences. In addition, the model includes a dynamically enhanced adaptive graph convolutional network module, which improves the capture of dynamic spatial correlations in traffic systems.

Compared with the above existing studies, our proposed MMHFormer offers the following advantages and distinctions: (1) A multi-source gated embedding layer adaptively fuses spatial Laplacian, temporal periodic, and occupancy features, effectively mitigating the limitations of single-source or uniformly weighted embeddings. (2) A hierarchical multi-view spatial attention module explicitly captures global, local, and dynamic-similarity dependencies, enabling comprehensive modeling of spatial heterogeneity. (3) A two-stage temporal attention mechanism jointly models global dependencies and node-specific variations, substantially improving robustness to diverse temporal dynamics.

3. Problem Formalization

Traffic flow is defined as $\mathcal{X} = \{X_1, X_2, \ldots, X_T\} \in R^{T\times N\times C}$, where $X_t \in R^{N\times C}$ represents the observed values at time step t for N nodes with C feature dimensions. Here, $C = 1$ corresponds to the traffic flow feature dimension. Our goal is to train a function $\mathcal{F}$ that predicts the sequence for the next Q time steps. This prediction is based on the past P time steps of traffic flow data $\mathcal{X}$. Therefore, the traffic prediction problem can be formulated as:

$$Y_{(t+1)}, \ldots, Y_{(t+Q)} = \mathcal{F}(\mathcal{X}_{(t-P+1)}, \ldots, \mathcal{X}_t) \tag{1}$$

4. Methodology

Figure 2 illustrates the framework of MMHFormer, which primarily consists of three modules: the multi-source gated embedding layer, the spatial–temporal encoder layer, and the output layer.

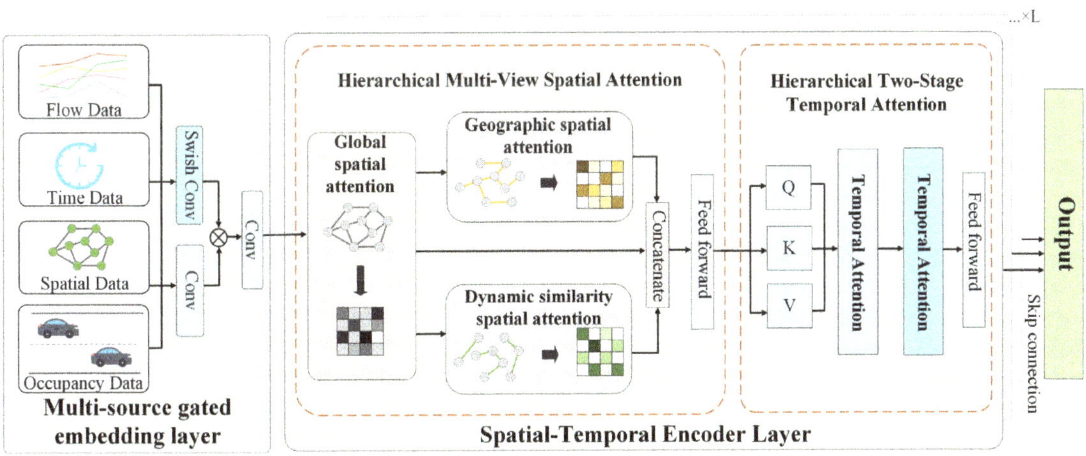

Figure 2. The overall architecture of the MMHFormer, which contains L identical layers.

To effectively capture the spatial–temporal features of traffic flow while considering road conditions. The multi-source gated embedding layer integrates multidimensional data inputs, including raw traffic flow, spatial Laplacian embeddings, temporal periodic embeddings, and traffic occupancy embeddings, using gated convolutions.

The spatial–temporal encoder layer employs a hierarchical multi-view spatial attention module. A global spatial attention mechanism captures global spatial dependencies between nodes. To enhance sensitivity to critical local information, a geospatial attention mechanism focuses on the local dynamic features of neighboring nodes while discarding connections to distant nodes. Additionally, dynamic similarity spatial attention is incorporated, allowing the model to ignore distance and extract long-range spatial features by capturing the dynamic similarity of traffic patterns between nodes.

To adapts to the varying traffic patterns across different nodes. In the hierarchical two-stage temporal attention module, the first stage captures global temporal dependencies, while the second stage refines the query focus and re-applies temporal attention to emphasize traffic pattern differences between nodes.

The output layer employs skip connections and convolutional layers to transform the outputs into the final dimensions required for prediction. In this section, we provide a detailed description of the MMHFormer architecture.

4.1. Multi-Source Gated Embedding Layer

At each time interval, traffic flow in different urban areas is influenced by various factors, including traffic flow in neighboring areas and external environmental conditions. For instance, congestion in one area can lead to a sudden decrease in traffic flow across the city. Similarly, during holidays, urban traffic flow often increases significantly compared to weekdays. Based on this observation, we propose a multi-source gated embedding layer that integrates multidimensional data to capture the spatial–temporal dependencies and the impact of road conditions on traffic flow.

Specifically, the raw traffic flow input X is transformed into $X_f \in R^{T\times N\times d_f}$ via a Linear layer and combined with three types of embeddings: spatial Laplacian embeddings to encode the spatial structural features of the road network, temporal periodic embeddings to capture the periodic variations in traffic flow, and traffic occupancy embeddings to reflect external environmental impacts. Finally, a gated convolutional network achieves efficient fusion of multi-source information, effectively integrating the spatial, temporal, and external influences on traffic flow.

Spatial–temporal embedding: The Laplacian matrix is used to learn the correlations between nodes in the road network, embedding the graph into Euclidean space to obtain the spatial embedding representation $X_{spe} \in R^{N\times d_{st}}$. Considering the periodicity of urban traffic flow, we introduce weekly and daily periodic embeddings [30], represented as $t_w(T)$ and $t_d(T) \in R^d$, where $w(T)$ and $d(T)$ convert time t into week and minute indices. Temporal embeddings X_w and $X_d \in R^{T\times d_{st}}$ are added to the spatial embeddings X_{spe} to obtain the final spatial–temporal representation $X_{st} \in R^{T\times N\times d_{st}}$.

Traffic occupancy embedding: Variations in traffic flow and occupancy across different nodes may reflect the characteristics of the roads in those regions (e.g., major or minor roads), providing the model with the ability to recognize differences between nodes. To achieve this, we introduce the traffic occupancy embedding mechanism. Initially, raw traffic occupancy data is processed to extract key features, and a linear layer is applied to generate the traffic occupancy embedding representation $X_o \in R^{T\times N\times d_o}$.

Information fusion: The traffic flow input representation X_f, spatial–temporal embedding X_{st}, and traffic occupancy embedding X_o are concatenated to produce the fused representation $X_e \in R^{T\times N\times d}$, where $d = d_f + d_{st} + d_o$ is the hidden dimension:

$$X_{fus} = concat(X_f, X_{st}, X_o) \tag{2}$$

$$X_e = Conv(Conv(X_{fus}) \odot silu(Conv(X_{fus}))) \tag{3}$$

$$X_e = X_e + PE(X_e) \tag{4}$$

The gated convolution mechanism dynamically adjusts feature weights based on different input data, enabling more comprehensive fusion. Here, *Conv* denotes 1×1 convolution, *PE* represents temporal positional encoding, *silu* denotes the Sigmoid Linear Unit, and $\odot$ represents the Hadamard product.

4.2. Hierarchical Multi-View Spatial Attention Module

As illustrated in Figure 3, global spatial attention is first introduced to capture the dependencies between traffic nodes and all other nodes. For the input feature representation $X_e \in R^{T\times N\times d}$, query, key, and value matrices are generated at each time step t through convolutional operations as follows:

$$
\begin{aligned}
Q_t^{(S1)} &= X_{e(t::)}W_Q, \quad K_t^{(S1)} = X_{e(t::)}W_K, \\
V_t^{(S1)} &= X_{e(t::)}W_V
\end{aligned}
\tag{5}
$$

Here, $W_Q, W_K, W_V \in R^{d\times d'}$ are learnable parameter matrices, where d' is the dimensionality of the queries, keys, and values. The attention score matrix $A^{(S1)}$ is computed as:

$$
A^{(S1)} = \frac{Q_t^{(S1)}K_t^{(S1)T}}{\sqrt{d'}}
\tag{6}
$$

The softmax function is applied to $A^{(S1)}$, generating attention weights for each node relative to all other nodes. The global spatial representation $X_s^{(Glo)}$ is then obtained:

$$
X_s^{(Glo)} = \text{softmax}(A^{(S1)})V_t^{(S1)}
\tag{7}
$$

In practical traffic scenarios, the dynamic features of a target node are often significantly influenced by its neighboring nodes, particularly during abrupt changes. To better capture the local dynamic characteristics of nodes, geospatial attention is introduced. This mechanism focuses on information from neighboring nodes while discarding connections with distant nodes. Based on the global spatial attention output $X_s^{(Glo)}$, a new query $Q_t^{(S2)}$ is generated at time step t using a 1×1 convolution and treated as $V_t^{(S2)}$. The secondary attention weight $A^{(S2)}$ is computed as:

$$
A^{(S2)} = \frac{Q_t^{(S2)}(Q_t^{(S1)})^T}{\sqrt{d'}}
\tag{8}
$$

To ensure that the attention mechanism focuses only on nearby nodes, a geospatial mask matrix M_{geo} is defined. The matrix is undirected, and only nodes with distances less than λ are considered for extracting important features. The geospatial representation $X_s^{(Geo)}$ is obtained as:

$$
X_s^{(Geo)} = \text{softmax}(A^{(S2)} \odot M_{geo})V_t^{(S2)}
\tag{9}
$$

Although some traffic nodes are not geographically adjacent, they may exhibit similar traffic patterns driven by shared traffic characteristics or demand patterns. Dynamic similarity spatial attention is introduced to dynamically adjust attention weights based on the similarity of traffic patterns between nodes, identifying potential correlations among non-adjacent nodes. We use a sliding 12-step time window to analyze traffic data and construct a dynamic similarity mask matrix M_{dyn} using the fast Fourier transform (FFT). Specifically, we apply FFT to the past 12 steps of each node to extract short-term periodicity and local oscillation patterns, and then compute the Euclidean distances between the resulting frequency-domain features. This matrix encodes the relationships between each node and its K-most similar nodes, with a weight of 1 assigned to the most similar nodes and 0 to others.

The choice of a 12 step window is driven by the forecasting task, which involves predicting the next 12 steps based on the previous 12 steps. A window of 12 steps corresponds to one hour of traffic data, offering a sufficient segment for capturing stable frequency components, while also being short enough to account for transient variations and peak-hour fluctuations.

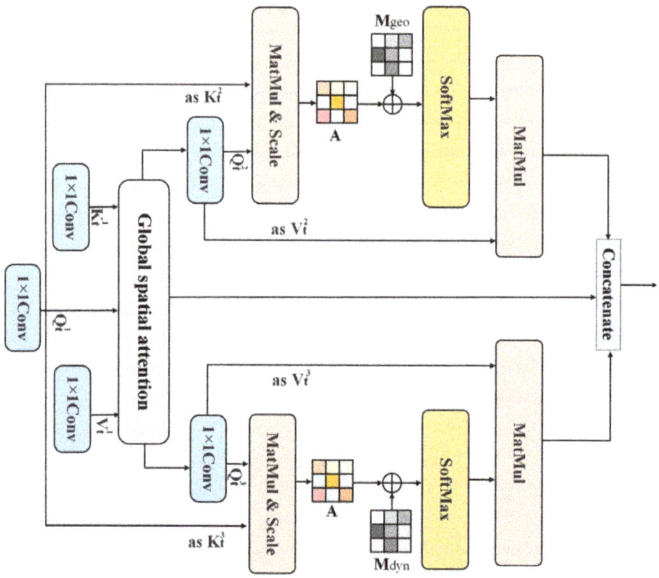

Figure 3. Hierarchical multi-view spatial attention module.

Based on $X_s^{(Glo)}$, a new query $Q_t^{(S3)}$ is generated at time step t using a 1×1 convolution and treated as $V_t^{(S3)}$. Under the dynamic similarity mask, the dynamic similarity attention weight $A^{(S3)}$ is calculated as:

$$A^{(S3)} = \frac{Q_t^{(S3)}(Q_t^{(S1)})^T}{\sqrt{d'}} \tag{10}$$

The dynamic similarity representation $X_s^{(Dyn)}$ is given by:

$$X_s^{(Dyn)} = \text{softmax}(A^{(S3)} \odot M_{dyn})V_t^{(S3)} \tag{11}$$

Finally, the model integrates global spatial, geospatial, and dynamic similarity attention mechanisms to model multi-perspective spatial dependencies. The aggregated spatial representation is expressed as:

$$X_s = \text{Concat}\left(X_s^{(\text{Dyn})}, X_s^{(\text{Geo})}, X_s^{(\text{Glo})}\right)W_F \tag{12}$$

$$X_s = \text{FFN}(\text{LN}(X_s)) + X_s \tag{13}$$

Here, $W_F \in R^{3d' \times d}$ is a learnable parameter. The feed-forward network (FFN) performs nonlinear transformations with two linear layers and a GELU activation function, while LN denotes the layer normalization operation.

4.3. Hierarchical Two-Stage Temporal Attention

4.3.1. Global Temporal Attention

In the first stage, to capture global dynamic temporal patterns, we first project the input tensor $X_s \in \mathbb{R}^{T\times N\times d}$ into temporal query, key, and value matrix using learnable 1×1 convolutions:

$$Q_t^{(T)} = X_{s(t,:,:)} W_Q^{(T)}, \quad K_t^{(T)} = X_{s(t,:,:)} W_K^{(T)}, \quad V_t^{(T)} = X_{s(t,:,:)} W_V^{(T)} \tag{14}$$

Here, $W_Q^{(T)}, W_K^{(T)}, W_V^{(T)} \in R^{d\times d}$ are learnable parameters.

Next, we compute the scaled dot-product between the queries and keys at each node to obtain the temporal attention scores. The output of the first-stage temporal attention module is then calculated by applying the attention scores to the values:

$$X_t^{(T1)} = \mathrm{softmax}(\frac{Q_t^{(T)} (K_t^{(T)})^\top}{\sqrt{d}}) V_t^{(T)}. \tag{15}$$

4.3.2. Spatially-Informed Temporal Attention

Time series data not only have long-term dependence, but also show complex cross-node changes due to the heterogeneity and dynamics of time series characteristics of each node. It is difficult to describe this heterogeneous evolution with a single time attention. To this end, as shown in Figure 4, we propose the spatially-informed temporal attention Module. Attention is combined with adaptive graph convolution to model the dynamic changes between nodes. Specifically, adaptive graph convolution dynamically adjusts the adjacency matrix by learning, fuses spatial topology information with the dynamic interaction between nodes, and can flexibly model the interdependence between different nodes. In the case of large cross-time changes between nodes, it can effectively distinguish and strengthen the timing pattern at critical moments. This mechanism enables the model not only to capture the dynamic evolution in time, but also to identify the temporal pattern differences between nodes, which further enhances the sensitivity to cross-node changes.

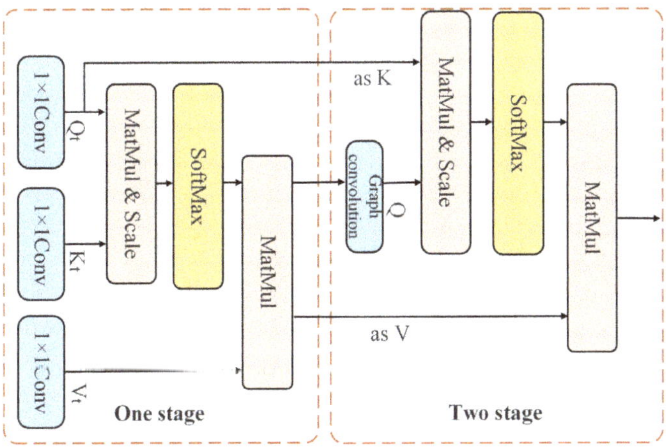

Figure 4. Cross-Node temporal attention module.

In the second stage, based on the output of the first stage, we capture temporal patterns that span multiple nodes by incorporating adaptive graph learning to infer latent inter-node dependencies. These learned dependencies are then used to compute the query matrix through a graph convolution step.

$$G = \text{softmax}(\text{ReLU}(E\,E^{\top})) \tag{16}$$

$$Q_t^{(T2)} = GX_t^{(T1)}W_Q^{(T2)} \tag{17}$$

Here, $W_Q^{(T2)} \in R^{d\times d}$ is a learnable projection matrix, and $E \in R^{N\times a}$ denotes the learnable node embeddings.

In this stage, the first stage key $K_t^{(T)}$ and output $X_t^{(T1)}$ serve as the key K and value V, respectively, to compute the secondary temporal attention weights. The weight matrix is normalized and applied to $X_t^{(T1)}$, which passes through a linear layer and an FFN to generate the final temporal context $X_t^{(T2)}$, completing the two stage aggregation of temporal features:

$$X_t^{(T2)} = \text{softmax}\left(\frac{Q_t^{(T2)}(K_t^{(T)})^{\top}}{\sqrt{d}}\right)X_t^{(T1)}. \tag{18}$$

$$X_t = \text{FFN}(\text{LN}(X_t^{(T2)})) + X_t^{(T2)} \tag{19}$$

4.4. Output Layer

To resize the spatial–temporal encoder output, we apply 1 × 1 convolution skip connections, summing each layer's output to obtain $X_h \in R^{T\times N\times d_{sk}}$.

$$\widehat{X} = Conv2(Conv1(X_h)) \tag{20}$$

For multi-step prediction, $Conv1$ and $Conv2$ adjusts X_h to the prediction window dimension, reducing error and enhancing model effectiveness.

5. Experiments

5.1. Datasets

The datasets used for the experiments in this paper include PeMS03, PeMS04, PeMS07, and PeMS08, which are all derived from the California State Traffic Management System (CSTMS) and collected by the California Department of Transportation (Caltrans) through its traffic monitoring network. The data cover highway traffic conditions across various regions, collected every 5 min by real-time sensors deployed on the roadways, providing multidimensional features such as traffic volume, speed, and roadway occupancy for time-efficient monitoring of traffic flow. These multidimensional features serve as inputs to the traffic flow prediction model, which effectively captures the traffic patterns and dynamic changes in each region, revealing regional traffic differences. Due to its large size and high frequency, the PeMS dataset has become a cornerstone of ITS research, providing a valuable testbed for data-driven traffic decision-making and system optimization. Specific information about each dataset is detailed in Table 1.

Table 1. The detailed information of datasets.

Dataset	Sensors	Edges	Time Range	Time Steps
PeMS03	358	547	September 2018–November 2018	26,208
PeMS04	307	340	January 2018–February 2018	16,992
PeMS07	883	866	May 2017–August 2017	28,224
PeMS08	170	295	July 2016–August 2016	17,856

5.2. Baselines

We compare MMHFormer with the following baseline models: traditional time series forecasting models, graph neural network-based models, and transformer-based models.

1. VAR [4]: VAR is based on the assumption that traffic flow follows an autoregressive pattern, meaning that future values can be predicted using past data in the series.
2. SVR [5]: SVR leverages the principles of Support Vector Machines (SVMs) to effectively model traffic flow in a nonlinear way, unlike traditional linear regression methods.
3. DCRNN [34]: DCRNN models traffic flow as a diffusion process on a directed graph, capturing spatial dependencies through bidirectional random walks. It also models temporal dependencies using a sequence-to-sequence structure combined with scheduled sampling.
4. GraphWaveNet [20]: GraphWaveNet automatically generates the graph adjacency matrix by adaptively learning node embeddings, which enables more accurate capturing of hidden spatial dependencies. Additionally, it utilizes stacked dilated causal convolutions to effectively handle long-term temporal dependencies.
5. AGCRN [21]: AGCRN introduces the NAPL and DAGG modules, which are designed to capture node-specific patterns and automatically infer interdependencies across different traffic series. This enables AGCRN to effectively capture fine-grained spatial and temporal correlations within traffic series data, enhancing its ability to model complex traffic dynamics.
6. STGCN [18]: STGCN is a model that combines graph convolutions to capture spatial dependencies between traffic nodes and 1D convolutions to model temporal dynamics, effectively handling the complex spatio-temporal correlations in traffic flow prediction.
7. MTGNN [35]: MTGNN exploits the underlying spatio-temporal dependencies by automatically learning the relationships between variables. The framework includes a graph learning layer, a graph convolution module, and a temporal convolution module, which adaptively learns the graph structure to capture spatial dependencies between variables and integrates multi-frequency temporal patterns to enhance prediction performance.
8. GMAN [36]: GMAN adopts an encoder-decoder structure, with both the encoder and decoder comprising multiple spatial–temporal attention blocks to model the complex spatio-temporal correlations of traffic systems. By incorporating a transformer attention mechanism, GMAN aims to reduce error propagation in long-term forecasting.
9. ASTGCN [19]: ASTGCN consists of three independent components that model the recent, daily, and weekly dependencies of traffic flow. Each component includes spatial–temporal attention mechanisms and spatial–temporal convolution modules to capture the dynamic spatio-temporal correlations and features of traffic data.
10. STFGNN [37]: STFGNN is a novel spatiotemporal fusion graph neural network for traffic flow prediction. Using data-driven "temporal graphs" to complement traditional spatial graphs, it effectively captures hidden spatiotemporal dependencies. STFGNN's fusion operation processes multiple spatial and temporal graphs in parallel, integrating with a gated convolution module to handle long sequences and capture richer dependencies.
11. STID [38]: By adding spatial and temporal identity information, this model addresses the issue of sample indistinguishability in the spatio-temporal dimensions, thereby enhancing the model's predictive capability.
12. GDGCN [39]: GDGCN systematically explores spatial, temporal, and feature dimensions of data by combining parameter-sharing and independent modules. It designs

a novel temporal graph convolution block to process the dynamic relationships of historical time slices in graph form. Additionally, a dynamic graph constructor is introduced to model time-specific spatial dependencies and the dynamic interaction relationships between different time slices.

13. PDFormer [30]: PDFormer addresses the limitations of current GNN-based models in static modeling, short-range spatial dependencies, and the neglect of propagation delays. PDFormer introduces a dynamic spatial self-attention module to capture dynamic spatial dependencies and employs both geographic and semantic graph mask matrices to simultaneously capture short-range and long-range dependencies.
14. DDGformer [33]: DDGformer captures the directional and relative positional relationships in traffic data using a direction- and distance-aware self-attention module. It also uses a dynamically enhanced adaptive graph convolution network to capture dynamic patterns in traffic systems.
15. STGAFormer [32]: STGAFormer effectively integrates both local and global dynamic spatio-temporal features and employs a distance-based self-attention module to capture critical features between different regions. The model incorporates multidimensional inputs, including traffic flow attributes, periodicity, proximity adjacency matrices, and adaptive adjacency matrices, to better capture the spatio-temporal characteristics of traffic flow.

5.3. Experimental Settings

In this experiment, data from the past 12 time steps (one hour) are used to predict the traffic flow for the next 12 time steps, based on current mainstream traffic flow prediction methods. The dataset is divided into training, validation, and test sets in the ratio of 6:2:2 to fully evaluate the generalization ability of the model. The experiments were conducted on an NVIDIA RTX 4090 GPU (NVIDIA, Santa Clara, CA, USA). The batch size was set to 16 for PeMS03, PeMS04, and PeMS08, and to 6 for PeMS07, with 200 epochs of training to ensure sufficient model convergence and to avoid overfitting. The model's spatial–temporal encoder comprises 6 layers (L), with the hidden dimension (d) set to 64. The optimizer uses AdamW with an initial learning rate of 0.001 and combines the weight decay mechanism to effectively prevent overfitting. In addition, overfitting is further prevented during training by employing an early stopping criterion, which halts training once the validation performance stops improving.

5.4. Evaluation Metrics

To comprehensively assess the model performance, we introduced three commonly used evaluation metrics: mean absolute error (MAE), root mean square error (RMSE), and mean absolute percentage error (MAPE) . To ensure the accuracy of the results, we filtered the missing data.

$$MAE = \frac{1}{n}\sum_{i=1}^{n}|y_i - \hat{y}_i| \tag{21}$$

$$RMSE = \sqrt{\frac{1}{n}\sum_{i=1}^{n}(y_i - \hat{y}_i)^2} \tag{22}$$

$$MAPE = \frac{1}{n}\sum_{i=1}^{n}\left|\frac{y_i - \hat{y}_i}{y_i}\right| \times 100\% \tag{23}$$

Here, y_i and $\hat{y}_i$ denote the actual and predicted traffic flow values at node i for a given time step, respectively. The variable n represents the total number of nodes.

5.5. Comparison Results

Table 2 presents the performance of our proposed MMHFormer model compared to other baseline models on next-hour predictions across four real-world datasets. The best-performing results are highlighted in bold, while the second-best results are underlined. To validate the superiority of the model, most baseline results were obtained from official documentation and related studies [28]. Our model achieved the best performance across all metrics on the four datasets: PeMS03, PeMS04, PeMS07, and PeMS08.

To evaluate the long-term prediction capability of the model, we visualized its MAE, RMSE, and MAPE across the next 12 steps, as shown in Figure 5. Compared to other models, our model demonstrates smaller variations in all metrics throughout the prediction horizon, highlighting its superior stability and robustness. This consistent performance further confirms its suitability for long-term prediction tasks.

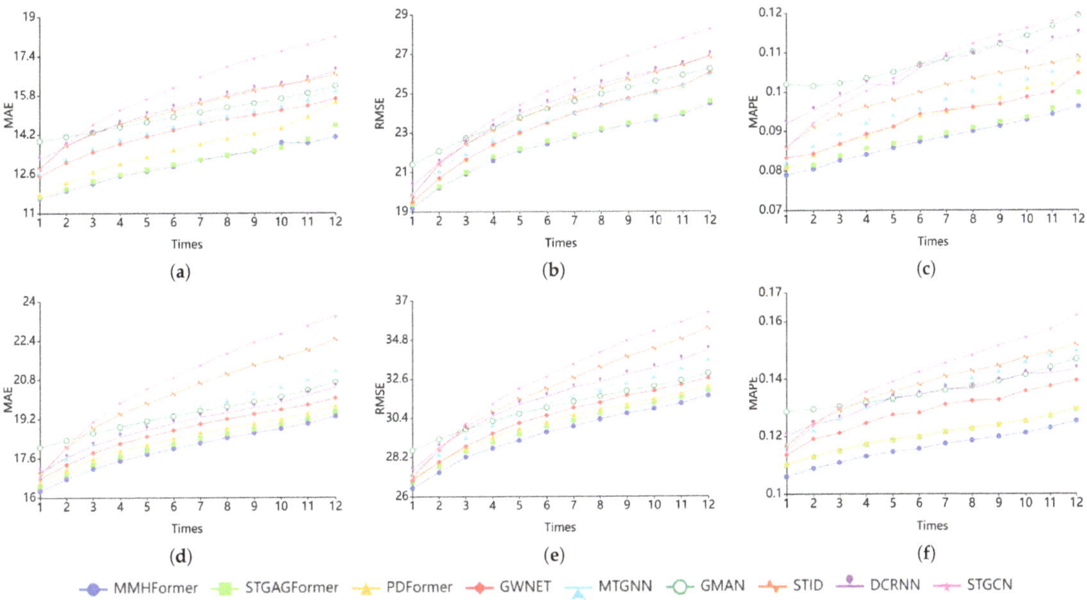

Figure 5. Comparison of single-step prediction on different datasets. (**a**) MAE on PeMS08, (**b**) RMSE on PeMS08, (**c**) MAPE on PeMS08, (**d**) MAE on PeMS04, (**e**) RMSE on PeMS04, (**f**) MAPE on PeMS04.

DDGformer, PDFormer, and STGAFormer are three representative attention-based models. However, they fuse features across different dimensions through simple addition during data input embedding. In contrast, MMHFormer employs a concatenation approach across dimensions, which not only preserves the complete representation of spatial–temporal features but also incorporates traffic occupancy as an additional data source, enabling a more accurate representation of real-world traffic conditions. Additionally, the introduction of a gating mechanism allows the model to dynamically adjust its focus on each dimension under different scenarios, thereby effectively retaining and utilizing information from each dimension.

Although PDFormer and STGAFormer are designed to address short-range and long-range spatial correlations, their focus is limited to local spatial relationships, failing to adequately capture global spatial features. In contrast, MMHFormer introduces two masking matrices to capture global, local, and long-distance dynamic similarity patterns of nodes simultaneously, enabling multi-view modeling of spatial features. Furthermore, in the temporal dimension, other models use single temporal attention to capture temporal features, which may be insufficient to adapt to significant temporal heterogeneity. In contrast, MMHFormer employs a hierarchical two-stage temporal attention mechanism to flexibly adjust to the temporal features of different nodes, dynamically adapting to varying traffic patterns.

To further demonstrate the effectiveness and practicality of MMHFormer, the computational complexity of the model is also analyzed in this section. The computational complexity of each module in STGAFormer's spatial–temporal encoder is analyzed as follows: the distance-based spatial self-attention module requires $O(4N^2 d_{\text{model}} + 2N^2)$ operations, the gated temporal self-attention module consumes $O(2T^2 d_{\text{model}} + T^2)$, and the position-wise feed-forward network contributes $O(NTd^2_{\text{model}})$. Consequently, the overall time complexity of STGAFormer's spatial–temporal encoder amounts to $O(l(4N^2 d_{\text{model}} + 2T^2 d_{\text{model}} + T^2 + 2N^2 + NTd^2_{\text{model}}))$, where l denotes the number of encoding layers, T represents the number of time steps, N indicates the number of sensors in the traffic network, and d_{model} corresponds to the hidden dimension size. In the MMHFormer framework, the hierarchical multi-view spatial attention module has a complexity of $O(6N^2 d_{\text{model}} + 3N^2)$. The hierarchical two-stage temporal attention mechanism requires $O(2T^2 d_{\text{model}} + 2T^2)$ operations. The feed-forward network maintains $O(NTd^2_{\text{model}})$ complexity, resulting in a total time complexity of $O(l(6N^2 d_{\text{model}} + 2T^2 d_{\text{model}} + 2T^2 + 3N^2 + NTd^2_{\text{model}}))$.

While MMHFormer demonstrates a moderately higher theoretical complexity compared to the best-performing baseline, STGAFormer, primarily due to its multi-view spatial attention design, this increased computational overhead is strategically justified. The hierarchical architecture enables more comprehensive spatial–temporal representation learning by simultaneously capturing global, local, and dynamic similarity patterns. The model remains practically feasible for real-world applications, as the polynomial complexity is manageable for typical urban traffic networks where N and T are constrained by physical infrastructure. The additional computational investment results in significant improvements in prediction accuracy and model interpretability, as shown by the experimental results across multiple benchmark datasets.

Table 2. Comparison Table of Model Performance. Bold: best; underline: second best.

Model	PeMS03			PeMS04			PeMS07			PeMS08		
	MAE	RMSE	MAPE (%)	MAE	RMSE	MAPE (%)	MAE	RMSE	MAPE (%)	MAE	RMSE	MAPE (%)
SVR [5]	27.40	26.46	44.51	28.66	44.59	19.15	32.97	50.15	15.43	23.25	36.15	14.71
VAR [4]	23.65	38.26	24.51	24.54	38.61	17.24	50.22	75.63	32.22	19.19	29.81	13.10
DCRNN [34]	17.99	30.31	18.34	21.22	33.44	14.17	25.22	38.61	11.82	16.82	26.36	10.92
GraphWaveNet [20]	19.12	32.77	18.89	39.66	31.72	17.29	26.39	41.50	11.97	18.28	30.05	12.15
AGCRN [21]	15.98	28.25	15.23	19.83	32.26	12.97	22.37	36.55	9.12	15.95	25.22	10.09
STGCN [18]	17.55	30.42	17.34	21.16	34.89	13.83	25.33	39.34	11.21	17.50	27.09	11.29
MTGNN [35]	15.85	26.23	15.55	19.08	31.56	12.96	20.82	34.09	9.03	15.40	24.93	10.17
ASTGCN [19]	17.34	29.56	17.21	22.93	35.22	16.56	24.01	37.87	10.73	18.25	28.06	11.64
STFGNN [37]	16.77	28.34	16.30	19.83	31.88	13.02	22.07	35.80	9.21	16.64	26.22	10.60
GDGCN [39]	14.66	**24.30**	**13.94**	18.44	29.79	12.52	20.15	33.21	8.50	14.82	23.87	9.35
GMAN [36]	16.87	27.92	18.23	19.14	31.60	13.19	20.96	34.10	9.05	15.31	24.92	10.13
STID [38]	15.33	27.40	16.40	18.29	29.82	12.49	19.54	32.82	8.25	14.20	23.49	9.28
PDFormer [30]	14.94	25.39	15.82	18.32	29.97	12.10	19.83	32.87	8.53	13.58	23.51	9.05
DDGFormer [33]	15.01	24.95	15.89	18.04	30.06	11.76	18.99	32.25	7.93	13.37	23.15	8.83
STGAFormer [32]	14.56	24.94	14.69	18.18	29.78	11.98	19.65	32.62	8.45	13.06	22.43	8.87
MMHFormer (ours)	**14.33**	24.80	14.68	**17.99**	**29.53**	**11.64**	**18.74**	**32.10**	**7.79**	**12.94**	**22.30**	**8.77**

5.6. Long-Range Forecasting

To further evaluate the long-horizon forecasting capability of different models, we conduct experiments on the PEMS04 and PEMS08 datasets under an extended prediction setting. Specifically, we increase the prediction horizons to 36, 48, and 60 steps and report the performance at each horizon as well as their average.

When the horizon is extended to 60 steps, models that perform competitively at short horizons—such as DCRNN, Graph WaveNet, and MTGNN—exhibit pronounced degradation in MAE, RMSE, and MAPE. In contrast, attention-based architectures such as GMAN and the Transformer-based PDFormer show greater robustness and consistently outperform the recurrent and convolutional baselines, highlighting the advantage of explicitly modeling global temporal dependencies for long-range traffic forecasting. The detailed numerical results are summarized in Table 3.

Compared with these baselines, MMHFormer achieves the lowest errors across all long horizons on both PEMS04 and PEMS08. On PEMS04, it reduces the average long-horizon MAE by about 4% compared with PDFormer, while on PEMS08 it yields relative improvements of around 8–9% at the most challenging 60-step horizon. These results demonstrate that the multi-source gated embedding, hierarchical multi-view spatial attention, and two-stage temporal attention enable MMHFormer to effectively capture long-range spatial–temporal dependencies, resulting in consistently superior long-horizon prediction performance.

Table 3. Performance on long-range traffic flow forecasting. Bold: best; underline: second best.

Dataset	Method	@Horizon 36			@Horizon 48			@Horizon 60			Average		
		MAE	RMSE	MAPE	MAE	RMSE	MAPE	MAE	RMSE	MAPE	MAE	RMSE	MAPE
PEMS04	DCRNN [34]	23.67	37.62	16.03%	24.47	38.86	16.81%	25.39	39.92	18.10%	22.79	36.19	15.70%
	STGCN [18]	26.85	41.78	18.32%	27.27	42.40	18.74%	28.90	44.54	19.65%	26.66	41.58	18.30%
	GraphWaveNet [20]	24.25	39.13	17.94%	24.32	39.29	18.01%	24.98	39.97	18.48%	24.26	39.20	17.78%
	MTGNN [35]	23.18	37.12	15.63%	23.78	37.80	16.17%	25.11	39.16	17.65%	22.37	35.88	15.25%
	AGCRN [21]	23.03	36.76	15.72%	23.47	37.65	16.38%	24.93	39.38	17.87%	22.29	35.58	15.72%
	GMAN [36]	22.45	36.91	16.07%	22.72	39.13	16.33%	23.20	40.48	17.00%	22.26	35.66	15.93%
	PDFormer [30]	22.18	35.57	15.26%	22.45	36.40	15.33%	27.64	37.73	15.91%	21.34	34.55	14.60%
	MMHFormer (ours)	**21.01**	**35.38**	**14.73%**	**21.49**	**35.19**	**14.95%**	**22.28**	**36.11**	**15.53%**	**20.43**	**33.34**	**14.43%**
PEMS08	DCRNN [34]	20.48	31.73	13.73%	21.38	33.03	14.56%	23.12	35.22	16.35%	20.02	30.93	13.58%
	STGCN [18]	27.25	41.28	17.02%	27.63	41.97	17.27%	28.80	43.55	18.24%	26.83	40.86	16.77%
	GraphWaveNet [20]	21.27	35.06	13.75%	21.54	35.76	13.96%	22.01	36.38	14.95%	21.00	34.63	13.63%
	MTGNN [35]	19.57	31.55	13.14%	20.45	32.74	14.19%	22.10	34.57	15.95%	18.86	30.36	12.52%
	AGCRN [21]	20.06	31.92	14.39%	20.49	32.71	14.56%	21.58	34.04	14.80%	19.10	30.56	12.99%
	GMAN [36]	17.69	30.69	13.75%	18.13	31.20	14.06%	18.95	32.73	14.52%	17.48	30.14	13.56%
	PDFormer [30]	17.61	28.98	12.22%	18.24	29.87	12.80%	19.52	31.40	14.12%	16.93	28.05	11.83%
	MMHFormer (ours)	**16.89**	**28.72**	**11.54%**	**17.29**	**29.73**	**11.96%**	**17.98**	**31.18**	**12.78%**	**16.26**	**27.49**	**11.11%**

5.7. Ablation Study

The MMHFormer model comprises three key modules: the multi-source gated embedding layer, the hierarchical multi-view spatial attention module, and the hierarchical two-stage temporal attention module. To validate the effectiveness of each component, we conducted ablation experiments on the PeMS08 dataset We compared MMHFormer against the following variants:

- MMHFormer w/o multi-source embedding: This variant removes all auxiliary spatial, temporal, and traffic occupancy embeddings, using only raw traffic flow data as input.
- MMHFormer w/o gated embedding: This variant retains all the embedding sources but removes the gating mechanism, instead fusing the features through simple addition.

- MMHFormer w/o hierarchical multi-view spatial attention: The multi-view spatial attention mechanism is replaced by a global spatial attention mechanism for direct spatial feature extraction.
- MMHFormer w/o hierarchical two-stage temporal attention: The second stage temporal attention structure is removed, and only the first stage temporal attention mechanism is retained.

Figure 6 illustrates the comparison results of these variants. We can observe that removing the multi-source embedding leads to a significant increase in MAE, RMSE, and MAPE, highlighting the importance of integrating diverse embeddings, such as spatial Laplacian, temporal periodic, and traffic occupancy, in effectively capturing spatial–temporal dependencies. Replacing the gated embedding mechanism with simple feature addition results in a slight performance decline, suggesting that the gating mechanism is crucial for dynamically adjusting feature contributions, thereby improving model accuracy.

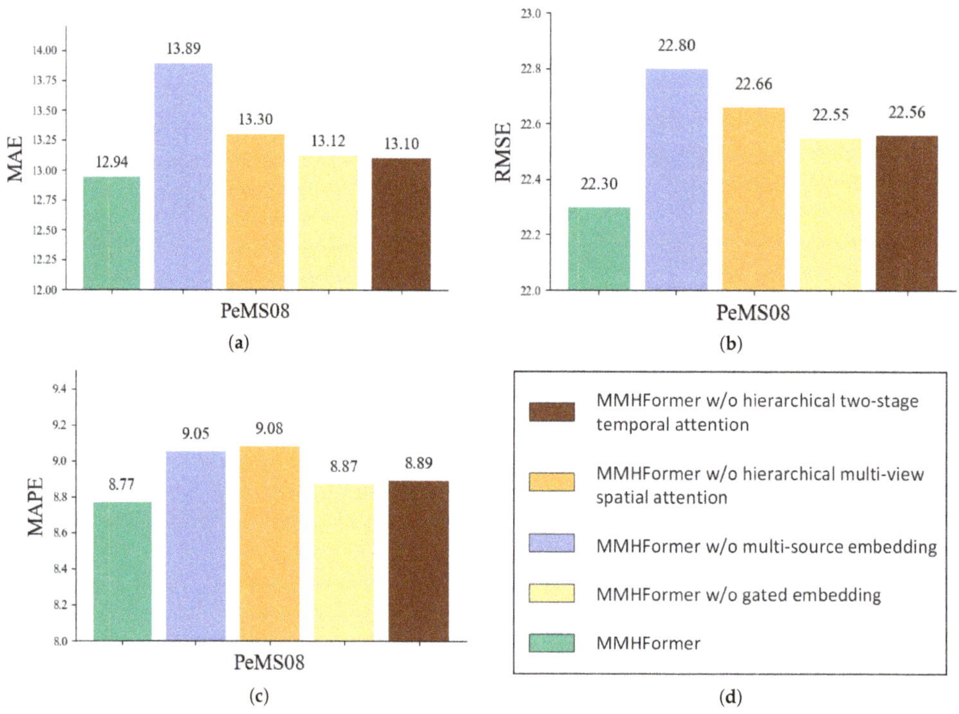

Figure 6. Ablation study of key designs in MMHFormer. (**a**) MAE on PeMS08, (**b**) RMSE on PeMS08, (**c**) MAPE on PeMS08, (**d**) Representations of connections between colors and variants.

The inclusion of the hierarchical multi-view spatial attention module enables MMHFormer to account for global spatial dependencies while also capturing geographic and dynamic similarity-based spatial dependencies. When this module is removed, the RMSE, MAE, and MAPE metrics increase, underscoring the importance of multi-view spatial information for modeling the spatial dependencies of traffic flow.

The hierarchical two-stage temporal attention module also proves highly impactful. Removing the second stage temporal attention structure results in increased RMSE, MAE, and MAPE values, demonstrating its effectiveness in capturing temporal dependencies within traffic flows. This module is particularly advantageous in addressing sudden traffic flow fluctuations and varying temporal patterns across nodes. The second-stage temporal attention mechanism, in particular, enhances the model's adaptability to diverse traffic patterns at different nodes.

5.8. Parameter Sensitivity Analysis

To further investigate the influence of different parameter settings on our proposed model for the traffic forecasting task, we conduct a parameter sensitivity analysis for MMHFormer. Specifically, we explored various values for each hyperparameter within predefined search spaces: [4–7] for the geographic distance threshold λ, and [6–9] for the number of nearest neighbors K based on dynamic similarity mask matrix. These hyperparameters are pre-trained using only the training set data, ensuring no information leakage during the training or inference processes. This analysis allowed us to evaluate the impact of different configurations on the performance of our MMHFormer model.

The results, as shown in Figure 7, reveal the following observations: (1) The geographic distance threshold $\lambda = 5$ best preserves spatial dependencies, enabling the model to capture relevant relationships without unnecessary complexity. Smaller values (e.g., $\lambda = 4$) place too much emphasis on local dependencies, resulting in a loss of broader spatial context, while larger values (e.g., $\lambda = 7$) lead to overfitting and reduce efficiency by including irrelevant or weak connections. (2) Increasing the number of nearest neighbors K based on FFT-calculated similarity improves model performance up to $K = 7$. Beyond this point, further increases provide minimal gains. Specifically, when K is smaller than 7 (e.g., $K = 6$), the model may fail to capture long-range dependencies adequately, leading to underfitting and a less accurate prediction of traffic flow. On the other hand, when K exceeds 7 (e.g., $K = 8$ or $K = 9$), the model becomes overly sensitive to noise, which can lead to overfitting and reduced generalization ability. A value of $K = 7$ strikes a balance, capturing both local and global patterns effectively while avoiding unnecessary complexity.

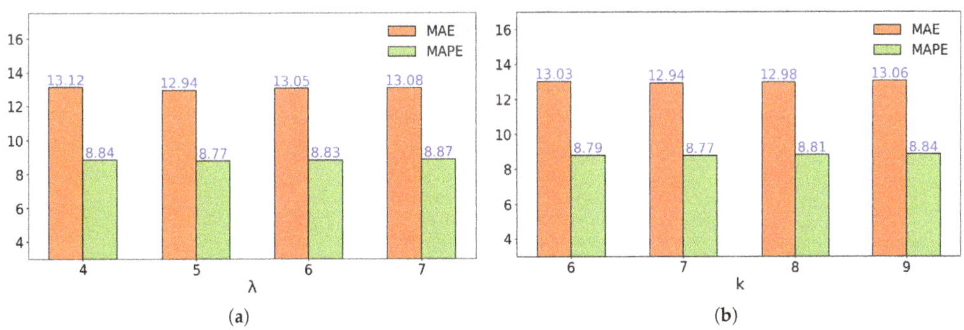

Figure 7. Experimental results of the hyperparameter study on the PeMS08 dataset. (**a**) Effect of geographic distance threshold λ on model performance; (**b**) Effect of number of nearest neighbors K on model performance.

5.9. Traffic Occupancy Embedding

To highlight the importance of traffic occupancy data as an input embedding, this study conducts an in-depth analysis of the daily variations in both traffic flow and traffic occupancy. Twenty specific sensors were selected from the PeMS08 dataset, and heatmaps of both traffic flow and traffic occupancy were generated. Figure 8 visually demonstrates that, during peak periods (such as 7–9 a.m. and 5–7 p.m.), both traffic flow and traffic occupancy reach peak values at multiple nodes simultaneously. These significant peak regions indicate strong synchronization between flow and road occupancy during peak hours. Moreover, data from Node 16 shows that, although traffic flow decreases during certain periods, traffic occupancy remains relatively high. This may be due to factors such as reduced speed and increased vehicle density, leading to localized congestion. In such cases, relying solely on traffic flow may not accurately reflect congestion conditions. Traffic occupancy, as a supplementary indicator, can assist the model in identifying abnormal features under varying congestion states. Additionally, variations in traffic flow and occupancy across different nodes may reflect the features of the roads in the region, providing the model with the ability to identify differences between nodes. Therefore, embedding traffic occupancy into the model helps capture the complex spatial–temporal dependencies of traffic flow, thereby improving the model's prediction accuracy and adaptability.

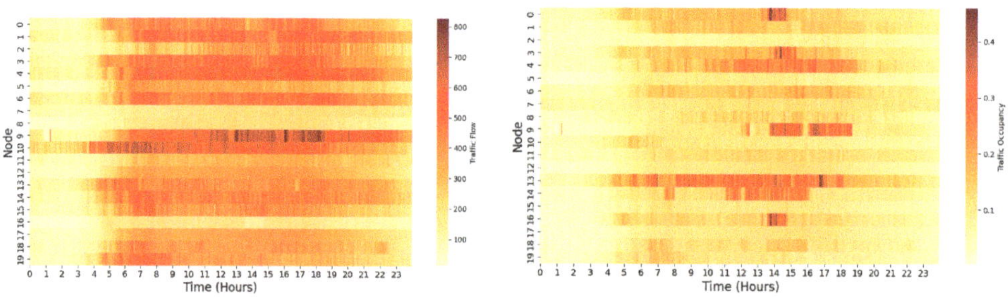

(**a**) (**b**)

Figure 8. Heat maps of traffic flow and traffic occupancy of MMHFormer on PeMS08 dataset. (**a**) Daily traffic flow; (**b**) Daily traffic occupancy.

5.10. Case Study

To gain insights into the model's spatial reasoning capabilities, We visualized the normalized attention weights of the geographic spatial attention during the morning peak (9:00–10:00) and the dynamic similarity spatial attention during the off-peak period (15:00–16:00). Figure 9 demonstrates that two attention mechanism captures distinct spatial patterns: During the morning peak (9:00–10:00), traffic congestion is typically higher on major arterial roads and key intersections, as people commute to work or school. The Geographic Spatial Attention mechanism captures this phenomenon by emphasizing the nodes (roads or intersections) with immediate neighbors that are highly trafficked, thus modeling the flow patterns in densely populated areas. As seen in Figure 9a, the attention heatmap predominantly highlights these localized clusters of road nodes, reflecting the high traffic density and the strong temporal dependencies at this time. The corresponding geospatial mask (Figure 9b) further refines this attention by suppressing connections that are distant or irrelevant to the immediate traffic flow. Dynamic similarity spatial attention (Figure 9c) reveals interesting long-range dependencies between nodes sharing similar traffic patterns, such as residential areas that experience synchronized morning outbound traffic flows. Similarly, by comparing the dynamic similarity attention heatmap (Figure 9c) with its mask matrix (Figure 9d), we observe that the mask strategically preserves connections between functionally similar nodes regardless of geographical distance, while filtering out irrelevant correlations. This enables the model to identify and leverage synchronized traffic behaviors across the network, capturing complex urban mobility patterns that transcend physical connectivity

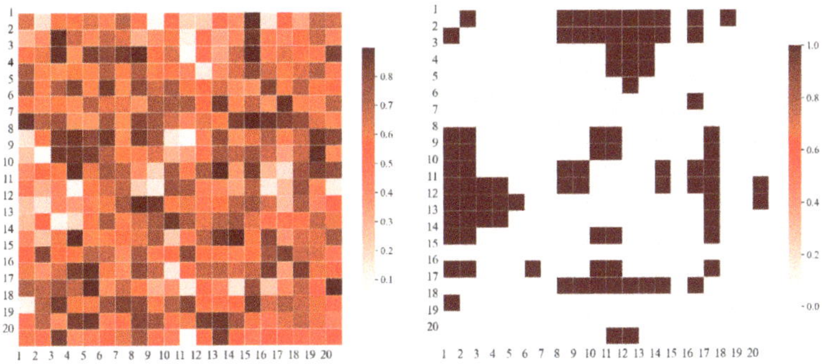

(**a**) (**b**)

Figure 9. *Cont.*

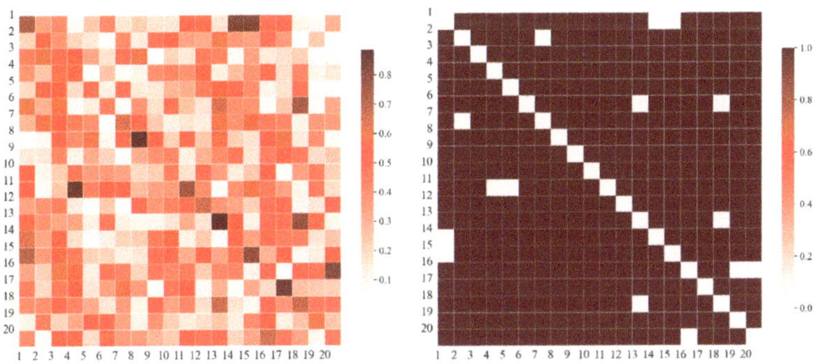

(**c**) (**d**)

Figure 9. Visualization of multi-view spatial attention weights on PeMS08 dataset. Darker colors in the heat maps denote higher attention weights, indicating stronger spatial dependencies. The mask's black regions represent not selected areas. (**a**) Heat maps of geographic spatial attention during the morning peak, (**b**) Visualization of the geospatial mask matrix, (**c**) Heat maps of dynamic similarity spatial attention during the off-peak period, (**d**) Visualization of the dynamic similarity mask matrix.

5.11. Forecasting Results and Visualization

After selecting two specific nodes from the PeMS08 dataset, we visualized the prediction results for the test set (horizon = 12), as shown in Figure 10a. The true traffic flow curve exhibits relatively smooth variations with a clear trend. In this scenario, the model performs exceptionally well, almost perfectly capturing the overall trend of traffic flow, especially at peak and trough points. Furthermore, as shown in Figure 10b, despite the larger and more frequent fluctuations in the actual traffic flow, the model successfully identifies various abrupt changes and responds quickly, closely approximating the true values even in high-frequency fluctuation intervals. This stable prediction of extreme fluctuations further highlights the model's superiority in handling complex and unstable traffic patterns.

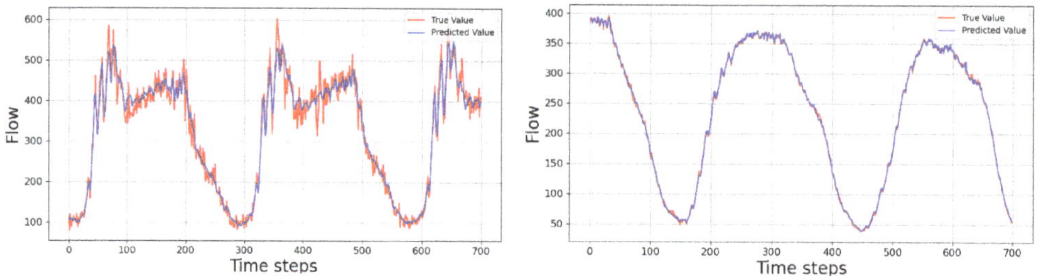

(a) (b)

Figure 10. Visualization of prediction results of MMHFormer on PeMS08 dataset. (**a**) Predicted traffic flow with smooth trends; (**b**) Predicted traffic flow with abrupt changes and fluctuations.

6. Conclusions

In this paper, we proposed MMHFormer, a novel multi-source and multi-view hierarchical Transformer model designed for traffic flow prediction, effectively tackling the challenges of complex spatial–temporal dependencies, dynamic variations, and external influences in traffic networks. MMHFormer incorporates a multi-source gated embedding layer to dynamically fuse multidimensional features, including spatial, temporal, and external conditions, enhancing the representation of traffic scenarios. Additionally, it employs a hierarchical multi-view spatial attention module to capture global, local, and similarity-based spatial dependencies, effectively addressing spatial heterogeneity. To further improve adaptability, the model leverages a hierarchical two-stage temporal attention mechanism, which models global temporal patterns while adapting to node-specific variations. Extensive experiments on four benchmark datasets (PeMS03, PeMS04, PeMS07, and PeMS08) demonstrate that MMHFormer consistently outperforms state-of-the-art methods across various evaluation metrics, including MAE, RMSE, and MAPE. The model also exhibits remarkable long-term prediction stability and effectiveness, making it a reliable tool for intelligent transportation systems. Future work will aim to enhance MMHFormer by incorporating a broader range of external contextual features, such as various weather conditions and traffic incidents, and extending its capabilities to support multimodal data inputs.

Author Contributions: Conceptualization, H.W. (Han Wu) and G.T.; methodology, H.W. (Han Wu) and G.T.; software, H.W. (Hao Wu); validation, H.W. (Han Wu), G.T., and Z.Q.; formal analysis, H.W. (Hao Wu); investigation, G.T.; data curation, M.Z.; writing—original draft preparation, H.W. (Han Wu); writing—review and editing, H.W. (Han Wu), G.T., and Z.Q.; visualization, M.Z.; supervision, M.Z.; project administration, G.T. All authors have read and agreed to the published version of the manuscript.

Funding: This research received no external funding.

Data Availability Statement: The data presented in this study are available on request from the corresponding author.

Conflicts of Interest: The authors declare no conflicts of interest.

References

1. Liu, H.; Zhu, C.; Zhang, D.; Li, Q. Multi-Scale Spatial-Temporal Recurrent Networks for Traffic Flow Prediction. *arXiv* **2023**, arXiv:2310.08138. [CrossRef]

2. Kim, K.; Jin, S.; Ko, S.; Choo, J. Stgrat: A spatio-temporal graph attention network for traffic forecasting. In Proceedings of the International Conference on Information and Knowledge Management, Online, 19–23 October 2020.
3. Williams, B.M.; Hoel, L.A. Modeling and Forecasting Vehicular Traffic Flow as a Seasonal ARIMA Process: Theoretical Basis and Empirical Results. *J. Transp. Eng.* **2003**, *129*, 664–672. [CrossRef]
4. Lu, Z.; Zhou, C.; Wu, J.; Jiang, H.; Cui, S. Integrating granger causality and vector auto-regression for traffic prediction of large-scale WLANs. *KSII Trans. Internet Inf. Syst. (TIIS)* **2016**, *10*, 136–151. [CrossRef]
5. Dhiman, H.S.; Deb, D.; Guerrero, J.M. Hybrid machine intelligent SVR variants for wind forecasting and ramp events. *Renew. Sustain. Energy Rev.* **2019**, *108*, 369–379. [CrossRef]
6. Hochreiter, S. The Vanishing Gradient Problem During Learning Recurrent Neural Nets and Problem Solutions. *Int. J. Uncertain. Fuzziness Knowl. Based Syst.* **1998**, *06*, 107–116. [CrossRef]
7. Chung, J.; Gulcehre, C.; Cho, K.; Bengio, Y. Empirical Evaluation of Gated Recurrent Neural Networks on Sequence Modeling. *arXiv* **2014**, arXiv:1412.3555. [CrossRef]
8. Xu, J.; Song, R.; Wei, H.; Guo, J.; Zhou, Y.; Huang, X. A fast human action recognition network based on spatio-temporal features. *Neurocomputing* **2021**, *441*, 350–358. [CrossRef]
9. Defferrard, M.; Bresson, X.; Vandergheynst, P. Convolutional Neural Networks on Graphs with Fast Localized Spectral Filtering. In Proceedings of the Advances in Neural Information Processing Systems, Barcelona, Spain, 5–10 December 2016; Volume 29.
10. Zhao, L.; Song, Y.; Zhang, C.; Liu, Y.; Wang, P.; Lin, T.; Deng, M.; Li, H. T-GCN: A Temporal Graph Convolutional Network for Traffic Prediction. *IEEE Trans. Intell. Transp. Syst.* **2020**, *21*, 3848–3858. [CrossRef]
11. Huang, R.; Li, P. Hub-hub connections matter: Improving edge dropout to relieve over-smoothing in graph neural networks. *Knowl. Based Syst.* **2023**, *270*, 110556. [CrossRef]
12. Zheng, G.; Chai, W.K.; Duanmu, J.L.; Katos, V. Hybrid deep learning models for traffic prediction in large-scale road networks. *Inf. Fusion* **2023**, *92*, 93–114. [CrossRef]
13. Hamed, M.M.; Al-Masaeid, H.R.; Said, Z.M.B. Short-Term Prediction of Traffic Volume in Urban Arterials. *J. Transp. Eng.* **1995**, *121*, 249–254. [CrossRef]
14. Sun, Y.; Leng, B.; Guan, W. A novel wavelet-SVM short-time passenger flow prediction in Beijing subway system. *Neurocomputing* **2015**, *166*, 109–121. [CrossRef]
15. Sun, B.; Cheng, W.; Goswami, P.; Bai, G. Flow-aware WPT k-nearest neighbours regression for short-term traffic prediction. In Proceedings of the 2017 IEEE Symposium on Computers and Communications (ISCC), Heraklion, Greece, 3–6 July 2017; pp. 48–53. [CrossRef]
16. Liu, Y.; Zheng, H.; Feng, X.; Chen, Z. Short-term traffic flow prediction with Conv-LSTM. In Proceedings of the 2017 9th International Conference on Wireless Communications and Signal Processing (WCSP), Nanjing, China, 11–13 October 2017; pp. 1–6. [CrossRef]
17. Lv, Z.; Xu, J.; Zheng, K.; Yin, H.; Zhao, P.; Zhou, X. LC-RNN: A deep learning model for traffic speed prediction. In Proceedings of the 27th International Joint Conference on Artificial Intelligence, IJCAI'18, Stockholm, Sweden, 13–19 July 2018; pp. 3470–3476.
18. Yu, B.; Yin, H.; Zhu, Z. Spatio-Temporal Graph Convolutional Networks: A Deep Learning Framework for Traffic Forecasting. In Proceedings of the 27th International Joint Conference on Artificial Intelligence, IJCAI'18, Stockholm, Sweden, 13–19 July 2018; pp. 3634–3640. [CrossRef]
19. Guo, S.; Lin, Y.; Feng, N.; Song, C.; Wan, H. Attention Based Spatial-Temporal Graph Convolutional Networks for Traffic Flow Forecasting. *Proc. AAAI Conf. Artif. Intell.* **2019**, *33*, 922–929. [CrossRef]
20. Wu, Z.; Pan, S.; Long, G.; Jiang, J.; Zhang, C. Graph WaveNet for Deep Spatial-Temporal Graph Modeling. *arXiv* **2019**, arXiv:1906.00121. [CrossRef]
21. BAI, L.; Yao, L.; Li, C.; Wang, X.; Wang, C. Adaptive Graph Convolutional Recurrent Network for Traffic Forecasting. In Proceedings of the Advances in Neural Information Processing Systems, Virtual, 6–12 December 2020; Volume 33, pp. 17804–17815.
22. Guo, K.; Hu, Y.; Sun, Y.; Qian, S.; Gao, J.; Yin, B. Hierarchical Graph Convolution Network for Traffic Forecasting. *Proc. AAAI Conf. Artif. Intell.* **2021**, *35*, 151–159. [CrossRef]
23. Ye, Y.; Xiao, Y.; Zhou, Y.; Li, S.; Zang, Y.; Zhang, Y. Dynamic multi-graph neural network for traffic flow prediction incorporating traffic accidents. *Expert Syst. Appl.* **2023**, *234*, 121101. [CrossRef]
24. Yin, X.; Zhang, W.; Jing, X. Static-dynamic collaborative graph convolutional network with meta-learning for node-level traffic flow prediction. *Expert Syst. Appl.* **2023**, *227*, 120333. [CrossRef]
25. Huang, X.; Ye, Y.; Yang, X.; Xiong, L. Multi-view dynamic graph convolution neural network for traffic flow prediction. *Expert Syst. Appl.* **2023**, *222*, 119779. [CrossRef]

26. Zhong, W.; Suo, Q.; Jia, X.; Zhang, A.; Su, L. Heterogeneous Spatio-Temporal Graph Convolution Network for Traffic Forecasting with Missing Values. In Proceedings of the 2021 IEEE 41st International Conference on Distributed Computing Systems (ICDCS), Washington, DC, USA, 7–10 July 2021; pp. 707–717. [CrossRef]
27. Xu, J.; Li, Y.; Lu, W.; Wu, S.; Li, Y. A heterogeneous traffic spatio-temporal graph convolution model for traffic prediction. *Phys. A Stat. Mech. Its Appl.* **2024**, *641*, 129746. [CrossRef]
28. Wu, M.; Lin, Y.; Jiang, T.; Weng, W. MHGNet: Multi-Heterogeneous Graph Neural Network for Traffic Prediction. In Proceedings of the ICASSP 2025—2025 IEEE International Conference on Acoustics, Speech and Signal Processing (ICASSP), Hyderabad, India, 6–11 April 2025; pp. 1–5. [CrossRef]
29. Guo, S.; Lin, Y.; Wan, H.; Li, X.; Cong, G. Learning Dynamics and Heterogeneity of Spatial-Temporal Graph Data for Traffic Forecasting. *IEEE Trans. Knowl. Data Eng.* **2022**, *34*, 5415–5428. [CrossRef]
30. Jiang, J.; Han, C.; Zhao, W.X.; Wang, J. PDFormer: Propagation Delay-Aware Dynamic Long-Range Transformer for Traffic Flow Prediction. *Proc. AAAI Conf. Artif. Intell.* **2023**, *37*, 4365–4373. [CrossRef]
31. Cai, J.; Wang, C.H.; Hu, K. LCDFormer: Long-term correlations dual-graph transformer for traffic forecasting. *Expert Syst. Appl.* **2024**, *249*, 123721. [CrossRef]
32. Geng, Z.; Xu, J.; Wu, R.; Zhao, C.; Wang, J.; Li, Y.; Zhang, C. STGAFormer: Spatial–temporal Gated Attention Transformer based Graph Neural Network for traffic flow forecasting. *Inf. Fusion* **2024**, *105*, 102228. [CrossRef]
33. Li, Y.; Xu, H.; Zhang, T.; Li, X.; Li, G.; Tian, W. DDGformer: Direction- and distance-aware graph transformer for traffic flow prediction. *Knowl. Based Syst.* **2024**, *302*, 112381. [CrossRef]
34. Li, Y.; Yu, R.; Shahabi, C.; Liu, Y. Diffusion Convolutional Recurrent Neural Network: Data-Driven Traffic Forecasting. *arXiv* **2018**, arXiv:1707.01926. [CrossRef]
35. Wu, Z.; Pan, S.; Long, G.; Jiang, J.; Chang, X.; Zhang, C. Connecting the Dots: Multivariate Time Series Forecasting with Graph Neural Networks. In Proceedings of the 26th ACM SIGKDD International Conference on Knowledge Discovery & Data Mining, KDD '20, Virtual, 6–10 July 2020; pp. 753–763. [CrossRef]
36. Zheng, C.; Fan, X.; Wang, C.; Qi, J. GMAN: A Graph Multi-Attention Network for Traffic Prediction. *Proc. AAAI Conf. Artif. Intell.* **2020**, *34*, 1234–1241. [CrossRef]
37. Li, M.; Zhu, Z. Spatial-Temporal Fusion Graph Neural Networks for Traffic Flow Forecasting. *Proc. AAAI Conf. Artif. Intell.* **2021**, *35*, 4189–4196. [CrossRef]
38. Shao, Z.; Zhang, Z.; Wang, F.; Wei, W.; Xu, Y. Spatial-Temporal Identity: A Simple yet Effective Baseline for Multivariate Time Series Forecasting. In Proceedings of the 31st ACM International Conference on Information & Knowledge Management, CIKM '22, Atlanta, GA, USA, 17–21 October 2022; pp. 4454–4458. [CrossRef]
39. Xu, Y.; Han, L.; Zhu, T.; Sun, L.; Du, B.; Lv, W. Generic Dynamic Graph Convolutional Network for traffic flow forecasting. *Inf. Fusion* **2023**, *100*, 101946. [CrossRef]

applied sciences

MDPI

Article

A Time and Frequency Domain Based Dual-Attention Neural Network for Tropical Cyclone Track Prediction

Fancheng Meng [1,2], Xiran Xiong [3] and Liling Zhao [1,2,*]

[1] School of Automation, Nanjing University of Information Science & Technology, Nanjing 210044, China; 202312490071@nuist.edu.cn

[2] Jiangsu Collaborative Innovation Center on Atmospheric Environment and Equipment Technology, Nanjing 210044, China

[3] Reading Academy, Nanjing University of Information Science & Technology, Nanjing 210044, China; 202383710070@nuist.edu.cn

* Correspondence: zhaoliling@nuist.edu.cn

Abstract

Due to the influence of various dynamic meteorological factors, accurate Tropical Cyclone (TC) track prediction is a significant challenge. However, current deep learning based time series prediction models fail to simultaneously capture both short-term and long-term dependencies, while also neglecting the change in meteorological environment pattern associated with TC motion. This limitation becomes particularly pronounced during sudden turning in the TC track, resulting in significant deterioration of prediction accuracy. To overcome these limitations, we propose LFInformer, a hybrid deep learning framework that integrates an Informer backbone, a Frequency-Enhanced Channel Attention Mechanism (FECAM), and a Long Short-Term Memory (LSTM) network for TC track prediction. The Informer backbone is underpinned by ProbSparse Self-Attention in both the encoder and the causally masked decoder, prioritizing the most informative query–key interactions to deliver robust long-range modeling and sharper detection of turning signals. FECAM enhances meteorological inputs via discrete cosine transforms, band-wise weighting, and channel-wise reweighting, then projects the enhanced signals back into the time domain to produce frequency-aware representations. The LSTM branch captures short-term variations and localized temporal dynamics through its recurrent structure. Together, these components sustain high accuracy during both steady evolution and sudden turnin. Experiments based on the JMA and IBTrACS 1951–2022 Northwest Pacific TC data show that the proposed model achieves an average absolute position error (APE) of 72.39 km, 117.72 km, 145.31 km and 168.64 km for the 6-h, 12-h, 24-h and 48-h forecasting tasks, respectively. The proposed model enhances the accuracy of TC track predictions, offering an innovative approach that optimally balances precision and efficiency in forecasting sudden turning points.

Keywords: tropical cyclone; track prediction; informer; LSTM

1. Introduction

TC are among the most destructive natural disasters globally [1], particularly in the Northwest Pacific region, where TC frequently occur and cause severe impacts on the economy, society, and ecosystems of coastal areas. The intense storm surges, heavy rainfall, and strong winds associated with TC often result in significant loss of life and property damage. For example, in 2017, China experienced eight TC landfalls, affecting 5.879 million people and causing approximately 5 billion USD in economic losses [2]. Therefore, accurate TC track prediction is crucial not only for minimizing losses, but also for providing timely warnings to the government and the public, and offering a scientific foundation for post-disaster response and mitigation [3].

The evolution of TC track is influenced by a variety of factors, including ocean temperature, air pressure, topography, wind speed, and other complex meteorological environmental factors [4]. The interactions among these factors result in highly nonlinear and complex TC track. However, traditional TC track prediction methods rely on statistical models (CLIPER) [5] and numerical weather prediction (NWP) [6] models. Statistical models typically base their predictions on the relationship between historical data and meteorological factors, using regression analysis and pattern recognition techniques. While they provide quick results, their accuracy is significantly limited when handling complex nonlinear relationships and long-term predictions [7]. However, NWP models can more accurately simulate meteorological conditions, but they come with higher computational complexity and often exhibit large errors in medium- to long-term track predictions [8].

With the rapid increase in meteorological observation data, traditional prediction methods may face challenges in efficiently handling such large-scale spatiotemporal data, which has led deep learning methods to gradually emerge as a promising research direction in the field of TC track prediction. In particular, recurrent neural networks (RNNs) [9] and their variants, such as long short-term memory networks (LSTM) [9] and gated recurrent units (GRU) [10], excel at handling long-term dependencies in sequential data, leading to significant advancements in TC track prediction [11]. For example, Lian, J et al. [12] proposed a TC track prediction model that combines RNN and attention mechanisms, demonstrating the effectiveness of deep learning in handling complex meteorological data. However, traditional RNNs still face challenges such as long-term dependencies and gradient vanishing [13], which limit their performance in practical applications. To address these issues, LSTM networks have been widely applied to TC track prediction [14]. By introducing gating units, LSTM effectively overcomes the gradient explosion and vanishing problems in RNNs, and performs more stably in learning from long time-series data [15]. For example, Gao, S et al. [16] trained an LSTM network on TC data from mainland China and proposed a new TC track prediction method, significantly improving prediction accuracy. Tong, B et al. [17] proposed a spatiotemporal model based on ConvLSTM, which combines the advantages of convolutional neural networks and LSTM to better handle large-scale meteorological data. However, LSTM models still face issues such as high computational resource requirements and slow training speeds, especially when handling long time-series data, which limits their effectiveness [18]. With the development of deep learning technologies, the Informer model [19], as a new framework for processing sequential data, has begun to demonstrate its potential in the field of meteorological forecasting. The Informer model, by introducing Self-Attention Mechanisms and Sparse Attention computation, is capable of efficiently processing long time-series data, particularly suited for large-scale meteorological datasets. Compared to traditional LSTM and GRU models, which struggle with capturing long-term dependencies due to their limited memory and vanishing gradient issues, Informer excels at modeling long-term sequential dependencies, significantly improving both prediction accuracy and computational efficiency.

Recent TC forecasting studies have also explored more domain-tailored deep learning architectures. For example, Transformer-based models have been developed for joint TC track–intensity prediction [20]. Beyond TC-specific models, GNN-based global weather forecasting systems have demonstrated strong medium-range skill and can support tropical cyclone tracking [21]. In addition, climate-oriented Transformer foundation models provide a unified pretrained backbone that can be fine-tuned for diverse weather and climate prediction tasks [22].

In summary, many current deep learning approaches for TC track prediction face challenges in jointly learning multiple temporal features, hindering their ability to simultaneously capture both short-term dynamic changes and long-term development trends of TC track progression. thereby affecting their ability to accurately predict the overall trend of the TC. Furthermore, several traditional approaches to TC track prediction often overlook the periodic change characteristics of meteorological environment factors such as wind speed and air pressure that are related to TC track development, which affects their ability to accurately predict sudden turning points in the TC's track. We use the term "periodic change" to denote recurring meteorological environment change patterns rather than strict cycles; a precise definition and modeling rationale are provided in Section 2.3. Additionally, LSTM has advantages in handling short-term temporal dependencies, and the Informer model can effectively address the computational complexity issues in long-sequence predictions. However, there is scant research on combining the two to enhance the ability to capture the complex dynamic change in TC characteristics. To address this, this paper proposes a novel TC track prediction model, LFInformer, with the following main contributions:

- A novel neural network LFInformer: Integrating Informer, FECAM and LSTM. The model can simultaneously focus on overall development trends of TC track turning and the local dynamic features.
- TC track turning critical time point attention block: Introducing ProbSparse Self-Attention mechanism into Informer encoder, this mechanism has the ability to filter out critical time points and focus precise attention on points carrying turning point information, effectively improving the ability to predict TC track critical turning points.
- Meteorological environment change pattern perception block: This block is mainly composed of Frequency-Enhanced Channel Attention mechanis, enabling our model to find physical processes at different time scales in the raw data based on DCT, and focus on rapidly changing meteorological factors during TC trajectories. This provides crucial pattern features for predicting sudden turnings.

2. Materials and Methods

2.1. LFInformer Architecture

To improve the accuracy and stability of TC track prediction, this paper proposes a new Informer architecture—LFInformer, an efficient Transformer-style architecture [23]. Its core objective is to combine short-term and long-term temporal capturing abilities and enhance the capture of the influence of periodic change in meteorological factors change patterns on track development. The structure of LFInformer is shown in Figure 1, which mainly consists of the Informer encoder, Informer decoder, LSTM network, FECAM block and fully connected layer (FCL). The block order is designed to first obtain a compact long-range temporal representation via the Informer (with distilling/downsampling), and then refine it using FECAM and LSTM for periodic pattern enhancement and short-term dynamics modeling before the final regression layer.

The proposed model input is divided into two parts: the Informer encoder input and the Informer decoder input.In the encoder section, the input consists of the predefined step-length data of the TC's track, with each time step processed through position encoding to ensure the model can comprehend the order and dependencies of the time series. The decoder operates under a causal constraint. It only consumes the historical prefix of the target sequence together with the encoder context; no future observations are used In this way, the model can simultaneously utilize both preceding and subsequent data to predict the turning in the TC's track, thereby improving the accuracy and stability of the TC track forecast. Furthermore, the core components of the model include two main parts: the Informer encoder and the Informer decoder. The Informer encoder employs a Probabilistic Sparse Self-Attention Mechanism to capture the global dependency characteristics of the input sequence through efficient sparse computation, while also using Self-Attention distilling to extract key information and compress data redundancy. The Informer decoder section utilizes a Masked ProbSparse Self-Attention Mechanism to generate high-quality decoding outputs, ensuring no information leakage during the prediction process. Additionally, the decoder uses a multi-head attention mechanism to combine the encoder's output features with the decoder's contextual information, thereby enhancing its ability to predict future turning in the TC's track.

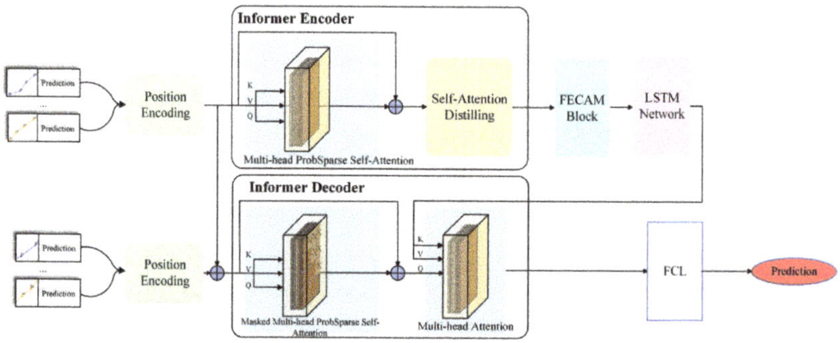

Figure 1. Workflow of the proposed method.

Furthermore, to enhance the model's ability to capture the complex periodic dynamic features of meteorological factors patterns during the TC's development and turning, the FECAM block is introduced. This module combines frequency domain enhancement techniques and channel attention mechanisms to extract the periodic change in meteorological environment change patterns in the TC's track, such as the impact of the periodic change in environmental factors like wind speed and air pressure on the track trend, thereby enhancing the model's sensitivity to periodic change and nonlinear relationships in time series data. The FECAM block receives the output from the Informer encoder, and the output features of the FECAM block are fed into the LSTM network to further capture and process short-term dependencies and local dynamic change, ensuring that the model can effectively handle the short-term fluctuations and trend turning in the TC's track. Finally, the output of the Informer decoder is mapped to the target space through a fully connected layer, generating the predicted results for the TC's track. The LFInformer model fully combines the Informer framework's global modeling capability for long sequences, LSTM's short-term dependency modeling ability, and FECAM's periodic dynamic feature enhancement mechanism, providing an efficient and reliable solution for accurate TC track forecasting.

2.2. Informer Encoder

The Informer encoder is designed to improve computational efficiency in long-term time series forecasting, while enhancing the model's ability to capture both global dependencies and local features.Compared to Transformer architectures proposed prior to Informer, the Informer encoder incorporates a Probabilistic Sparse Self-Attention Mechanism, integrates Conv1d to enhance feature extraction, and employs a progressive downsampling strategy to reduce sequence length, thereby preserving critical information while lowering computational cost. Figure 2 illustrates the complete architecture of the Informer encoder, which consists of several key components: input embedding, ProbSparse Self-Attention Blocks, Conv1d, Maxpooling for downsampling, and feature map generation.

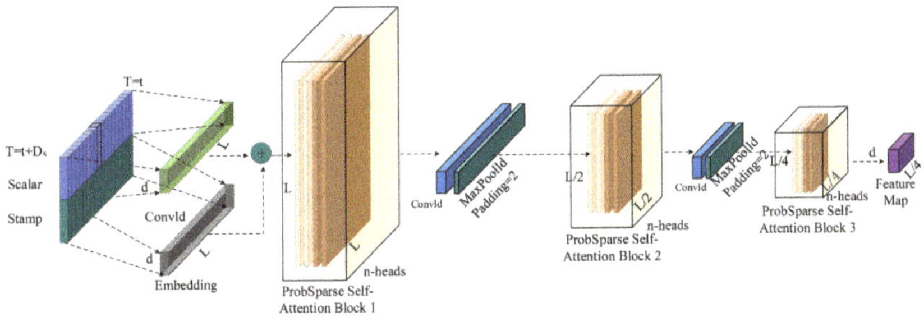

Figure 2. The structure of Informer encoder.

The input sequence is first processed by the input embedding layer, which maps the raw time series data into a high-dimensional feature space and incorporates temporal information to enhance the model's ability to capture temporal dependencies. Subsequently, the data passes through multiple ProbSparse Self-Attention Blocks, where Sparse Attention matrices are computed to capture global dependencies. In addition, the one-dimensional convolution layer is employed to further enhance the representation of local information, providing more stable temporal features. To reduce computational complexity, the encoder applies Maxpooling layers to downsample the sequence, gradually compressing the data stream while reducing redundant information, and ultimately generating feature maps for use by the decoder.

2.2.1. Input Embedding

The input embedding method of Informer enables the time series data to incorporate not only basic feature information, but also fully leverage time stamp and time difference information. Let the input time series be:

$$X = \{x_1, x_2, \ldots, x_T\}, x_t \in \mathbb{R}^d \tag{1}$$

where T denotes the number of time steps, and d represents the dimensionality of the input features. To enhance the model's understanding of temporal information, each time step x_t is augmented with a corresponding time scalar S_t and timestamp P_t, resulting in the final input representation H_0 as:

$$H_0 = X + \text{Conv1d}(P_t) + \text{Conv1d}(S_t) \tag{2}$$

The Conv1d is used to map temporal information into the same dimensional space, ensuring that the model can capture time-dependent features. In this way, the Informer encoder leverages not only the numerical features of the data but also incorporates temporal information, thereby enhancing its ability to handle irregular time intervals.

2.2.2. ProbSparse Self-Attention Blocks

The Self-Attention Mechanism in Transformers, while powerful for feature extraction, suffers from quadratic computational complexity ($O(T^2)$) in processing long sequences. Recent studies reveal that self-attention weights demonstrate inherent sparsity patterns [24], suggesting opportunities for efficiency optimization. The Sparse Transformer leverages spatial correlations to reduce computation, while the LogSparse Transformer employs exponential step sizes to capture temporal patterns [25]. More recently, the Longformer extends these approaches with configurable sparsity patterns [26]. However, current methods face two critical limitations: their heuristic designs lack theoretical grounding, and their uniform treatment of attention heads overlooks the multi-head mechanism's diverse representational capacities. These shortcomings result in both computational redundancy and suboptimal performance, particularly in long-sequence temporal forecasting tasks.

To address this issue, the model in this paper employs a ProbSparse Self-Attention Mechanism that focuses on the most informative query positions, significantly reducing computational complexity while maintaining the model's predictive capability.

The core idea of the ProbSparse Self-Attention Mechanism is to quantify the dependence distribution discrepancy between query vectors and key vectors using a sparsity scoring function, and select the attention terms that contribute the most to the prediction. Given the input sequence $X \in R^L$, its Q (Query), K (Key), and V (Value) [27] matrices are generated through linear transformations:

$$Q = XW^Q, K = XW^K, V = XW^V \tag{3}$$

where W^Q, W^K, and W^V are learnable parameters. For each query vector $q_i \in Q$, the sparsity scoring function is defined as:

$$M(q_i, K) = \max_j \left(\frac{q_i^T k_j}{\sqrt{d_k}} \right) - \frac{1}{L} \sum_{j=1}^{L} \frac{q_i^T k_j}{\sqrt{d_k}} \tag{4}$$

This function measures the concentration of the dependence distribution of q_i over the key set K. A higher score indicates that q_i has significant dependence on a few k_j. Let L be the sequence length. Based on the sparsity scoring $M(q_i, K)$, the Self-Attention Mechanism employs a Top-U selection strategy, retaining only the top $U = 5 \log L$ highest-scoring terms in each row to construct the Sparse Attention matrix A_{prob}. The calculation is accelerated through random projection approximation. The process involves randomly sampling $m = L \log L$ candidate key vectors from K, then calculating the similarity between q_i and the candidate keys to approximate $M(q_i, K)$, and finally selecting the top U keys with the highest scores to generate Sparse Attention weights.

$$A_{prob,i,j} = \frac{\exp\left(\frac{q_i^T k_j}{\sqrt{d_k}} \right)}{\sum_{l \in Top-U} \exp\left(\frac{q_i^T k_l}{\sqrt{d_k}} \right)} \tag{5}$$

Only the top U important terms are retained in the attention matrix, and the scores of the remaining positions are set to zero. Subsequently, the Sparse Attention matrix performs a weighted sum on the value matrix V, generating the global representation $\mathbb{Z}$ for the time step:

$$\mathbb{Z} = \mathbb{A}_{prob} \mathbb{V} \tag{6}$$

This ProbSparse Self-Attention Mechanism effectively captures the dependencies between important time steps in TC track prediction, such as moments when significant changes occur in wind speed or pressure, or critical time points when the TC track undergoes major shifts.

2.2.3. MaxPooling

Although the ProbSparse Self-Attention Mechanism effectively reduces redundant attention computations, the generated feature representations may still contain some redundant information. To address this, the Informer encoder incorporates a Self-Attention distillation operation, which refines the feature representations by pruning the temporal dimension, significantly improving computational efficiency. The Self-Attention distillation extracts local features through Conv1d and handles nonlinear relationships with an ELU activation function [28]. As shown in Figure 2 Then, MaxPooling with a stride of 2 is applied to downsample the input temporal dimension [29,30], halving the input length. The computation process can be expressed as:

$$X_{j+1}^{t} = MaxPool\Big(ELU(Conv1d([X_j^t]_{AB}))\Big) \tag{7}$$

where $[\cdot]_{AB}$ represents the output of Multi-Head ProbSparse Self-Attention. The kernel size of the 1D convolution is 3, and the stride of the max pooling is 2, reducing the temporal dimension length from $T \rightarrow T/2$ to optimize memory. In the distillation operation, the encoder adopts a pyramid structure to gradually reduce the input temporal dimension while ensuring that the output hidden representations retain global contextual information. The time step length of each layer's output is progressively reduced, and the final hidden representation of the Z_{final} is generated by stacking the outputs of each layer:

$$Z_{final} = Concat(Z_{stack1}, Z_{stack2}, \ldots, Z_{stackN}). \tag{8}$$

2.3. FECAM Block

In this paper, we do not use "periodic" to mean a strict, perfectly repeating cycle. Rather, we refer to recurring short-term temporal patterns in environmental fields (e.g., wind and pressure) over time scales of hours to days. Concretely, the track tendency often responds in a similar way when a sequence of changes reappears—for example, a gradual strengthening of easterly steering flow followed by a brief relaxation can repeatedly nudge the track poleward–eastward. To this end, we endow the model with climate–environmental pattern perception: FECAM operates in the frequency domain to highlight the time scales where such repeatable patterns live and uses channel attention to emphasize those bands that are most predictive of track deflection. In short, the model learns mappings from Δwind/Δpressure patterns over a recent window to the subsequent track tendency, and can reuse this knowledge when similar patterns recur later in the sequence.

FECAM is a novel attention mechanism block introduced in this paper, aimed at significantly enhancing the ability to capture frequency-domain features in time series prediction tasks. Traditional attention mechanisms primarily focus on capturing temporal dependencies in sequential data. However, in many practical applications, especially in tasks like TC track prediction, which depend on the periodic change of meteorological factors environment change patterns, the frequency-domain information of environmental factors plays a crucial role. Traditional methods often overlook the periodic change of these meteorological environmental factors in the frequency domain, making it difficult for the model to accurately capture the long-term trends and dynamic turning of the TC track.

The innovation of FECAM lies in the introduction of DCT [31] to replace the traditional Fourier Transform (FT), effectively overcoming the Gibbs phenomenon [32,33] that may occur in Fourier Transform. This approach avoids the interference of high-frequency noise with the model by eliminating the associated artifacts. FECAM can more accurately capture the periodic change of meteorological factors patterns in the TC track, such as the temporal variations in wind speed, pressure, and other factors. By extracting these frequency-domain features, FECAM not only enhances the model's sensitivity to frequency-domain information but also strengthens its ability to understand complex dynamic turning, thereby optimizing the TC track prediction performance.

In the TC track prediction task, the input multivariate time series includes longitude, latitude, time, wind speed, and pressure. As shown in Figure 3, the FECAM aims to explicitly extract and utilize periodic patterns hidden in these time series features.

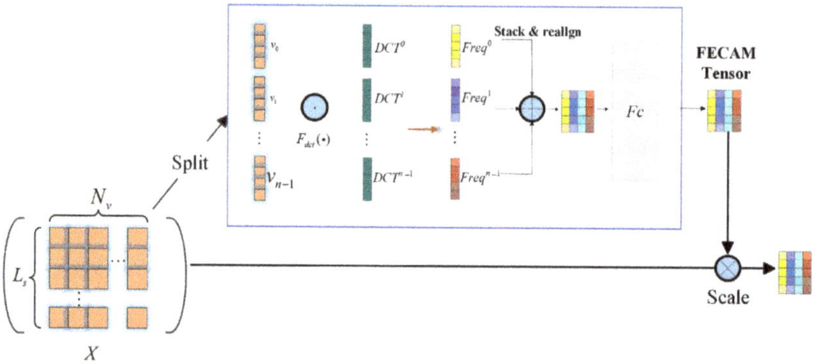

Figure 3. The structure of frequency enhanced channel attention mechanism.In this diagram, the color coding represents different data states and operations: the orange squares denote the input feature segments after splitting; the teal vertical blocks indicate the DCT operations; the distinct colors (yellow, purple, reddish-brown) represent different frequency components extracted from the input; the light grey block corresponds to the Fully Connected (Fc) layer; and the multi-colored grids illustrate the stacked and realigned frequency-enhanced features.

The input to FECAM is the feature tensor $X \in \mathbb{R}^{N_v \times L_s}$, where N_v denotes the number of variables and L_s denotes the sequence length. First, the module splits the feature tensor along the variable channel dimension, treating each variable's sequence $V_i \in \mathbb{R}^{1 \times L_s}$ independently. Each of these sequences is then transformed into the frequency domain using the DCT, which has stronger energy compaction properties than the traditional Fourier Transform and avoids boundary effects such as the Gibbs phenomenon.

The frequency representation of each input V_i is computed as follows:

$$Freq^i = DCT_j(V_i) = \sum_{j=0}^{L_s-1} V_{i,j}\, B_j^i \tag{9}$$

where B_j^i represents the j-th basis function of the DCT for variable i, and L_s is the length of the frequency vector. The DCT operation extracts both the low-frequency components and high-frequency components from the input sequences.

Once the DCT is applied to all n variables, the resulting frequency vectors are stacked together to form a comprehensive frequency-domain representation:

$$Freq = DCT(V) = stack([Freq^0, Freq^1, \cdots, Freq^{n-1}]) \tag{10}$$

This stacked frequency tensor is then fed into a fully connected layer to generate attention weights for each variable channel, allowing the model to emphasize features with strong periodic signals. The output attention weights are used to scale the original feature tensor X, enhancing the model's capacity to capture periodic and trend-related dependencies in TC dynamics.

Based on the frequency domain tensor, FECAM further optimizes feature representation through a channel attention mechanism. Specifically, FECAM first calculates the global feature vector for each channel via Global Average Pooling (GAP):

$$g = GAP(F) = \frac{1}{L}\sum_{i=1}^{L} F_i \tag{11}$$

where $g \in \mathbb{R}^N$ represents the global feature description for each channel. Then, the channel attention mechanism generates channel weights through a two-layer fully connected network:

$$bw_{channel} = \sigma(W_2 \cdot \delta(W_1 \cdot g)) \tag{12}$$

where W_1 and W_2 are learnable weight parameters, $\delta(\cdot)$ is the ReLU activation function, and $\sigma(\cdot)$ is the Sigmoid activation function. The channel weights $w_{channel} \in \mathbb{R}^N$ represent the relative importance of different variables and dynamically adjust the contribution of variables such as wind speed and pressure to the TC track prediction. FECAM optimizes the frequency domain features by adjusting the channel weights, generating enhanced feature representations:

$$Z_{FECAM} = w_{channel} \odot F \tag{13}$$

where $\odot$ denotes element-wise multiplication. In this way, FECAM captures both the periodic patterns and the dynamic interactions between channels, providing more accurate input for the subsequent LSTM module.

Figure 4 illustrates how FECAM leverages frequency-domain information to complement time-domain attention. Starting from the environmental time series (Figure 4a), each channel is transformed from the time domain to the frequency domain so that energy is separated by time scale; the DCT spectrum (Figure 4b) makes slow trends (low frequencies) and recurring short-term motifs (higher frequencies) explicit, with the frequency index inversely related to period. On this spectrum, FECAM learns data-driven channel–frequency weights A (Figure 4c) that emphasize scales which historically precede track deflection while de-emphasizing irrelevant bands. Applying these weights yields the reweighted spectrum $A \odot F$ (Figure 4d), suppressing noise and amplifying informative scales. The enhanced signal is then projected back to the time domain via IDCT $(A \odot F)$ (Figure 4e, blue), which can be interpreted as a scale-aware version of the original input (gold). Finally, the emphasized band produces a stable time-domain signature that cues a turn tendency (Figure 4f) and is less sensitive to phase shifts than position-based, time-domain attention. In short, FECAM attends to scales × channels rather than positions, providing a denoised, scale-selective representation that is fused with the main branch for track prediction.

FECAM combines the dual advantages of frequency domain enhancement and channel attention mechanisms. In the TC track prediction task, this module is capable of capturing the periodic in meteorological factors environment change patterns during the development of the TC track. At the same time, FECAM dynamically adjusts the relative weights between variables, enabling the model to efficiently extract key information even in the presence of complex multivariable interactions. Through this feature optimization, FECAM effectively compensates for the encoder's limitations in modeling the periodic change of TC environmental factors patterns.

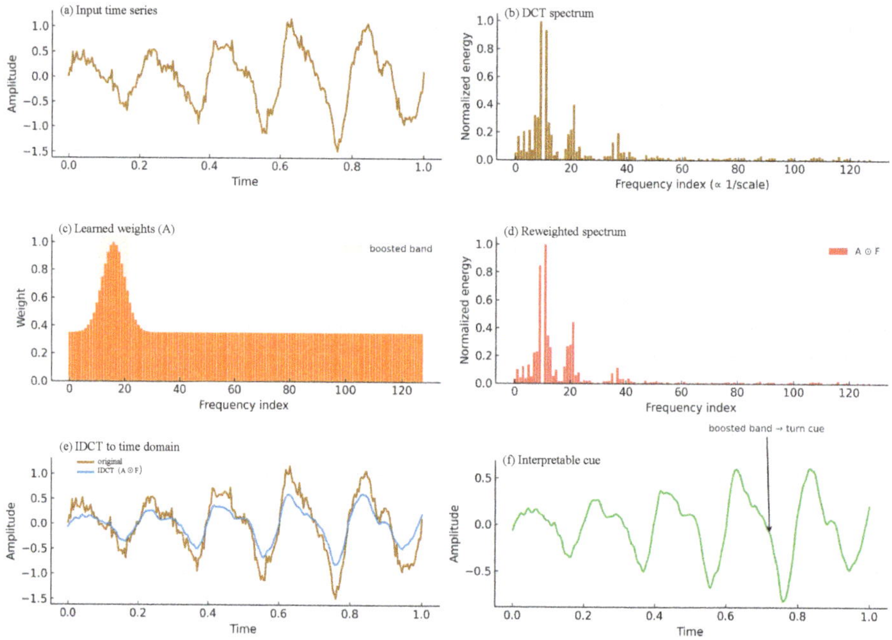

Figure 4. FECAM frequency-domain processing with panels (**a**–**f**). In panel (**a**), the orange line represents the input time series. In panel (**b**), the yellow bars indicate the magnitude of the DCT spectrum frequencies. In panel (**f**), the green line illustrates the extracted interpretable cue highlighting the periodic patterns.

2.4. LSTM Network

In TC track prediction, the dynamic turn in the track is not only influenced by long-term trends but also exhibits significant short-term dependencies. In particular, moments of sudden shifts in the TC track require a more focused learning of short-term dependencies. Although the Informer encoder and FECAM block have captured the global features of the time series and the periodic change of environmental meteorological factors patterns, their performance in capturing short-term dependencies still shows some limitations. Specifically, for the rapid change of variables such as wind speed and pressure at local time steps in the TC track, relying solely on the Informer encoder and frequency domain analysis may fail to capture the fine-grained dynamic information. To address this, the model introduces an LSTM network after FECAM, focusing on enhancing short-term dependencies and local dynamic features, thereby further improving the model's ability to accurately predict the TC track.

LSTM is a classic variant of RNN that effectively addresses the gradient vanishing and explosion issues in traditional RNN for long-sequence tasks by introducing gating mechanisms. In the TC track prediction task, the LSTM module is placed after the FECAM block to model the short-term dependencies and dynamic changes of environmental factors in the time series. By processing input features step by step, the LSTM can extract local dynamic features of variables such as wind speed and pressure along the track, while preserving key long-term dependency information.

As shown in Figure 5, the input to the LSTM includes the feature X_t at the current time step, the hidden state h_{t-1} , and the memory state C_{t-1} from the previous time step, with X_t being the output feature from the FECAM block. The output consists of the hidden state h_t at the current time step and the updated memory state C_t. This mechanism ensures the flow of information between consecutive time steps in the time series while dynamically updating the key feature representations.

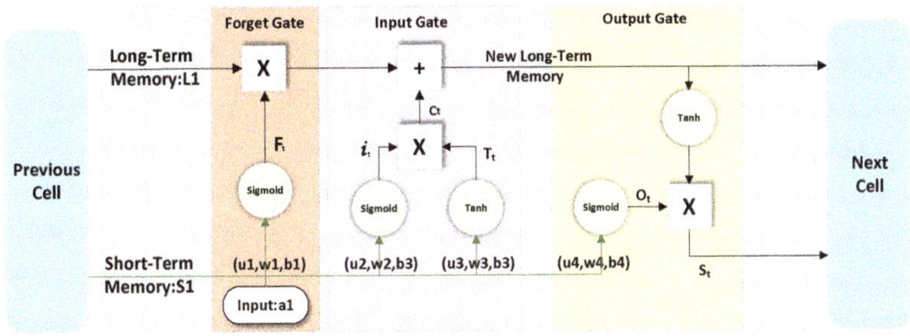

Figure 5. The cell structure of LSTM.

In the internal processing flow, the LSTM consists of three main gating mechanisms [34]: the forget gate, the input gate, and the output gate. The forget gate is responsible for selectively retaining or discarding the memory state from the previous time step, the input gate controls the writing of new information, and the output gate determines the update of the hidden state. The specific calculations are as follows:

Forget Gate: Determines which information in the previous time step's memory state should be retained or discarded:

$$f_t = \sigma\left(W_f X_t + U_f h_{t-1} + b_f\right). \quad (14)$$

where f_t is the output of the forget gate, and σ is the Sigmoid activation function.

Input Gate: Selects the new information to be written into the memory state at the current time step:

$$\mathrm{i_t} = \sigma\ (\mathrm{W_i X_t} + \mathrm{U_i h_{t-1}} + \mathrm{b_i}) \quad (15)$$

Simultaneously, it computes the candidate memory information:

$$\tilde{c}_t = \tanh(\mathrm{W_c X_t} + \mathrm{U_c h_{t-1}} + \mathrm{b_c}) \quad (16)$$

Memory State Update: The memory state for the current time step is updated through the forget and input gates:

$$\mathrm{c_t} = \mathrm{f_t} \odot \mathrm{c_{t-1}} + \mathrm{i_t} \odot \tilde{\mathrm{c}}_t \quad (17)$$

where $\odot$ denotes element-wise multiplication.

Output Gate: Determines the output of the hidden state, generating the time step feature based on the current memory state:

$$\mathrm{o_t} = \sigma\ (\mathrm{W_o X_t} + \mathrm{U_o h_{t-1}} + \mathrm{b_o}) \quad (18)$$

$$\mathrm{h_t} = \mathrm{o_t} \odot \tanh(\mathrm{c_t}) \quad (19)$$

Through this series of gating operations, LSTM captures short-term dynamic features in time series while effectively memorizing long-term dependencies, preventing interference from redundant or irrelevant information. The primary role of LSTM is to model the local dynamic characteristics in the TC track, such as the impact of wind speed and pressure change at specific time steps on track deviation. Combined with the frequency-domain enhanced features provided by the FECAM block, LSTM further strengthens the temporal dependencies in the time series, generating accurate feature representations for the subsequent decoder module.

2.5. *Informer Decoder*

The primary task of the Informer decoder is to combine the global features output by the encoder with the historical information of the target sequence, generating predictions for the future TC track step by step. In the TC track prediction task, the decoder captures the short-term dependencies and the interaction between global features through a dynamic attention mechanism, ensuring the temporal continuity and spatial accuracy of the predictions.

The input to the decoder consists of two parts: first, the global feature representations extracted by the Informer encoder, and second, the known historical values of the target sequence. After position encoding, the historical part of the target sequence is combined with the encoder features, and through multi-layer dynamic modeling, the decoder generates the predicted results. Unlike the encoder's input, the decoder guides the future predictions using the historical values of the target sequence, further enhancing the model's accuracy. The core components of the decoder include the Masked ProbSparse Self-Attention mechanism and the Multi-head Attention Mechanism. Together, they ensure that the decoder captures short-term dynamic turning while also incorporating the global context from the encoder during the prediction process.

2.5.1. Masked ProbSparse Self-Attention Block

The Masked ProbSparse Self-Attention Mechanism models the historical part of the target sequence. This mechanism ensures that each time step can only access its preceding historical information through the mask operation, preventing future information leakage. The decoder uses a masked matrix M to constrain the self-attention calculation:

$$M_{i,j} = \begin{cases} 0, & if\ j \leq i \\ -\infty, & otherwise \end{cases} \tag{20}$$

The mask ensures that during the prediction at time step i, the model can only access the historical information up to time step $i-1$, as shown in the Masked ProbSparse Self-Attention Block in Figure 6.

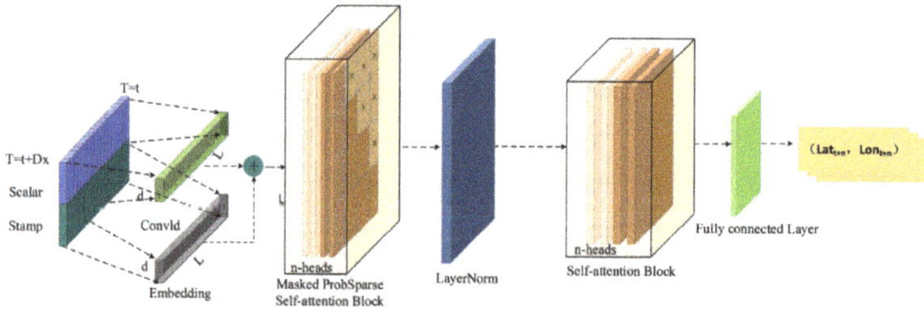

Figure 6. The structure of Informer decoder.

The attention weight calculation method is similar to the one used in the encoder, where Sparse Attention weights are computed under the mask constraint:

$$A_{masked} = Softmax\left(Mask\left(\frac{Q_{dec}K_{dec}^{T}}{\sqrt{d_k}}\right)\right) \tag{21}$$

where Q_{dec},K_{dec} are the query and key matrices for the decoder input, and A_{masked} is the sparsified attention weight matrix. Using the $Top - U$ selection strategy, the top U most important items are retained for each row. For example, when predicting time step $t + 1$, the model can only rely on information from time step t and earlier. Specifically, the mask operation sets the attention weights for future time steps to negative infinity, ensuring that after the Softmax operation, these weights become zero.

2.5.2. Layer Normalization

The main purpose of Layer Normalization is to address the issue of Internal Covariate Shift, which arises due to change in the input distribution of the network layers during training. Its core idea is to normalize the activations of all neurons in a given layer for each sample, ensuring that the input distribution of each layer remains stable, thus accelerating training convergence and improving the model's generalization ability.

After the Masked ProbSparse Self-Attention, the decoder introduces Layer Normalization (LayerNorm) to stabilize training and improve the model's convergence speed. The calculation of LayerNorm is as follows:

$$H' = \frac{H-\mu}{\sigma}\gamma + \beta \tag{22}$$

where H represents the input features, μ and σ are the mean and standard deviation of the features, and γ and β are learnable parameters. Through Layer Normalization, the Informer decoder can reduce internal covariate shift, thereby enhancing the stability of training.

2.5.3. Self-Attention Block

After Layer Normalization, the data enters the Self-Attention Block, further enhancing feature representation. On the basis of modeling the historical values of the target sequence, the decoder uses the encoder-decoder Multi-head Attention mechanism to combine the short-term dynamic features of the target sequence with the global features from the encoder. Specifically, the query matrix Q comes from the decoder, and the key matrix K and value matrix V come from the encoder's global feature representation:

$$H_{context} = Softmax\left(\frac{Q_{decoder}K_{encoder}^{\top}}{\sqrt{d_k}}\right) \cdot V_{encoder} \tag{23}$$

This mechanism captures the time-step relationships between the target sequence and the encoder features, generating a context representation that integrates both short-term and long-term features. After the self-attention computation, the decoder's output features pass through a Fully Connected Layer (FC) to map the high-dimensional features to the final prediction result, i.e., the future latitude and longitude of the TC track:

$$(Lat_{t+n}, Lon_{t+n}) = FC(H') \tag{24}$$

3. Results

In this section, we mainly present the experimental dataset and the comparison results of various metrics for TC track prediction to validate the performance of the proposed model. The experiments were conducted on a workstation with an Intel® Core™ i7-6498 DU CPU@2.50 GHz and 256 GB of main memory. All deep learning methods were implemented using Python 3.6 and several open-source machine learning libraries, including scikit-learn 0.20.3 and TensorFlow 2.1.

3.1. Dataset

This experiment utilizes a comprehensive TC track dataset covering the Western North Pacific (WNP) region. The dataset is a collaborative effort between the Japan Meteorological Agency (JMA) [35] and the International Best Track Archive for Climate Stewardship (IBTrACS) [36]. The dataset contains records of all TC tracks from 1951 to 2022, including detailed TC track and associated meteorological factors. The dataset covers the genesis, development, movement, and dissipation processes of thousands of TCs, comprising over 60,000 timestamped records. Each record includes key features such as the TC center's latitude and longitude, maximum wind speed, minimum sea-level pressure (MSLP), storm classification, storm radius, and wind intensity. The records are sampled at 6-h intervals, capturing the critical time points throughout the lifecycle of each TC.

Specifically, each record includes the following features: TCID, name, timestamp, center longitude and latitude, minimum pressure, maximum wind speed, wind circles Radius (Radius30, Radius50, and Radius70), movement speed, movement direction, and TC classification.

To evaluate the performance of the proposed model, as shown in Table 1, the dataset was randomly divided into training and testing sets, with 90% of the TC records used for training and the remaining 10% for testing. After removing records with missing or incomplete data, a total of 1748 TCs were used for training, and 195 TCs were reserved for testing.

Table 1. The detailed experiment dataset.

Parameters	Value
Dataset size	Year 1951–2022; total 1943 records
Training set	1748 records
Testing set	195 records
Train/Test	9:1
Temporal dimension	6 h

To ensure the statistical reliability of the results and mitigate any bias introduced by a single random split, as shown in Table 2, we employed a 10-fold cross-validation strategy. The entire dataset (1943 TCs) was partitioned into ten subsets. In each round, one subset was utilized for testing while the remaining nine served as the training set. The final performance metrics reported are the averages across these ten independent folds.

Table 2. The 10-fold cross-validation results (APE, unit: km) of LFInformer at different forecast horizons.

Forecast	6 h	12 h	48 h
Fold 1	72.39	117.72	168.64
Fold 2	73.52	119.45	170.15
Fold 3	71.85	116.30	167.22
Fold 4	72.90	118.10	169.50
Fold 5	71.45	116.93	167.80
Fold 6	73.01	117.83	169.24

Table 2. *Cont.*

Forecast	6 h	12 h	48 h
Fold 7	72.74	119.31	166.45
Fold 8	73.97	116.61	168.72
Fold 9	71.63	119.87	167.42
Fold 10	72.12	116.87	168.95
Mean ± Std	72.42 ± 0.73	117.69 ± 1.12	168.66 ± 1.08

3.2. *Performance Measurements*

The objective evaluation metrics used in this study include Mean Absolute Error (MAE), Root Mean Square Error (RMSE), Mean Absolute Percentage Error (MAPE), and Average Positioning Error (APE). MAE represents the average of the absolute differences between the predicted values and the observed values. It is calculated as follows:

$$\mathrm{MAE} = \frac{\sum_{i=1}^{n} |P_i - O_i|}{n} \tag{25}$$

where n is the number of records in the test set, P_i is the predicted value, and O_i is the observed value, A larger MAE indicates poorer model performance.

RMSE is the square root of the ratio between the squared deviations of predicted values from the actual values and the number of observations. It is defined as:

$$\mathrm{RMSE} = \sqrt{\frac{\sum_{i=1}^{n} (P_i - O_i)^2}{n}} \tag{26}$$

MAPE represents the average percentage error of the predictions relative to the actual values. It takes into account both the magnitude of the prediction error and its proportion relative to the actual track. A MAPE value of 0 indicates a perfect model, while values greater than 1 suggest poorer performance. It is calculated as:

$$\mathrm{MAPE} = \frac{1}{n} \sum_{i=1}^{n} \left| \frac{P_i - O_i}{O_i} \right| \tag{27}$$

APE is calculated using the formula below, where R is the radius of the Earth, $\mathrm{Lat_{real}}$, $\mathrm{Lon_{pred}}$ represent the actual latitude and longitude of the tropical cyclone, and $\mathrm{Lat_{real}}$, $\mathrm{Lon_{pred}}$ represent the predicted position. APE measures the distance between the actual and predicted positions, with the unit being kilometers. A smaller APE indicates better model performance.

$$\begin{aligned} \mathrm{APE} = R \times \arccos\big(&\sin(\mathrm{Lat_{real}}) \times \sin(\mathrm{Lat_{pred}}) \\ &\times \cos(\mathrm{Lon_{real}} - \mathrm{Lon_{pred}}) \\ &+ \cos(\mathrm{Lat_{real}}) \times \cos(\mathrm{Lat_{pred}})\big) \times \pi/180 \end{aligned} \tag{28}$$

3.3. *Experimental Results*

3.3.1. Ablation Experiments

To evaluate the performance of each major network component designed in this study for TC track prediction, we conducted ablation experiments with different module combinations. The evaluation metrics include MAE, RMSE.

Table 3 presents the results of the ablation study. Taking the 6-h TC prediction experiment as an example, the prediction performance improves significantly as more modules are added to the model configuration. When the LSTM module is used alone, it can handle short-term dependencies in TC track. However, the MAE and RMSE values for both longitude and latitude predictions remain relatively high, indicating that the model has considerable errors in capturing the dynamic turning in TC track. In contrast, the Informer model demonstrates significant advantages in the 6-h TC track prediction task due to its powerful ability to capture global dependencies. Notably, the prediction performance shows clear improvements in RMSE, indicating that the Informer model is more effective in capturing the long-term trends of TC track. By combining the Informer and LSTM modules, the prediction accuracy is further improved, particularly with significant reductions in MAE and RMSE values. This demonstrates that the integration of both modules effectively compensates for the limitations of individual models in capturing short-term and long-term dependencies. When the LSTM module is combined with the FECAM block, the model's capability to handle the periodic change of meteorological factors patterns in TC track is further enhanced, resulting in lower RMSE values. This highlights FECAM's role in improving the model's sensitivity to the cyclical fluctuations of the meteorological environment. Ultimately, the combination of the Informer, LSTM and FECAM achieves the best performance across all evaluation metrics, especially with notable improvements in metrics.

In summary, with the continuous optimization of the model architecture and the enhancement of its components, the accuracy of TC track prediction has been significantly improved. Our proposed model (Informer + LSTM + FECAM) achieves the best prediction performance across all evaluation metrics. The results presented in Table 3 validate the effectiveness of each component in the LFInformer model. As more modules are integrated, the model performance improves consistently, fully demonstrating the synergistic effects among the components and providing strong support for further enhancing prediction accuracy. In addition, the visual comparison in Figure 7 highlights the path prediction differences across module combinations. An inset around the turning point highlights the differences, where other module combinations show larger deviations. The proposed model aligns best with the actual track, visually validating its superior performance.

Table 3. Taking 6-h typhoon track prediction as an example, we conduct a comparative evaluation of different module combinations and the evaluation based on MAE (°) and RMSE (°) metrics, bold values indicate the best performance for each metric.

Methods	Longitude (°E)		Latitude (°N)	
	MAE	RMSE	MAE	RMSE
LSTM	0.721	0.836	0.784	0.891
Informer	0.657	0.713	0.742	0.818
Informer + LSTM	0.634	0.695	0.724	0.790
LSTM + FECAM	0.597	0.673	0.651	0.776
LFInformer	**0.555**	**0.623**	**0.637**	**0.758**

3.3.2. Comparison Experiment

This section presents a performance comparison between the LFInformer model and traditional deep learning methods (such as RNN, GRU, LSTM, and Transformer) in TC track prediction. To comprehensively evaluate the predictive capabilities of each model, we selected multiple forecasting time steps (6 h, 12 h, 24 h, and 48 h) and calculated the Absolute Position Error (APE) for each method. The main objective of this experiment is to verify whether the LFInformer model can achieve more accurate TC track predictions than traditional methods across different time scales, particularly in longer-term forecasts.

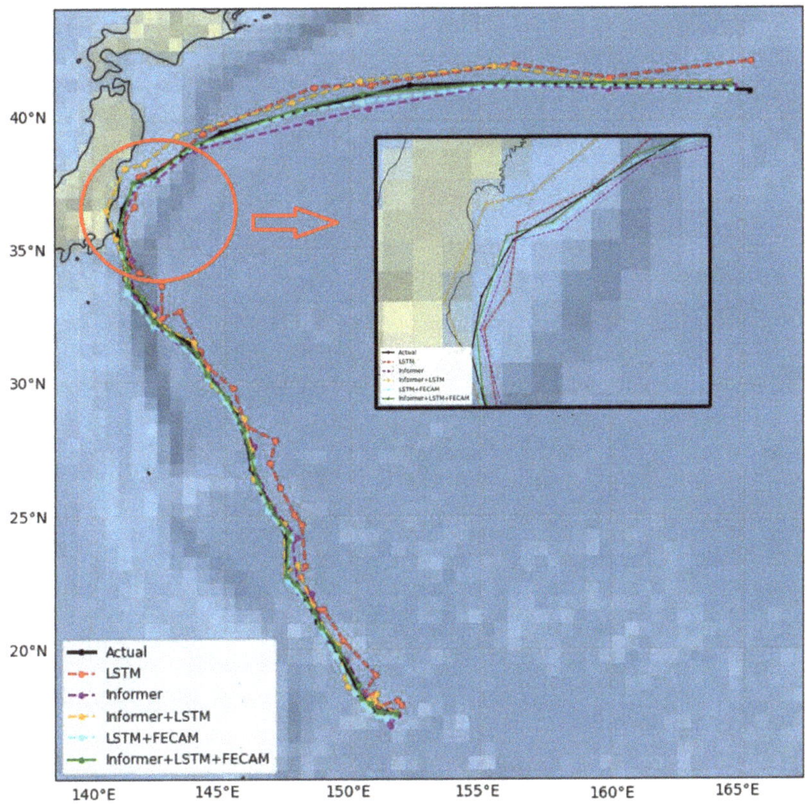

Figure 7. The visualization of ablation experiments.The **red circle** highlights the key turning point region of the typhoon track, and the **red arrow** indicates the corresponding zoomed-in view shown in the inset box. The inset allows for a detailed comparison of different models' performances in this critical area.

As shown in Table 4, the LFInformer model achieved APE values of 72.39 km and 117.72 km for the 6-h and 12-h forecasts, respectively, which are significantly lower than those of other models. For instance, the APE of the RNN model was 106.76 km for the 6-h prediction, while GRU, LSTM, and Transformer recorded 94.81 km, 89.68 km, and 86.73 km, respectively. These results indicate that LFInformer demonstrates outstanding performance in short-term predictions, effectively capturing local variations in TC track. As the prediction horizon increases, although all models experience a rise in prediction error, LFInformer maintains relatively stable performance. In the 48-h forecast, LFInformer achieves an APE of 168.64 km, which is slightly lower than that of LSTM and Transformer, showcasing its advantage in long-term track prediction.To further illustrate the model's prediction performance, Figure 8 presents the APE of LFInformer at each forecast step, showing its stable accuracy across time and its ability to maintain low errors during sudden track changes.

Furthermore, we analyze and compare the multi–lead-time errors. Table 5 reports the APE at 12, 24, and 48 h for the proposed model and the traditional baselines. At short and medium leads, our errors are still higher than recent operational forecasts, reflecting the advantage of operations in real-time data assimilation and multi-source predictors. At 48 h, however, the proposed model attains 168.64 km, already close to operational skill and lower than all learning baselines, indicating clear potential for longer-range track prediction.

Table 4. APE metric comparison: LFInformer versus deep learning methods for TC prediction in the future of 6 h, 12 h, 24 h, and 48 h, bold values indicate the best performance for each metric.

Methods	6 h	12 h	24 h	48 h
RNN	106.76 km	152.91 km	186.49 km	178.61 km
GRU	94.81 km	148.47 km	173.68 km	174.63 km
LSTM	89.68 km	150.61 km	172.53 km	172.25 km
Transformer	86.73 km	135.12 km	172.02 km	173.46 km
LFInformer	**72.39** km	**117.72** km	**145.31** km	**168.64** km

In summary, LFInformer demonstrates high prediction accuracy across all forecast horizons for TC track prediction, with particularly superior performance in short-term forecasts.

Table 5. APE metric comparison: LFInformer versus deep learning methods for TC prediction in the future of 12, 24, and 48 h, bold values indicate the best performance for each metric.

Methods	12 h	24 h	48 h
China Meteorological Administration	**70.37** km	93.88 km	166.52 km
Japan Meteorological Agency	74.13 km	99.11 km	167.23 km
Joint Typhoon Warning Center	76.54 km	**93.39** km	**165.31** km
LFInformer	117.72 km	145.31 km	168.64 km

Table 6 presents a comparison of the objective evaluation metrics—MAE, RMSE, and MAPE—across the RNN, GRU, LSTM, Transformer, and LFInformer models. Taking the 6-h forecast of tropical cyclone positions as an example, the best results are highlighted in bold. As shown in Table 6, the LFInformer model consistently outperforms the other models across all three metrics. Specifically, LFInformer achieves a MAE of 0.555, RMSE of 0.623, and MAPE of 0.585, which are significantly lower than those of RNN and GRU. Compared to LSTM and Transformer, LFInformer also shows notable advantages in MAE and RMSE, indicating its superior ability to capture the dynamic variations in TC track and reduce prediction errors.

Additionally, LFInformer excels in the MAPE metric, with a value of 0.585, which is significantly lower than that of LSTM and Transformer. This demonstrates that LFInformer has a strong ability to capture relative errors in TC track predictions. LFInformer not only performs excellently in short-term predictions but also consistently maintains lower error values in medium- and long-term predictions, with a particularly notable improvement in MAPE.

Based on the analysis of these experimental results, LFInformer outperforms traditional deep learning methods across multiple evaluation metrics, demonstrating its superiority in the TC track prediction task. This confirms that by combining the long-sequence processing capabilities of Informer, the short-term handling capabilities of LSTM, and the modeling of periodic change through FECAM, the model can effectively improve the accuracy and stability in capturing TC track variations.

Taking the prediction of the TC's position for the next 48 h as an example, Table 7 presents a comparison of the RNN, GRU, LSTM, Transformer, and LFInformer models based on objective evaluation metrics such as MAE, RMSE, and MAPE, analyzing their predictive performance in both longitude and latitude directions, This indicates that while traditional models perform well in short-term predictions, their accuracy significantly decreases over longer time horizons, such as in the 48-h TC track prediction. Particularly in the latitude direction, the MAPE of Transformer and LSTM are 3.221 and 3.132, respectively, while LFInformer achieves a lower value of 2.276, demonstrating a significant improvement in prediction accuracy.

Table 6. Comparative analysis of LFInformer versus conventional deep learning approaches (RNN, GRU, LSTM, Transformer) in 6-h typhoon forecasting: evaluation based on MAE (°), RMSE (°), and MAPE (°) metrics, bold values indicate the best performance for each metric.

Methods	Dataset	Longitude			Latitude		
		MAE	RMSE	MAPE	MAE	RMSE	MAPE
RNN	Train	0.863	0.921	0.895	0.972	1.032	0.982
	Validation	0.899	0.952	0.912	1.049	1.052	1.025
	Test	0.852	0.950	0.891	1.010	0.991	0.997
GRU	Train	0.792	0.912	0.883	0.863	0.932	0.989
	Validation	0.812	0.962	0.897	0.892	1.023	1.043
	Test	0.800	0.948	0.876	0.878	0.942	0.997
LSTM	Train	0.721	0.836	0.866	0.784	0.838	0.829
	Validation	0.753	0.882	0.911	0.805	0.862	0.860
	Test	0.739	0.863	0.822	0.792	0.849	0.851
Transformer	Train	0.682	0.768	0.7426	0.791	0.831	0.815
	Validation	0.685	0.797	0.7796	0.826	0.853	0.832
	Test	0.672	0.784	0.7518	0.752	0.834	0.824
LFInformer	Train	0.563	0.635	0.589	0.640	0.796	0.754
	Validation	0.563	0.624	0.593	0.648	0.761	0.751
	Test	**0.555**	**0.555**	**0.585**	**0.637**	**0.758**	**0.746**

Table 7. Comparative analysis of LFInformer versus conventional deep learning approaches (RNN, GRU, LSTM, Transformer) in 48-h typhoon forecasting: evaluation based on MAE (°), RMSE (°), and MAPE (°) metrics, bold values indicate the best performance for each metric.

Methods	Dataset	Longitude			Latitude		
		MAE	RMSE	MAPE	MAE	RMSE	MAPE
RNN	Train	3.421	3.754	3.853	3.752	4.032	4.139
	Validation	3.125	3.532	3.673	3.589	3.822	3.394
	Test	3.121	3.452	3.616	3.552	3.731	3.824
GRU	Train	3.212	3.885	3.562	3.423	3.653	3.822
	Validation	3.107	3.414	3.445	3.226	3.348	3.857
	Test	3.198	3.323	3.437	3.223	3.741	3.755
LSTM	Train	2.995	3.218	3.525	3.315	3.136	3.845
	Validation	2.883	3.505	3.119	3.208	3.227	3.334
	Test	2.679	3.406	3.119	3.201	3.226	3.132
Transformer	Train	2.880	3.604	3.512	2.798	3.715	3.823
	Validation	2.474	2.998	3.307	2.692	3.209	3.617
	Test	2.368	2.493	3.204	2.389	3.512	3.221
LFInformer	Train	2.355	2.578	2.269	2.678	2.692	2.285
	Validation	2.249	2.373	2.256	2.475	2.689	2.278
	Test	**2.142**	**2.283**	**2.557**	**2.271**	**2.383**	**2.276**

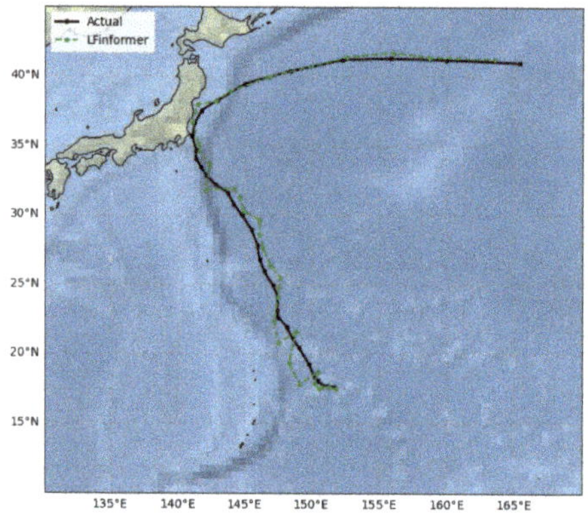

Time	APE (km)	Time	APE (km)
T1	27.54 km	T19	95.10 km
T2	14.91 km	T20	92.98 km
T3	18.22 km	T21	134.74 km
T4	61.35 km	T22	112.33 km
T5	48.74 km	T23	65.46 km
T6	96.59 km	T24	105.92 km
T7	289.47 km	T25	25.77 km
T8	202.03 km	T26	71.26 km
T9	79.13 km	T27	14.05 km
T10	208.45 km	T28	55.07 km
T11	89.51 km	T29	15.00 km
T12	3.12 km	T30	113.65 km
T13	48.53 km	T31	148.40 km
T14	127.08 km	T32	35.87 km
T15	97.10 km	T33	50.09 km
T16	49.13 km	T34	107.57 km
T17	72.09 km	T35	159.28 km
T18	141.17 km		

(**a**)

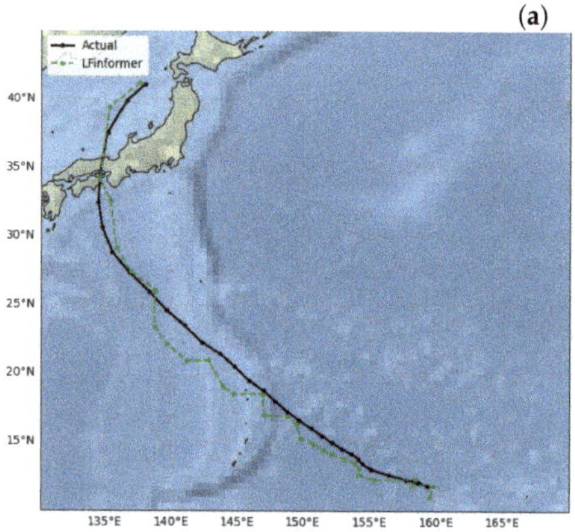

Time	APE (km)	Time	APE (km)
T1	86.89 km	T17	152.65 km
T2	122.22 km	T18	222.86 km
T3	61.62 km	T19	266.41 km
T4	228.67 km	T20	155.55 km
T5	89.89 km	T21	287.38 km
T6	91.90 km	T22	271.75 km
T7	82.15 km	T23	292.35 km
T8	76.11 km	T24	239.16 km
T9	87.90 km	T25	205.76 km
T10	93.25 km	T26	197.38 km
T11	115.95 km	T27	89.94 km
T12	122.03 km	T28	42.89 km
T13	147.75 km	T29	390.05 km
T14	127.57 km	T30	121.76 km
T15	154.56 km	T31	31.93 km
T16	198.92 km		

(**b**)

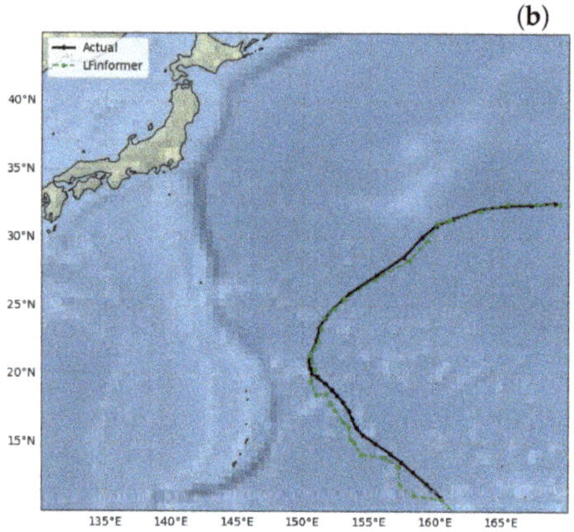

Time	APE (km)	Time	APE (km)
T1	205.24 km	T18	20.33 km
T2	174.86 km	T19	3.96 km
T3	189.54 km	T20	5.05 km
T4	186.37 km	T21	14.30 km
T5	123.86 km	T22	10.52 km
T6	129.95 km	T23	5.94 km
T7	158.35 km	T24	4.70 km
T8	124.56 km	T25	49.79 km
T9	167.32 km	T26	33.23 km
T10	127.41 km	T27	50.26 km
T11	163.97 km	T28	46.76 km
T12	122.91 km	T29	20.55 km
T13	121.45 km	T30	33.31 km
T14	78.54 km	T31	7.58 km
T15	141.56 km	T32	17.01 km
T16	53.31 km	T33	43.39 km
T17	42.51 km	T34	35.96 km

(**c**)

Figure 8. APE visualization of 12-h track predictions: (**a**) TC "Shanshan" (2018), (**b**) TC "Cimaron" (2018), (**c**) TC "Halong" (2019).

Specifically, LFInformer exhibits lower MAE, RMSE, and MAPE values in both the longitude and latitude directions compared to other models. In particular, in terms of MAPE, LFInformer achieves the lowest values of 2.557 and 2.276 in longitude and latitude, respectively, while the MAPE of other models is generally higher than that of LFInformer, indicating LFInformer's outstanding performance in reducing relative errors. This further demonstrates that LFInformer, by integrating the advantages of LSTM and Informer and introducing the FECAM, can effectively improve the accuracy of TC track prediction, especially in long-term prediction tasks.

Through comparative analysis, LFInformer demonstrated stronger resistance to interference and better performance in the 48-h prediction task. Compared to deep learning models such as LSTM and Transformer, it provides more accurate predictions. Experimental results show that LFInformer not only effectively improves prediction accuracy but also maintains reasonable control over computational complexity, making it suitable for long-term TC track prediction tasks in practical applications.

Beyond prediction accuracy, we further evaluated the model's efficiency by measuring the average inference time per sample. As shown in Table 8, the LFInformer achieves an inference latency of 5.89 ms. While this incurs a slight computational cost compared to the baseline Informer (4.85 ms) attributed to the integration of the FECAM and LSTM components, it remains substantially faster than the standard Transformer (12.56 ms). This validates the design choice of using an Informer backbone to ensure high efficiency without compromising the enhanced predictive capability.

3.3.3. Visualization Experiment

To further demonstrate the performance of the proposed model in TC track prediction, particularly in predicting track mutation points, we used visualization to compare the predictions of LSTM, GRU, LFInformer, and the observed cyclone track for 6-h, 12-h, and 48-h forecasts. Figures 9–11 show the predicted track for TCs "Shanshan" (2018) and "Cimaron" (2018).

Table 8. Comparison of Average Inference Time of Different Models.

Model	Average Inference Time (ms)
RNN	3.12
GRU	3.45
LSTM	4.08
Transformer	12.56
Informer	4.85
LFInformer	5.89

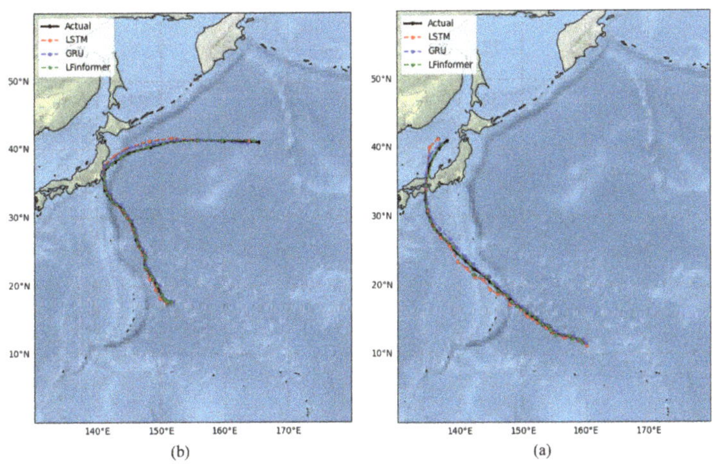

Figure 9. 6 h forecast results (markers denote 6-h intervals): (**a**) Track prediction of TC "Shanshan" (2018). (**b**) Track prediction of TC "Cimaron" (2018).

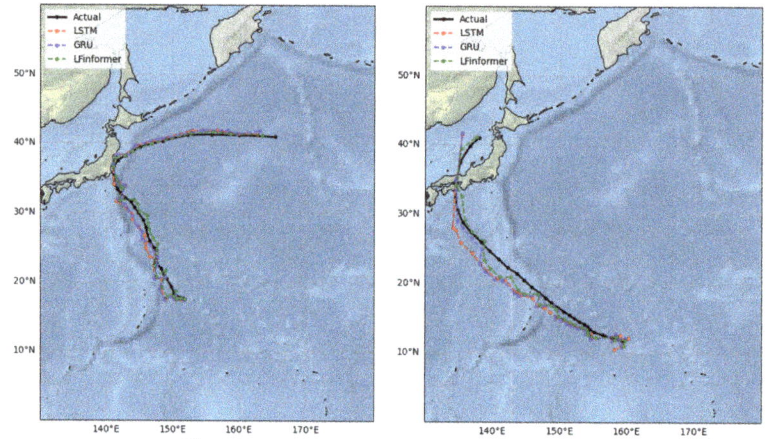

Figure 10. 12 h forecast results (markers denote 6-h intervals): (**a**) Track prediction of TC "Shanshan" (2018). (**b**) Track prediction of TC "Cimaron" (2018).

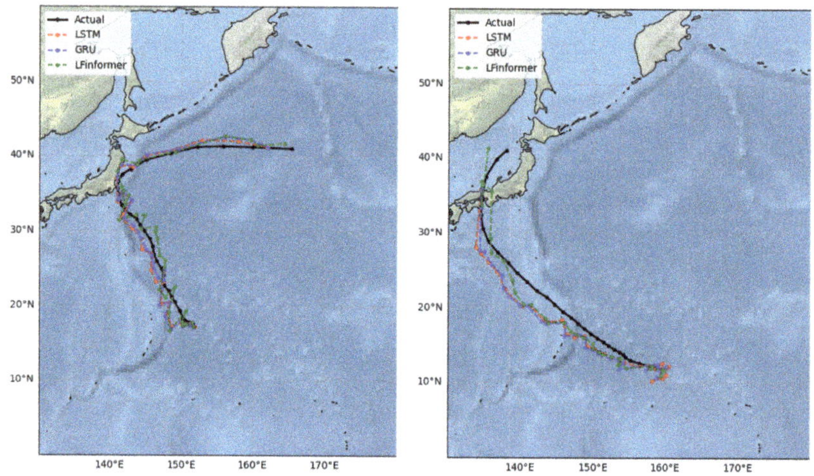

Figure 11. 48 h forecast results (markers denote 6-h intervals): (**a**) Track prediction of TC "Shanshan" (2018). (**b**) Track prediction of TC "Cimaron" (2018).

In the 6-h prediction results, all models broadly reproduced the synoptic-scale evolution of TC trajectories; in particular, for TCs "Shanshan" and "Cimaron", The predicted tracks closely matched the observed paths, whereas LFInformer achieved the lowest overall displacement error, indicating more accurate short-term TC track prediction and better ability to capture the sudden track changes in the TC track.

In the 12-h prediction results, although all models showed some deviation in the predicted track, LFInformer still managed to capture the TC track's trend more effectively. For TC "Cimaron", LFInformer stood out, showing the smallest deviation from the actual track, and for TC "Shanshan", LFInformer demonstrates superior performance in medium-range forecasts, capturing localized dynamic turns in TC tracks more accurately than competing models.

For the 48-h prediction results, all models started to show significant deviations in their predicted tracks, especially for TC "Shanshan", where the predictions from other models deviated notably from the actual track. However, LFInformer still performed relatively stably, particularly in the case of TC "Cimaron", where LFInformer's predicted track had the smallest difference from the observed track. This demonstrates LFInformer's strong stability and error saturation in long-term time series predictions.

3.3.4. Research on Critical Turning Points

The above results validate that the LFInformer model, through the synergistic design of the FECAM and the ProbSparse Self-Attention mechanism, significantly improves its ability to predict sudden turning in TC track. To further demonstrate the model's attention to track turning points, three representative cases were selected, all of which historically exhibited pronounced track deviations: TC "Hagibis" (2019) [37], TC "Lekima" (2019) [38], and TC Maria (2018) [39]. Among them, the turning point of "Lekima" occurred over the ocean, that of "Hagibis" occurred just before landfall, and that of Maria occurred after landfall. As shown in Figure 12, the proposed model exhibits notable advantages in predicting track turning points, which are marked by red pentagrams.

For example, in the northwestward turning event of TC "Lekima" shown in Figure 12a, the unexpected breakdown of the subtropical high led to a sudden turning in the TC's movement direction. Through the collaborative mechanism of FECAM and the ProbSparse Self-Attention, the LFInformer model is able to detect subtle turning in the ridge line of the subtropical high in advance. As a result, the predicted track closely follows the actual track near the turning point, demonstrating strong robustness. In contrast, LSTM and GRU fail to reflect the shift in TC trends, as their predictions continue to follow a northward inertial track due to their neglect of the periodic change of atmospheric circulation. This case highlights that the ProbSparse Self-Attention mechanism in the proposed model effectively selects the most relevant time steps from long historical sequences through a sparsity strategy, thereby enhancing prediction accuracy. In the case of TC "Hagibis" shown in Figure 12b, the TC's track exhibited a significant northeastward turn during the forecast period due to the sudden intensification of the westerly jet stream [40]. The predicted track of the proposed LFInformer model (blue) almost completely overlaps with the actual track (red) at this turning point, demonstrating accurate detection of the sudden turning of track. This performance is attributed to the model's internal frequency-domain enhancement module, which amplifies the periodic features of the westerly trough, and the ProbSparse Self-Attention mechanism, which emphasizes critical turning points. In contrast, LSTM and GRU, limited by their short-term memory characteristics, failed to effectively respond to the long-term evolution of the circulation system, resulting in continued northwestward deviation after the turning point and significant departure from the actual track. In the case of TC "Maria" shown in Figure 12c, the TC experienced a turning event after landfall due to the influence of inland convection, affecting pressure, wind speed, and wind direction. The proposed model still managed to accurately track and predict this turning point.

Figure 13 compares the predicted tracks of LSTM, GRU, and the proposed LFInformer with the observed track for TC "Lionrock". As a canonical abrupt re-curvature event, "Lionrock" provides a stringent test of a model's sensitivity to turning cues along the track. The cyclone exhibited an abrupt northward re-curvature over the northern South China Sea. While all baselines roughly follow the pre-turn southwesterly leg, they fail to respond promptly to the regime shift at the second turning point. Following the turning point, the LSTM and GRU models tend to retain a west–northwest heading for several steps, which manifests as a left-sided cross-track bias with respect to the observed track. By contrast,

LFInformer changes heading earlier and keeps closer to the observed segment after the turning, yielding a visibly smaller cross-track error. In the inset around the second turning point. The enlargement highlights that LFInformer begins its northward adjustment within one–two time steps of the observed turn and then stays nearly on the right of the observed track, avoiding the overshoot seen in the baselines. This behavior is consistent with our design: the FECAM block enhances frequency-domain cues associated with the evolving steering flow, and the ProbSparse self-attention selectively focuses on the most informative historical steps, enabling earlier and more robust detection of the turning signal.

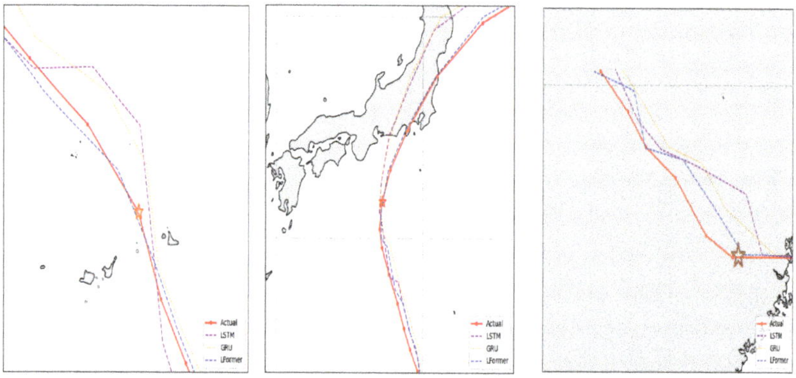

(a) (b) (c)

Figure 12. (**a**) Local map of track turning points for TC "Lekima" (2019) (**b**) Local map of track turning points for TC "Hagibis" (2019) (**c**) Local map of track turning points for TC "Maria" (2018). The red pentagrams symbols represent the turning points of the TC track.

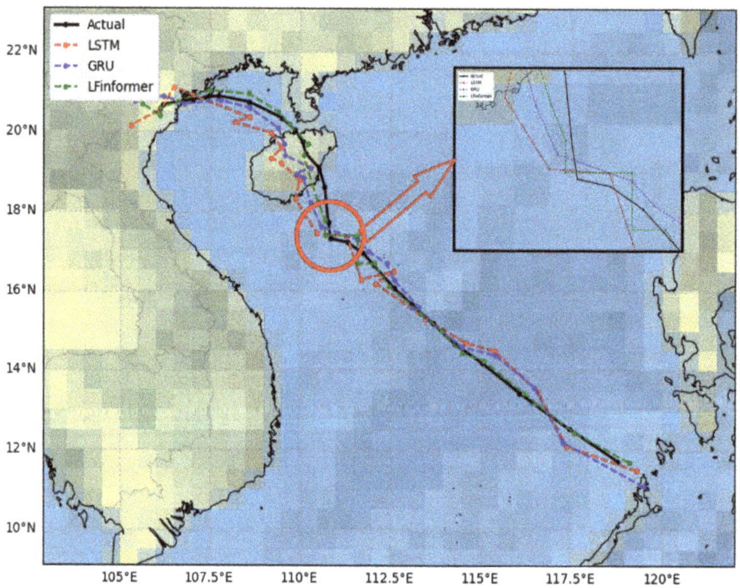

Figure 13. Case study of TC "Lionrock" (2021): model comparison for the abrupt northward recurvature. The **red circle** highlights the key turning point region of the typhoon track, and the **red arrow** indicates the corresponding zoomed-in view shown in the inset box. The inset allows for a detailed comparison of different models' performances in this critical area.

The above experiments demonstrate that the proposed LFInformer performs excellently in predicting complex TC movement track. The predicted tracks of the model closely match the actual track at turning points, further validating the effectiveness of the FECAM and ProbSparse Self-Attention mechanisms in joint modeling across frequency and time domains. The synergistic effect of the Frequency-domain Enhanced Channel Attention Mechanism and the ProbSparse Self-Attention mechanism provides a reliable theoretical foundation and technical support for accurately forecasting sudden turning in TC tracks.

4. Conclusions

In this study, in light of the limitations faced by current deep learning models in TC prediction—such as single-feature fusion and the difficulty of balancing short-term and long-term track forecasting, particularly in predicting sudden turning of track.we propose a hybrid deep learning model based on the Informer architecture, integrated with both FECAM block and LSTM modules, to enable accurate TC track prediction. The Informer module captures long-term dependencies. The LSTM focuses on short-term temporal features. FECAM, using frequency-domain enhanced channel attention, models the track's sensitivity to periodic meteorological environment patterns change. Overall, the proposed model efficiently handles complex TC time series data and achieves accurate track prediction.

This study conducted extensive experiments based on the JMA and IBTrACS dataset, including 6-h, 12-h, 24-h, and 48-h track predictions. The experimental results demonstrate that, compared with traditional models, the proposed model exhibits superior performance in short-term, mid-term, and long-term track predictions, especially in capturing the global trends and local dynamic turning of TC track. Ablation studies further validate the importance of the LSTM and FECAM block within the overall framework, indicating that the synergy among modules effectively enhances the comprehensive performance of the model.

Nevertheless, Several limitations should be noted. Firstly, the current model focuses solely on TC track prediction, without addressing other key TC characteristics such as wind speed and atmospheric pressure. Secondly, for TCs with significant intensity fluctuations, the model's predictions still exhibit noticeable deviations. In addition, the forecasting time range can be further extended to fully leverage the model's long-sequence prediction capability, and the ability to predict over longer periods remains to be further explored. Future research will extend LFInformer toward joint track–intensity forecasting by incorporating intensity indicators as additional prediction targets. Specifically, we will adopt a multi-task design with a shared Informer–FECAM–LSTM backbone and separate output heads for track and intensity, and enrich the inputs with intensity-related environmental factors to better represent the physical drivers of intensity changes. We will further investigate appropriate loss balancing between track and intensity objectives and evaluate intensity performance using standard MAE/RMSE metrics alongside track APE, aiming to improve long-term prediction of both TC track and characteristics.

Author Contributions: Conceptualization, F.M. and L.Z.; methodology, F.M. and L.Z.; software, F.M.; validation, F.M., X.X. and L.Z.; formal analysis, L.Z.; investigation, F.M. and X.X.; resources, L.Z.; data curation, F.M. and X.X.; writing—original draft preparation, F.M. and X.X.; writing—review and editing, X.X. and L.Z.; visualization, F.M. and X.X.; supervision, L.Z.; project administration, L.Z.; funding acquisition, L.Z. All authors have read and agreed to the published version of the manuscript.

Funding: This work was supported by the Shanghai Typhoon Research Foundation from Shanghai Typhoon Institute of China Meteorological Administration under Grant TFJJ202208.

Data Availability Statement: JMA data are available at the Japan Meteorological Agency RSMC archive JMA track archives (https://www.jma.go.jp/jma/jma-eng/jma-center/rsmc-hp-pub-eg/trackarchives.html) (accessed on 26 December 2024). IBTrACS data are available from NOAA NCEI IBTrACS (https://www.ncei.noaa.gov/products/international-best-track-archive) (accessed on 26 December 2024).

Conflicts of Interest: The authors declare no conflicts of interest.

References

1. Zhang, Q.; Wu, L.; Liu, Q. Tropical cyclone damages in China 1983–2006. *Bull. Am. Meteorol. Soc.* **2009**, *90*, 489–496. [CrossRef]
2. China Meteorological Administration. *China Climate Bulletin 2017*; China Meteorological Administration: Beijing, China, 2018. Available online: https://www.wmc-bj.net/publish/cms/2e8413abe9d443aba2bec7ed6b2946d2/index.html (accessed on 30 July 2025).
3. Barnes, M.D.; Hanson, C.L.; Novilla, L.M.; Meacham, A.T.; McIntyre, E.; Erickson, B.C. Analysis of media agenda setting during and after Hurricane Katrina: Implications for emergency preparedness, disaster response, and disaster policy. *Am. J. Public Health* **2008**, *98*, 604–610. [CrossRef] [PubMed]
4. Yanai, M. Dynamical aspects of typhoon formation. *J. Meteorol. Soc. Japan. Ser. II* **1961**, *39*, 282–309. [CrossRef]
5. Neumann, C.J. *An Alternate to the HURRAN Tropical Cyclone Forecast System*; NOAA Technical Memorandum NWS SR-62; NOAA National Weather Service: Fort Worth, TX, USA, 1972. Available online: https://repository.library.noaa.gov/view/noaa/3605 (accessed on 30 July 2025).
6. Al-Yahyai, S.; Charabi, Y.; Gastli, A. Review of the use of numerical weather prediction (NWP) models for wind energy assessment. *Renew. Sustain. Energy Rev.* **2010**, *14*, 3192–3198. [CrossRef]
7. Le Marshall, J.; Leslie, L.; Morison, R.; Pescod, N.; Seecamp, R.; Spinoso, C. Recent developments in the continuous assimilation of satellite wind data for tropical cyclone track forecasting. *Adv. Space Res.* **2000**, *25*, 1077–1080. [CrossRef]
8. Thu, H.N.; Thanh, N.P.T.; Thanh, H.V.; Thanh, H.P.; Tuan, L.T.; Thi, T.D.; Duy, T.T.; Phuong, H.N.T. Development of an R-CLIPER model using GSMaP and TRMM precipitation data for tropical cyclones affecting Vietnam. *Nat. Hazards* **2025**, *121*, 1241–1263. [CrossRef]
9. Sherstinsky, A. Fundamentals of recurrent neural network (RNN) and long short-term memory (LSTM) network. *Phys. D Nonlinear Phenom.* **2020**, *404*, 132306. [CrossRef]
10. Dey, R.; Salem, F.M. Gate-variants of gated recurrent unit (GRU) neural networks. In Proceedings of the 2017 IEEE 60th International Midwest Symposium on Circuits and Systems (MWSCAS), Boston, MA, USA, 6–9 August 2017; IEEE: Piscataway, NJ, USA, 2017; pp. 1597–1600. [CrossRef]
11. Lian, J.; Dong, P.; Zhang, Y.; Pan, J.; Liu, K. A novel data-driven tropical cyclone track prediction model based on CNN and GRU with multi-dimensional feature selection. *IEEE Access* **2020**, *8*, 97114–97128. [CrossRef]
12. Lian, J.; Dong, P.; Zhang, Y.; Pan, J. A novel deep learning approach for tropical cyclone track prediction based on auto-encoder and gated recurrent unit networks. *Appl. Sci.* **2020**, *10*, 3965. [CrossRef]
13. Noh, S.H. Analysis of gradient vanishing of RNNs and performance comparison. *Information* **2021**, *12*, 442. [CrossRef]
14. Kumar, J.S.; Venkataraman, V.; Meganathan, S.; Krithivasan, K. Tropical cyclone intensity and track prediction in the Bay of Bengal using LSTM-CSO method. *IEEE Access* **2023**, *11*, 81613–81622. [CrossRef]
15. Fei, H.; Tan, F. Bidirectional grid long short-term memory (bigridlstm): A method to address context-sensitivity and vanishing gradient. *Algorithms* **2018**, *11*, 172. [CrossRef]
16. Gao, S.; Zhao, P.; Pan, B.; Li, Y.; Zhou, M.; Xu, J.; Zhong, S.; Shi, Z. A nowcasting model for the prediction of typhoon tracks based on a long short term memory neural network. *Acta Oceanol. Sin.* **2018**, *37*, 8–12. [CrossRef]
17. Tong, B.; Wang, X.; Fu, J.; Chan, P.; He, Y. Short-term prediction of the intensity and track of tropical cyclone via ConvLSTM model. *J. Wind. Eng. Ind. Aerodyn.* **2022**, *226*, 105026. [CrossRef]
18. Kandadi, T.; Shankarlingam, G. Drawbacks of lstm algorithm: A case study. *Preprints* **2025**. [CrossRef]
19. Zhou, H.; Zhang, S.; Peng, J.; Zhang, S.; Li, J.; Xiong, H.; Zhang, W. Informer: Beyond Efficient Transformer for Long Sequence Time-Series Forecasting. *Proc. AAAI Conf. Artif. Intell.* **2021**, *35*, 11106–11115. [CrossRef]
20. Lin, Z.; Chu, J.E.; Ham, Y.G. Enhancing tropical cyclone track and intensity predictions with the OWZP-Transformer model. *npj Artif. Intell.* **2025**, *1*, 33. [CrossRef]

21. Lam, R.; Sanchez-Gonzalez, A.; Willson, M.; Wirnsberger, P.; Fortunato, M.; Alet, F.; Ravuri, S.; Ewalds, T.; Eaton-Rosen, Z.; Hu, W.; et al. Learning skillful medium-range global weather forecasting. *Science* **2023**, *382*, 1416–1421. [CrossRef] [PubMed]
22. Nguyen, T.; Brandstetter, J.; Kapoor, A.; Gupta, J.K.; Grover, A. ClimaX: A foundation model for weather and climate. *arXiv* **2023**. [CrossRef]
23. Vaswani, A.; Shazeer, N.; Parmar, N.; Uszkoreit, J.; Jones, L.; Gomez, A.N.; Kaiser, Ł.; Polosukhin, I. Attention Is All You Need. *Adv. Neural Inf. Process. Syst.* **2017**, *30*, 5998–6008. Available online: https://papers.nips.cc/paper/7181-attention-is-all-you-need (accessed on 18 January 2025).
24. Child, R.; Gray, S.; Radford, A.; Sutskever, I. Generating long sequences with sparse transformers. *arXiv* **2019**, arXiv:1904.10509. [CrossRef]
25. Li, S.; Jin, X.; Xuan, Y.; Zhou, X.; Chen, W.; Wang, Y.X.; Yan, X. Enhancing the locality and breaking the memory bottleneck of transformer on time series forecasting. *arXiv* **2019**.
26. Beltagy, I.; Peters, M.E.; Cohan, A. Longformer: The long-document transformer. *arXiv* **2020**, arXiv:2004.05150. [CrossRef]
27. Tsai, Y.H.H.; Bai, S.; Yamada, M.; Morency, L.P.; Salakhutdinov, R. Transformer dissection: A unified understanding of transformer's attention via the lens of kernel. *arXiv* **2019**, arXiv:1908.11775. [CrossRef]
28. Clevert, D.A.; Unterthiner, T.; Hochreiter, S. Fast and accurate deep network learning by exponential linear units (elus). *arXiv* **2015**, arXiv:1511.07289. [CrossRef]
29. Yu, F.; Koltun, V.; Funkhouser, T. Dilated residual networks. In Proceedings of the IEEE Conference on Computer Vision and Pattern Recognition, Honolulu, HI, USA, 21–26 July 2017; pp. 472–480. Available online: https://openaccess.thecvf.com/content_cvpr_2017/papers/Yu_Dilated_Residual_Networks_CVPR_2017_paper.pdf (accessed on 16 April 2025).
30. Gupta, A.; Rush, A.M. Dilated convolutions for modeling long-distance genomic dependencies. *arXiv* **2017**, arXiv:1710.01278. [CrossRef]
31. Xu, K.; Qin, M.; Sun, F.; Wang, Y.; Chen, Y.K.; Ren, F. Learning in the frequency domain. In Proceedings of the IEEE/CVF Conference on Computer Vision and Pattern Recognition, Seattle, WA, USA, 13–19 June 2020; pp. 1740–1749. Available online: https://openaccess.thecvf.com/content_CVPR_2020/papers/Xu_Learning_in_the_Frequency_Domain_CVPR_2020_paper.pdf (accessed on 18 April 2025).
32. Foster, J.; Richards, F. The Gibbs phenomenon for piecewise-linear approximation. *Am. Math. Mon.* **1991**, *98*, 47–49. [CrossRef]
33. Shizgal, B.D.; Jung, J.H. Towards the resolution of the Gibbs phenomena. *J. Comput. Appl. Math.* **2003**, *161*, 41–65. [CrossRef]
34. Shi, X.; Chen, Z.; Wang, H.; Yeung, D.Y.; Wong, W.K.; Woo, W.C. Convolutional lstm network: A machine learning approach for precipitation nowcasting. *Adv. Neural Inf. Process. Syst.* **2015**, *28*, 1. Available online: https://proceedings.neurips.cc/paper_files/paper/2015/file/07563a3fe3bbe7e3ba84431ad9d055af-Paper.pdf (accessed on 18 April 2025).
35. Kitamoto, A.; Hwang, J.; Vuillod, B.; Gautier, L.; Tian, Y.; Clanuwat, T. Digital typhoon: Long-term satellite image dataset for the spatio-temporal modeling of tropical cyclones. *Adv. Neural Inf. Process. Syst.* **2023**, *36*, 40623–40636. Available online: https://proceedings.neurips.cc/paper_files/paper/2023/file/7fc36bce5de315751001981baaf4751a-Paper-Datasets_and_Benchmarks.pdf (accessed on 18 April 2025).
36. Knapp, K.R.; Kruk, M.C.; Levinson, D.H.; Diamond, H.J.; Neumann, C.J. The international best track archive for climate stewardship (IBTrACS) unifying tropical cyclone data. *Bull. Am. Meteorol. Soc.* **2010**, *91*, 363–376. [CrossRef]
37. Jeon, J.; Tomita, T. Investigating the effects of super typhoon HAGIBIS in the Northwest Pacific Ocean using multiple observational data. *Remote Sens.* **2022**, *14*, 5667. [CrossRef]
38. Zhou, C.; Chen, P.; Yang, S.; Zheng, F.; Yu, H.; Tang, J.; Lu, Y.; Chen, G.; Lu, X.; Zhang, X.; et al. The impact of Typhoon Lekima (2019) on East China: A postevent survey in Wenzhou city and Taizhou city. *Front. Earth Sci.* **2022**, *16*, 109–120. [CrossRef]
39. Bao, X.; Wu, L.; Zhang, S.; Li, Q.; Lin, L.; Zhao, B.; Wu, D.; Xia, W.; Xu, B. Distinct raindrop size distributions of convective inner-and outer-rainband rain in Typhoon Maria (2018). *J. Geophys. Res. Atmos.* **2020**, *125*, e2020JD032482. [CrossRef]
40. Yanase, W.; Araki, K.; Wada, A.; Shimada, U.; Hayashi, M.; Horinouchi, T. Multiple dynamics of precipitation concentrated on the north side of Typhoon Hagibis (2019) during extratropical transition. *J. Meteorol. Soc. Japan. Ser. II* **2022**, *100*, 783–805. [CrossRef]

applied sciences

MDPI

Article

Enhancing Demand Forecasting Using the Formicary Zebra Optimization with Distributed Attention Guided Deep Learning Model

Ikhalas Fandi * and Wagdi Khalifa

Institute of Graduate Research and Studies, University of Mediterranean Karpasia, Mersin 33010, Turkey
* Correspondence: 210634084@std.akun.edu.tr

Abstract

In the modern era, demand forecasting enhances the decision-making tasks of industries for controlling production planning and reducing inventory costs. However, the dynamic nature of the fashion and apparel retail industry necessitates precise demand forecasting to optimize supply chain operations and meet customer expectations. Consequently, this research proposes the Formicary Zebra Optimization-Based Distributed Attention-Guided Convolutional Recurrent Neural Network (FZ-DACR) model for improving the demand forecasting. In the proposed approach, the combination of the Formicary Zebra Optimization and Distributed Attention mechanism enabled deep learning architectures to assist in capturing the complex patterns of the retail sales data. Specifically, the neural networks, including convolutional neural networks (CNNs) and recurrent neural networks (RNNs), facilitate extracting the local features and temporal dependencies to analyze the volatile demand patterns. Furthermore, the proposed model integrates visual and textual data to enhance forecasting accuracy. By leveraging the adaptive optimization capabilities of the Formicary Zebra Algorithm, the proposed model effectively extracts features from product images and historical sales data while addressing the complexities of volatile demand patterns. Based on extensive experimental analysis of the proposed model using diverse datasets, the FZ-DACR model achieves superior performance, with minimum error values including MAE of 1.34, MSE of 4.7, RMS of 2.17, and R^2 of 93.3% using the DRESS dataset. Moreover, the findings highlight the ability of the proposed model in managing the fluctuating trends and supporting inventory and pricing strategies effectively. This innovative approach has significant implications for retailers, enabling more agile supply chains and improved decision making in a highly competitive market.

Keywords: inventory optimization; demand forecasting; deep learning; formicary zebra optimization; convolutional neural networks

1. Introduction

Inventory management involves making decisions for tactically managing and controlling the inventory level that are essential to meet customer demand at the right time and at the best cost for improving the production and delivery of the end products [1]. Inventory control attempts to meet customer demand at a predetermined service level. By assuming deterministic demand, most existing studies strive to minimize the total of projected ordering and inventory carrying costs [2]. Nevertheless, in practice, owing to market fluctuations, consumer demand becomes exceedingly unpredictable [3]. In a market where competition is fierce, precise demand forecasting is essential for streamlining inventory control, cutting expenses, and increasing income. Accurate and timely forecasting of future demand help retailers match marketing, distribution, and production to anticipated consumer demands, improving both operational effectiveness and customer satisfaction [4]. In general, having either excessive or insufficient inventory results in inefficient management of the stock inventory and negative outcomes affecting the upstream supply chain [5]. For instance, having too much inventory can lead to an oversupply scenario, as the quantity of stored goods has significantly surpassed market needs [6]. In this scenario, the related inventory expense becomes elevated, as many items will need to be kept for an extended period, resulting in slow or inadequate resource turnover for businesses [7]. Furthermore, having an inadequate inventory result in insufficient coverage of customer demands, which can gradually lead to product shortages, diminished customer trust, and potentially reduced profits [8]. Despite the difficulties in attaining precise demand forecasting amid market uncertainties, the abundance of historical data and the application of big data analytics have enhanced the accuracy of demand forecasting [9].

Numerous previous studies have supported the derived version of the inventory model to account for uncertainty and variability. The Economic Order Quantity (EOQ) model is developed using Monte Carlo simulation to account for demand variability and time-sensitive product quality [10]. In view of this problem, another study examined the lead time and defective products as uncertainty factors in addition to demand by formulating the Min–Max inventory method [11]. An additional approach to address uncertainty in inventory involves the use of predictive techniques, such as employing ensemble deep learning for optimizing order-up-to-level inventory in demand forecasting [12]. Nevertheless, none of the studies have taken into account not only the uncertainties in supply and demand but also the uncertainty of discount events offered by the supplier at any given time. Specifically, this discount event is viewed as an uncertain or fluctuating price that has become increasingly prevalent lately [13]. Studies on demand forecasting are crucial for many industries, but the retail industry presents greater challenges than others. The primary cause of this is the lack of precise knowledge about the factors that will influence demand. Seasonality and trend are two other factors that heavily impact sales in the apparel sector [14]. Sales in the retail sector typically happen quickly. Customers visit other stores to satisfy their demands if they are unable to find the thing they desire in the store. For the stores, this translates into lower sales. As a result, it is critical to anticipate client needs precisely [15]. Customer loss may result from higher stock expenses, stock transfers to subsequent years, and a break from fashion if production exceeds real demand. When there is less production than there is demand, there is a decline in sales and a bad effect on the reputation of the brand, as customers are unable to find the products. Additionally, this may have a detrimental impact on brand belief and client loyalty. Specifically, the retail fashion and apparel sector is known for its dynamic and fast-paced environment, which is fueled by shifting consumer tastes, seasonal fashions, and shifting market dynamics [16,17]. Consumers these days want their products supplied quickly and like to choose them from a vast array of alternatives. It is difficult for fashion retailers to quickly satisfy customer needs [18]. As a result, it is critical that sellers of fashion apparel act quickly and effectively when it comes to restocking inventory by forecasting future demand patterns [19].

Deep learning is a branch of machine learning that makes use of multi-layered artificial neural networks and has shown impressive results in identifying complex patterns and connections in big datasets [20,21]. Retailers can use sophisticated algorithms to derive valuable insights from a variety of data sources, such as past sales, product features, industry trends, and outside variables, by utilizing deep learning models [13]. Predicting future

 https://doi.org/10.3390/app16021039

demand and sales of fashion items with accuracy is still a major challenge for both business and academia. Studying the intricacy of the fashion industry and managerial techniques that would enable products to be created, manufactured, and delivered on schedule is essential in order to address this issue [14]. By utilizing deep learning techniques to forecast demand in the retail clothing and fashion industry and investigating the possibilities of deep learning models, including transformer models, convolution neural networks (CNNs) [3], and recurrent neural networks (RNNs) [15], it is possible to increase the precision and dependability of demand forecasts by addressing the shortcomings of conventional forecasting techniques. Predicting future demand and sales of fashion items with accuracy is still a major challenge for both business and academia [22]. It is essential to solve this challenge. The fashion industry is trying to manage these quick changes more effectively by implementing an agile supply chain in order to solve the complexity originating from these variables [16,17]. By presenting a revolutionary strategy for creating an intelligent demand forecasting system employing cutting-edge deep learning techniques and historical product images, this study solves the shortcomings of the traditional forecasting system [2]. However, it is crucial to understand that there are challenges associated with the demand forecasting models used in the fashion and apparel retail industry. Fashion cycles are seasonal and can be very volatile most of the time. The external factors are economic factors and social media trends that cannot be overlooked, and the data from the fashion and apparel retail industry may include historical sales, current trends, as well as consumer data. Models fail to consider fluctuating trends in consumer preferences and other characteristics.

Therefore, this research offers significant improvement to demand forecasting in fashion and apparel retail using FZ with a DACR model. Such an integration of FZ with deep learning architectures leverages the inherent adaptability of FZ to optimize the feature extraction and sequential learning strength of deep learning architectures. In this study, FZ is used to select the most relevant search space that helps in improving the efficiency of the model. The key contribution of the proposed model is explained as follows.

Formicary Zebra Optimization (FZ): The FZ algorithm is a novel optimization algorithm inspired by the foraging behavior of zebras and the path search capability of ants to assist in obtaining the optimal solution. Based on the patterns of collected and shared collective intelligence, the FZ algorithm balances the trade-off between exploration–exploitation, leading to redundancy across other domains. Furthermore, the incorporation of unique traits improves the foraging and decision making concepts to solve high-dimensional optimization problems and provides improved performance.

FZ-DACR: In the proposed model, the combination of the convolutional neural network (CNN) and the recurrent neural network (RNN) model offers the benefits of extracting the context and hierarchical patterns present in the input data. Specifically, the proposed model is trained over the features extracted from product images and historical sales data. Furthermore, the distributed attention mechanism allows the model to focus on the important regions in the sequence data representations using the attention weights, improving forecasting performance.

The study is coherently organized, with Section 2 primarily concentrating on contemporary techniques and providing an overview of the strategies and obstacles faced. The recommended method for demand forecasting and the DACR model's mathematical form is explained in detail in Section 3. The DACR technique's outcomes are examined in Section 4. A thorough examination of the study's findings is provided as conclusion in Section 5.

2. Literature Review

The goal of Chandadevi Giri et al.'s research [2] was to use ML clustering and classification algorithms to propose a novel sales forecasting approach for fashion products. The

https://doi.org/10.3390/app16021039

clustering and classification model has been trained using real fashion retail data. This model has the potential to be useful for handling several supply chain planning activities in the fashion garment business. The results of the study show that the fashion retailing business can forecast the sales of a new item by using images and the item's past data. However, it was possible to improve the image database in subsequent work to increase model accuracy.

Simsir Fuat and Ilker Guven [8] examine the impact of product variety and artificial intelligence on sales forecasting in the retail apparel sector, intending to minimize errors. Within this framework, models of artificial intelligence, including support vector machines (SVM) and artificial neural networks (ANN), have been developed and conclusions drawn from the datasets that attain a high performance and efficiently decrease the over-fitting issues. Sales decline and brand prestige suffer when there is insufficient supply compared to demand because customers cannot locate the products.

For apparel products, demand forecasting and inventory prediction were first presented by Tran Thi Bich Chau et al. [11]. The ARIMA (Autoregressive Integrated Moving Average) model's demand was predicted in their study. A policy framework for production and forecasting models was proposed to ensure optimal inventory levels and enhance enterprise supply capacity. Additionally, the next goal is to identify the distribution path for the whole distribution system.

Giri Chandadevi et al. [3] present a revolutionary deep learning and non-linear neural network regression technique for fashion product sales forecasting. For estimating the amount of products that are to be sold in the future, the study appears to be promising. The study's primary weakness stems from the usage of a smaller dataset. This restriction should be overcome in future work by choosing a sizable image dataset to enhance model performance. This strategy may help fashion retailers and designers in the big data and data mining era.

Majd Kharfan et al. [10] used machine learning techniques to provide a data-driven forecasting solution for recently launched seasonal products. Forecasting has an impact on a company's entire supply chain activities. It is influenced by a variety of implicit product attributes in addition to sales patterns. To improve the process of demand forecasting for recently launched products in the fashion retailing industry, the study suggests a data-driven methodology based on machine learning techniques. Decision makers can use the suggested methodology to anticipate demand for recently introduced products even in the absence of previous data for fashion supply chains. Still, there are unanswered questions regarding the features to be used and the chosen machine learning algorithms for further research.

The use of modular neural networks for demand forecasting with seasonal climate predictions was first suggested by Smit Marvaniya et al. [18]. The challenge of climate-aware demand forecasting in the seasonal context involves employing a sub-neural network architecture to efficiently learn joint latent representations of historical data, known inputs, and climate predictions. This model effectively overcomes the existing problems and attains high performance. Additionally, there is a need to create procedures to assess the uncertainty associated with demand projections and transmit it from ensemble forecasts at different granularities.

Muhammad Yasir et al.'s [5] main objective was to forecast the textile and clothing sectors' output. GLS and single-layer perceptron models were used on the time-series dataset of a textile apparel company. The model has high potential to forecast the textile and clothing sectors' output with high accuracy. However, time-series datasets are rarely available; this is a problem. The same issue was present in this investigation. Future research must be based on large datasets in accordance with this limitation of unavailability. It is also necessary to include more macro-level (exogenous) elements in the forecasting models.

Sushil Punia et al.'s [15] research presents a novel forecasting technique that integrates deep learning with an advantage over previous forecasting techniques. The suggested

 https://doi.org/10.3390/app16021039

method may represent intricate correlations of both temporal and regression types. For time-series forecasting challenges, the more recent and sophisticated neural networks, including spiking neural networks or convolutional neural networks (CNNs), were used. CNN handles large amounts of data with notable success. However, the overfitting issue persists in the model, affecting overall performance.

Javad Feizabadi [19] introduced a hybrid demand forecasting method grounded in machine learning, such as ARIMAX and a neural network. The method was applied and evaluated in the context of a functional product and a steel manufacturer. However, the major limitation of the research is the use of a single dataset, limiting the generalization ability of the model.

Marco A. Villegas [20] presented a new model selection approach that combines different criteria using a support vector machine (SVM). This methodology will be exciting for scenarios with highly volatile demand. The methodology is very interesting because it allows for changing the model when it does not fit the data sufficiently well, thereby reducing the risk of misusing modeling techniques in the automatic processing of large datasets. However, the model is limited in terms of slow learning if new trends and outside forces are involved.

Rathipriya et al. [23] introduced the Demand Forecast Model (DFM), which assists pharmaceutical firms in achieving high efficiency in the global market. Specifically, the DFM model employed diverse shallow and deep neural networks to improve demand forecasting. In addition, the DFM model effectively suggested sales and marketing strategies considering trends or seasonal effects associated with diverse groups of pharmaceutical products with dissimilar characteristics. From the overall analysis, DFMs utilizing shallow neural networks efficiently predicted the future demand of pharmaceutical products.

Chandriah and Naraganahalli [24] implemented the RNN/LSTM with a modified Adam optimizer that effectively predicted the demand for spare parts. To improve the RNN/LSTM model's performance, the weights were optimally tuned utilizing the Modified-Adam algorithm. Further, the experimental validation demonstrated that RNN/LSTM with a Modified-Adam results in minimal errors compared to other baseline techniques utilized for the comparison.

Zohdi et al. [25] utilized the machine learning algorithms that addressed the bias in statistical methods for the demand prediction. Furthermore, the extreme learning machine was introduced for improving intermittent demand prediction. Ultimately, the innovation of the model was analyzed by validating the algorithm with other machine learning models, including the decision tree, gradient boosting, K-nearest neighbors, and multi-layer perceptron, for its accuracy and performance in comparison to other approaches.

Most of the existing techniques prevalent in demand prediction are limited in providing precise demand forecasting. Specifically, existing methods face difficulty in handling complex patterns hidden in the attributes, and modeling their subordinate and complementary relationships adds additional complexity. Existing models need to be trained with large sets of data for them to have proper functionality, limiting the ability of demand forecasting to predict demand for products that are new or in high-demand variability. Due to the limited feature representations, the model had difficulties in extracting the complex and dynamic relations in the historical sales data. Furthermore, the existing models are observed to be complex to fit with large data for automatic processing and slow to learn, resulting in subpar performance for analyzing new trends. So, the proposed research utilizes the FZ-DACR model with an advanced architecture that extracts the complex relationships present in the historical sales data, utilizing the effective feature representations, minimizing computational complexity, and improving the generalization performance.

https://doi.org/10.3390/app16021039

Challenges

- Some of the common issues that fashion retailers have to face are missing data, incorrect data, and noisy data. Lack of sufficient capacity for existing frameworks in handling such problems could also be attributed to their weaknesses [2].
- Numerous types of relationships can exist between various products, and some of them include a subordinate relationship, a complementary relationship, and a cannibal relationship. Maybe the currently available models are not so prescriptive in capturing these interdependencies [8].
- Many fashion items have relatively low durability, which dramatically affects the availability of historical data for such forecasting. Current models need to be trained with large sets of data for them to have proper functionality; this severely limits the capability of demand forecasting to predict demand for products that are new or in high-demand variability [8].
- Certain contemporary models have difficulties in extracting complex and dynamic relationships due to limited feature representations [11].
- Demand is another factor that influences fashion and may be shaped by several factors, such as the climate, changes in the economy, social media, and endorsement of celebrities. There are often external factors resulting in poor forecasts of results; in fact, historical models may not include such factors fully [3].
- In the SVM model, the model faced difficulty to fit with the large data for automatic processing, thereby limited in terms of slow learning and resulting in subpar performance for the new trends and outside forces involved in the forecasting [20].

However, the proposed FZ-DACR model addresses the challenges in the existing techniques with the application of the FZ algorithm, which improves the decision making concepts and solves the high-dimensional optimization problem, resulting in improved performance. Furthermore, the combination of the CNN and RNN algorithms assists in extracting the context and hierarchical patterns present in the input historical sales data. Furthermore, the proposed model is trained over the features extracted from product images and historical sales data, which assist in improving the generalization performance. Additionally, the distributed attention mechanism highlights the important region in the sequence data representations using the attention weights, improving the forecasting performance. Moreover, hyperparameter tuning using the proposed FZ algorithm offers the optimal configuration for minimizing computational complexity while improving demand forecasting performance.

3. Proposed Methodology for Demand Forecasting

Existing demand forecasting models face limitations, including poor performance due to fluctuating fashion trends, limited resources, incorrect feature selection, overfitting, high computation time, and poor generalizability. The proposed research offers an effective methodology that incorporates advanced pre-processing, feature extraction, and a segmentation process for effective feature extraction that influences accurate demand prediction. A hybrid FZ optimization is developed to overcome the existing limitations, like poor convergence, false errors, and complexity issues. Furthermore, the combination of a distributed attention mechanism with deep learning models improves the training process for demand forecasting while simultaneously reducing errors. The proposed FZ-DACR model utilized three different fashion datasets, such as dress, skirt, and T-shirt datasets, to train the model (as shown in Figure 1). Each dataset includes diverse consumer reviews and product images, which enhance the generalizability of the model on unseen data.

https://doi.org/10.3390/app16021039

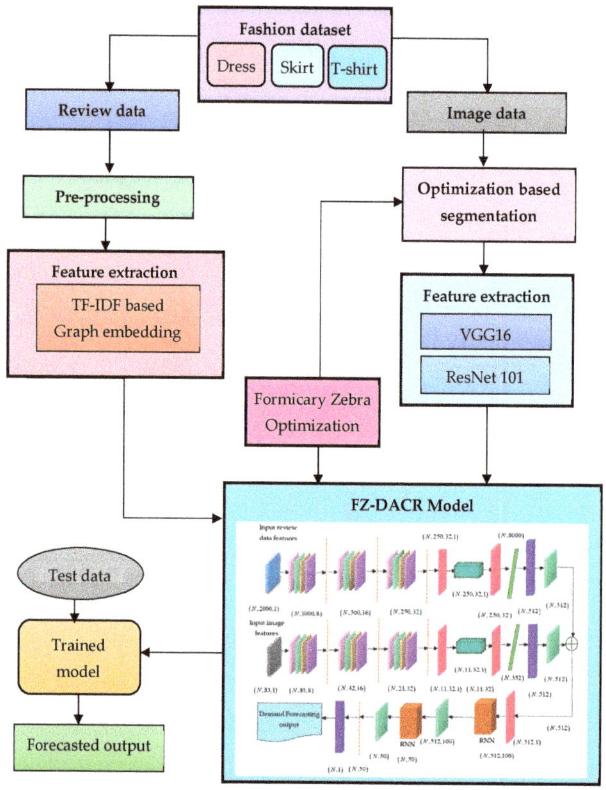

Figure 1. Architecture for the proposed FZ-DACR demand forecasting model.

The input data comprise price information, product qualities, historical sales indicators, and other relevant details in image and text formats. Initially, the review data input are processed through a pre-processing and feature extraction stage to guarantee data consistency and quality. Similarly, image input is processed in segmentation and feature extraction procedures. A key component of the process is feature extraction, which includes TF-IDF (Term Frequency–Inverse Document Frequency) and graph-based feature extraction techniques for review data in order to examine the links between fashion items and capture word importance inside of brand names and product descriptions.

In contrast, the input image is applied with the Formicary Zebra Optimization-based segmentation process for separating the fashion objects from the background. Furthermore, images are fed into the pre-trained deep neural networks, such as VGG16 and ResNet 101, for feature extraction, which extracts texture, form, and fine features. Hence, the extracted feature vectors received from the review data and image data are forwarded to the complex deep learning architecture, known as DACR. The proposed DACR model comprises a CNN model, a Zero Attention mechanism, and an RNN model for extracting the input features for accurate demand forecasting. Additionally, the proposed model integrates a distributed attention mechanism that enables it to concentrate on pertinent aspects from many data sources. The extracted features received from review and image data through the CNN model are concatenated and sent to the RNN layer. Furthermore, Formicary Zebra Optimization adaptively optimizes the FZ-DACR model parameters, boosting demand forecasting performance and generating accurate results. Furthermore, the method's outcome is the creation of demand

https://doi.org/10.3390/app16021039

forecasts for fashion items, which give traders useful information for improving supply chain decisions, pricing tactics, and inventory management.

3.1. Input Data for Demand Forecasting

The demand forecasting model for the fashion industry utilizes input data, such as fashion images, reviews, news, etc. In this research, three different fashion datasets, dress, skirt, and T-shirt datasets, are combined into a single Flipkart Grid Software Development Challenge 2.0 dataset [26], which is then used to train and test the proposed FZ-DACR model for accurate demand forecasting. The input image data are collected through scraping from the original dataset. This dataset also includes product reviews stored as time series data, which are mathematically presented below.

$$T = \sum_{n=1}^{u} T_e \tag{1}$$

where T denotes the database, T_e represents the input data, and n denotes the number of classes.

3.2. Pre-Processing for Review Data

The source dataset contains a time series of data that includes historical sales characteristics with reviews, URLs, brand items, discounted prices, MRP, stars, and ratings, providing the basics for forecasting the demand for fashion products. The input review data are applied to the pre-processing phase, where the text cleaning process is utilized to remove symbols, punctuations, stop words, and distracting elements from the text. The approach ensures that the textual data derived from consumer reviews are pre-processed and provides a reliable set of data to aid in more precise demand forecasting. The pre-processed input review data T_e^* with dimension $(N, 22, 224, 3)$ are mathematically represented as

$$T = \sum_{n=1}^{u} T_e^* \tag{2}$$

where n denotes the number of classes and T_e^* represents the pre-processed review data.

3.3. Feature Extraction for Review Data

The pre-processed review data T_e^* are forwarded into a combined feature extraction technique to obtain different aspects of the information contained in the reviews. This involves the TF-IDF graph embedding techniques, which include the cosine similarity, cosine distance, and linear integration. All of these techniques give a different view of the data and thus improve the reliability of feature extraction.

Term Frequency–Inverse Document Frequency (TF-IDF)-Based Graph Embedding

To determine a word's statistical significance in a document, the TF-IDF approach is utilized based on a keyword extraction technique in which the text is converted into vector values.

Its primary premise is that a word's significance to a text is positively connected with how often it appears in the document and negatively correlated with how often it appears in the corpus of documents. The term frequency is measured by the frequency of a word occurring in the document, mathematically expressed as

$$TF(y, d) = n_{yd} / \sum_{n} n_{yd} \tag{3}$$

https://doi.org/10.3390/app16021039

where d represents the document, y represents the word, and n_{yd} represents the number of times y word appears in the document.

Inverse Document Frequency (IDF) is used to reduce the weights of words that appear in multiple documents in the corpus, which is defined as

$$IDF(y,d) = \log \frac{|D|}{|\{n_y \in d\}|} \quad (4)$$

where $\{n_y \in d\}$ represents the documents that have the word y and $|D|$ is the total number of documents in the corpus. We typically add 1 to the denominator to account for words that do not appear in the corpus, which results in a 0 denominator. The final calculation of TF-IDF is defined as

$$TF - IDF(y,d,D) = TF(y,d) \times IDF(y,d) \quad (5)$$

The TF-IDF score is then computed for all of the terms in the documents, and the keywords can be obtained by choosing the terms with higher TF-IDF values. The TF-IDF-based graph embeddings function as a process of lowering data dimensions, which works well because it retains the inherent features and connections of the initial data. These transformations enable techniques, such as cosine similarity, cosine distance, and linear kernels, to improve the graph embeddings when used in demand forecasting. Therefore, TF-IDF-based graph embedding can be used to gain a better perspective on the structure and relationships of the data and therefore improve the comprehensibility of the prediction results.

(i) Cosine Similarity

Cosine similarity measures the similarity of the two vectors, generally texts or a particular segment of texts, based on the cosine of the angle formed by the two vectors in a high-dimensional space. Cosine similarity can be estimated as the ratio of the dot product of the two vectors B and C and the magnitudes of the vectors. This is where a vector score of 1 means that the two vectors are parallel, conveying the same information, while a score of 0 means that the two vectors are perpendicular, containing non-related information, and a score of -1 means that the two vectors are quite opposite and contain completely different information.

$$CS(B,C) = \frac{B \cdot C}{||B||\,||C||} \quad (6)$$

where $||B||$ and $||C||$ are the magnitudes of vectors B and C, respectively.

(ii) Cosine Distance

Cosine distance computes the cosine of the angle between two vectors, which can refer to documents or text segments, that are different from one another by using the complement of cosine similarity. It is defined as the value -1 cosine of the angle formed by the vectors. This metric implies that they can easily measure how orthogonal the two orientations of the two vectors are in a high-dimensional space. Cosine distance eliminates the dependence on the vector magnitudes while keeping the angle between vectors as the parameter for dissimilarity and, thus, allows for reliable measurement of the intrinsic dissimilarity of textual content.

$$CD(B,C) = 1 - CS(B,C) \quad (7)$$

(iii) Linear Kernels

https://doi.org/10.3390/app16021039

The linear kernels measure similarity based on the dot product of two input vectors B and C, which generates a single numerical value. The linear kernel K is mathematically expressed as

$$K(B, C) = B \cdot C \tag{8}$$

Finally, the graph embedding feature vector y_{hybrid} is generated by concatenating the findings of cosine similarity, cosine distance, and linear kernels. This hybrid vector is mathematically defined as

$$y_{hybrid} = \beta CS(B, C) + \gamma CD(B, C) + \delta K(B, C) \tag{9}$$

In Equation (9), β, γ, and δ represent the weights assigned according to the importance of each feature type. Graph embedding combines multiple features, including the cosine similarity, cosine distance, and linear kernel, into a unified feature vector with a dimension $(N, 83)$. This improves the representation of a term or document in different machine learning tasks.

3.4. Optimization-Based Segmentation for Image Data

The input image is gathered from the dataset using a scraping method and applied to the segmentation process. In this research, the segmentation process utilizes a thresholding method to transform the original input image into a binary format by separating the segmented image from the background. This binary thresholding transforms the original pixel intensity values, spanning a range of 0 to 255, into a binary representation of either 0 or 1. Furthermore, the segmentation process is enhanced by the integration of formicary zebra optimization. This FZ optimization utilizes the foraging and path-finding strategies to segment images or fashion items from their backgrounds. Therefore, the segmentation process is optimized and captures the important features of fashion items, like diverse image complexities or patterns, including the texture, color, and shape. Moreover, the segmented image is enhanced by the subsequent feature extraction techniques, which are explained in the next section.

3.5. Feature Extraction for Image Data

After the segmentation process, the segmented image is entered into the feature extraction phase, where two pre-trained deep neural networks are utilized to extract significant features that may influence demand forecasting. Here, two networks, such as VGG16 and ResNet101, are employed to learn multiple levels of abstraction, like edges, texture, shapes, and much more complex structures, from the input images.

In this research, we utilize both VGG16 and ResNet101 to take advantage of the strengths of both models to produce a better feature representation when dealing with image data. Different models, like VGG16 and ResNet101, are particularly good at highlighting different aspects of image features because of their structural differences. Compared with VGG19, which is more complex and inconsistent, VGG16 is good at preserving delicate texture and structure information. On the other hand, ResNet101, with a deeper architecture and residual connections, is able to learn more complex and abstract features. This integration benefits the ability of the system to capture a larger variety of visual features, thereby improving other related tasks, such as classification, clustering, and regression.

3.5.1. VGG16

VGG 16 is a type of CNN model that has proven to be simple but very effective in solving image classification problems. VGG16 was developed by the Visual Geometry

https://doi.org/10.3390/app16021039

Group (VGG) with a simple architecture that mainly consists of 16 layers, mostly of small 3×3 convolutional layers. These filters are used subsequently in all of the various layers of the network and then followed using max-pooling layers, and the final layers are fully connected layers. Small filter size and the homogeneous structure of the model help VGG16 extract detailed textures in the images. Although it is much shallower compared to today's architecture, VGG16 excels in pattern and shape recognition because of its feature extraction depth. It is also computationally efficient and easy to implement due to its simplicity of design, which has made it one of the most used solutions for many image recognition problems. The significant features extracted through the VGG16 model are denoted as f_{VGG18} with a dimension $(N, 1000)$.

3.5.2. ResNet101

ResNet101, or residual network with 101 layers, is a deep CNN model that seeks to overcome a major problem with very deep networks known as the vanishing gradient problem. ResNet101 was developed by researchers at Microsoft, where residual connections are integrated that enable the network to learn identity mappings. These residual connections allow the network to train more effectively by making sure that gradients can pass through the network without loss as the depth of the network increases. ResNet101 mainly consists of numerous residual blocks; each of these blocks contains convolutional layers succeeded by batch normalization and ReLU activation. This design helps ResNet101 pick up a variety of features from hierarchies, from the edges of the image to the objects in the picture. Due to its depth and remaining connections, ResNet101 is very efficient in the identification of complex patterns and relationships in image data. The input segmented image is extracted through the ResNet101 model and generates the feature map, represented as $f_{RESNET101}$ with dimension $(N, 1000)$.

The final output of the feature extraction technique for image data is produced by fusing the feature maps received from the pretrained model, represented as $f_c(I)$. This final feature vector is generated by concatenating the feature maps received from the VGG16, represented as f_{VGG18}, and the ResNet101 model is represented as $f_{RESNET101}$, which is mathematically expressed as

$$f_c(I) = f_{VGG18} \oplus f_{RESNET101} \tag{10}$$

where $\oplus$ denotes the concatenation operation. Hence, the feature vector $f_c(I)$ is generated with a dimension of $(N,\ 2000)$.

3.6. Distributed Attention-Based Deep Learning Model

The feature final vectors generated from the review data y_{hybrid} and image $f_c(I)$ are allowed into the proposed FZ-DACR model for demand forecasting (as shown in Figure 2). A distributed attention-based deep learning approach employs a CNN model integrated with a Zero Attention mechanism. This architecture is organized into two parallel branches dedicated to processing text data and image data, respectively. Furthermore, the RNN model is incorporated to combine textual and visual data to improve the prediction. Moreover, FZ optimization is integrated to tune the parameters to achieve accurate prediction and increase the convergence rate while reducing the false errors. The proposed FZ-DACR model starts by learning the data of the reviews and images separately through the distribution attention method.

https://doi.org/10.3390/app16021039

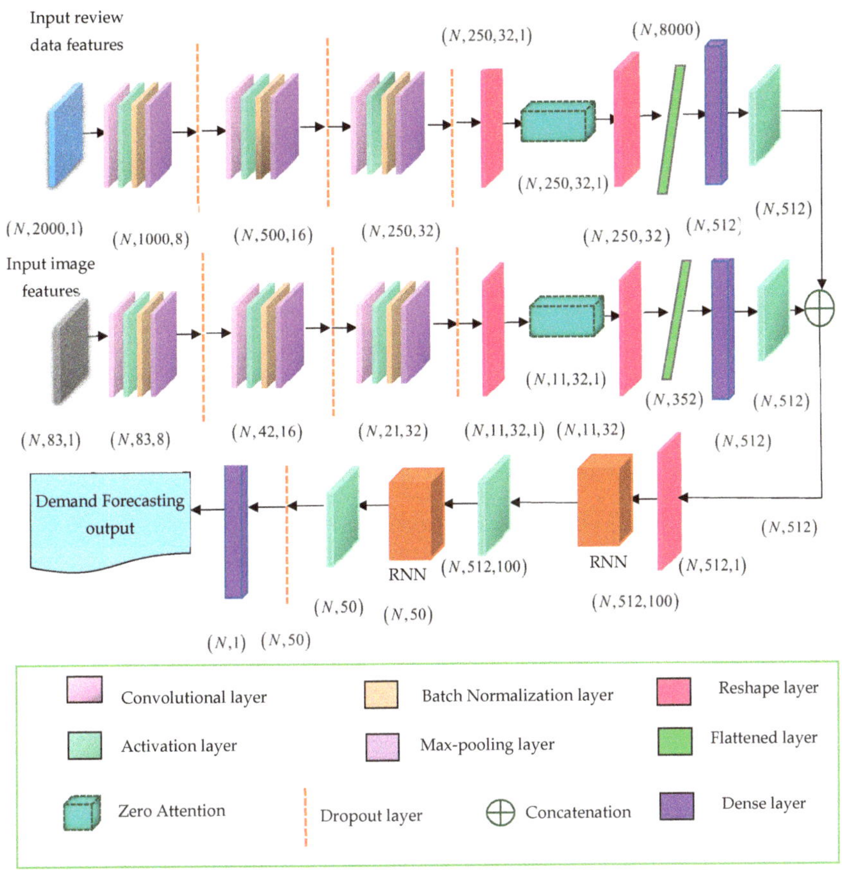

Figure 2. Architecture for the proposed FZ-DACR model.

At first, the feature vector received from review data with a dimension $(N, 83)$ is applied to the convolutional network for further processing. The input feature vector is transmitted through multiple layers, such as convolutional layers, linear activation function, batch normalization, max-pooling, reshape, Zero Attention mechanism, flatten layer, and dense layer in a deep CNN model separately to learn context and hierarchical patterns. In each convolutional layer, there are two or more filters (or kernels) that scan over the input feature map and perform element-wise multiplication or addition to generate several feature maps. These filters are capable of detecting various low-level patterns like edges, texture, and shapes in earlier layers and higher levels of abstractions in deeper layers. By using several layers of convolution, the network forms a deep, fully connected feature space. Thus, hierarchical feature extraction is vital to identify the details of the data from a combined review and image data set, which in turn helps the model to make an appropriate demand forecast. The architecture of the proposed FZ-DACR model is shown in Figure 2.

The convolutional layers calculate the dot product of the weights and the bias of the input received from the review data, which is expressed as

$$Conv = f\left(w * y_{hybrid} + b\right) \tag{11}$$

https://doi.org/10.3390/app16021039

where w and b represent the weights and bias of the input y_{hybrid} vector, which is then forwarded to the linear activation function, where the dimensionality is not changed, and sent to the next layer. Then, the batch normalization layer is employed to speed up the learning process while reducing overfitting. Furthermore, the pooling layer is applied to decrease the spatial extent of the feature maps, or, more precisely, to decrease the dimensions of the maps that were obtained, but the main objective here is to save all important data and decrease the computational load. Here, the CNN model is incorporated with a Zero Attention mechanism, which learns the informative spatial regions and complex patterns. Hence, the attention mechanism extracts important features from the input reviews and images while neglecting the irrelevant features effectively. Furthermore, the inclusion of a flatten and reshape layer reduces the dimensionality of the input. Then, the output received from the reshape layer is sent to the dense layer. A dense layer is also known as a fully connected layer, where multiple neurons are connected through the layers, which learns input and produces accurate predictions with dimension $(N, 512)$ to the consequent layers. The combination of the CNN model with the attention mechanism enhances the accuracy of demand forecasting results.

Similarly, the feature vector received from image data $(N,\ 2000)$ is processed through a separate CNN learning network. The feature vector is processed using a distributed attention mechanism and a CNN model to extract more detailed patterns and higher-level, hierarchical features, mirroring the processing applied to the review data.

Finally, the processed output generated for both review data and image data using the CNN model has the same dimension $(N, 512)$. Furthermore, these generated features are concatenated with a dimension $(N, 512)$ and then fed to the RNN model. Here, the combined feature vector is denoted as V with dimension $(N, 512)$ formed by concatenating the individual feature vectors received from the review-data-based CNN and image-data-based CNN model, expressed as

$$V = v_{review} \oplus v_{image} \tag{12}$$

where v_{review} represents the features based on review data, v_{image} represents the features based on image data, and $\oplus$ denotes the concatenation operation. Furthermore, the concatenated output $(N, 512)$ is fed into the reshape layer, where the dimension of the data is formed $(N, 512, 1)$, which is applied to the RNN model.

The RNN deals with temporal sequences and is important for understanding the temporal variation in sales. Moreover, the RNN model is incorporated with the Formicary Zebra Optimization by tuning the weights and bias parameters of the model to achieve accurate demands while minimizing false errors. In this context, the final output of the dense layer is the final one, which enables the organization to make accurate demand estimations. Finally, the predicted output is generated from the second RNN's hidden state and employed with a linear activation function, dropout layer, and dense layer to produce the final output, which may represent the predicted demand for fashion items with dimension $(N, 1)$. This multi-hierarchical combined strategy utilizes the potency of both CNNs and RNNs in addition to an attention mechanism to forecast demand and assist retailers in managing stock and meeting consumers' demands.

3.6.1. Distributed Attention Mechanism

In this research, a Zero Attention mechanism is utilized to extract the review and image data individually throughout the deep learning network. Here, the Zero Attention mechanism is computed by utilizing the zero-channel attention and zero spatial attention mechanisms to reduce the spatial dimensions of the input features. This aims to improve

https://doi.org/10.3390/app16021039

model performance using an attention mechanism without introducing any additional parameters, which calculates the element-wise summation to merge the pooled features and leverages spatial relationships among features to create a spatial attention map. The concatenation of features received from zero spatial attention S_{att} and zero channel attention C_{att} is represented as

$$Z_{att} = f(C_{att}) + f(S_{att}) \tag{13}$$

3.6.2. Recurrent Neural Network (RNN)

The RNN model extracts the temporal dependencies and sequences within the data, receiving its input from the reshape layer output with a dimension of $(N, 512, 1)$. The hidden state calculation h_t known in RNN is expressed as

$$h_t = \sigma(W_h I_t + X_h h_{t-1} + b_h) \tag{14}$$

where the hidden state at time t is denoted as h_t, the weight matrices are denoted as W_h and X_h, the bias vector is denoted as b_h, and the linear activation function is denoted as σ. Specifically, the output of the first RNN layer is integrated with FZ optimization, where the weights and bias of the RNN model are tuned for accurate demand prediction and to reduce false errors. Furthermore, the tuned data are applied to the linear activation function with dimension $(N, 512, 100)$ and also forwarded to the second RNN model.

$$z = W_o h_U + b_o \tag{15}$$

Here, W_o denotes the output weight parameters, b_o denotes the output bias parameters, and h_U is the final hidden state of the network at time step U. This concept also enables one to integrate inputs of different forms, namely, the images and the reviews, hence offering enhanced accuracy of the demand forecasting model.

3.7. Formicary Zebra Optimization

A new metaheuristic algorithm is developed from nature-inspired processes to address and overcome limitations found in existing algorithms, such as poor convergence, getting stuck in local optima, and the occurrence of false errors. Hence, the foraging behavior of the Zebra Optimization Algorithm (ZOA) [27] and the path search capability of Ant Colony Optimization (ACO) [28] are integrated into the proposed FZ optimization to overcome those previous limitations. Consequently, the FZ algorithm effectively navigates the large search space to find the optimal solution and offers the benefits of fewer parameter requirements while maintaining faster convergence. As a result, the FZ algorithm offers an efficient binary thresholding technique to facilitate accurate image segmentation. Specifically, the FZ algorithm operates through image data segmentation and effectively segments the relevant features from the background images of different fashion items. This simplifies the images and feeds the DACR model with the significant features of the image. In addition, the FZ algorithm is applied for optimal hyperparameter selection to improve the model's performance, thereby improving the overall demand forecasting performance. Thus, FZ optimization fine-tunes the weights and parameters of the DACR model to achieve high accuracy and reduce false errors while facilitating a high convergence rate. Altogether, these characteristics enhance the model's precision in providing accurate results in demand forecasting in the fashion industry.

Inspiration: FZ optimization is inspired by the efficient foraging characteristics of zebras and the path search capability of ants. The ACO algorithm comprises the collective intelligence and path-finding characteristics of ants, which use pheromone trails to discover

 https://doi.org/10.3390/app16021039

the shortest paths between food sources and their nest. FZ incorporates this mechanism to guide the search process towards optimal solutions. Additionally, the ZOA integrates the natural foraging and defense traits of zebras in the wild, including their coordinated foraging and defensive tactics against predators. By integrating these complementary behaviors, FZ enhances model effectiveness and robustness. The ACO ensures efficient exploration and exploitation of the search space, while the ZOA provides adaptability and resilience. Together, these characteristics improve the model's ability to generate accurate demand forecasting.

Initialization: The position of each solution in the search space determines the values for the decision variables. The population of the solution can be mathematically modeled using the matrix. The initial position of the solution in the search space is randomly assigned. The population matrix is specified below.

$$X = \begin{bmatrix} X_1 \\ \cdots \\ X_i \\ X_n \end{bmatrix}_{N \times M} \tag{16}$$

Here, X_1 means the first solution, X_i means the ith solution, N means the population, and M means the number of decision variables. The initial random solution X^t is initialized from the population matrix based on the weights and bias parameters.

Fitness Evaluation: In FZ, the fitness function is considered as the Mean Squared Error (MSE). The model is tuned for minimum MSE values to attain the optimal solution, which also increases the prediction accuracy. Thus, by minimizing the MSE, the FZ algorithm continuously refines the model's parameters to improve the method's ability to make an accurate demand forecast for the fashion industry.

$$Fit\left(X^{t+1}\right) = \min\left(MSE\left(X^{t+1}\right)\right) \tag{17}$$

Phase 1: Foraging strategy: If $\phi \geq 1$, where ϕ represents the deterministic factor. The pioneer solution and the leading solution are especially important in the foraging process of FZ, where the aim is to explore the search space and look for the best solutions. This leading character paves the way to other solutions, which are contingent on its guidance while mapping the search space. Here, the pioneer solution of zebra optimization and the following solution of ant optimization are incorporated to create an effective foraging strategy. This makes the pioneer solution effective in guiding the solution space, as it can identify the successful paths and use them for the group's benefit. It helps to create the average for the group to enhance to establish the position of the optimal solution. Besides supporting the effective exploration and exploitation solutions, it also makes the process strong, especially in the presence of noise and disturbances. This foraging strategy is expressed by incorporating the behavior of ants and zebras, which is updated as

$$X^{t+1} = 0.5\left[X_t + r_1 \cdot (z_p - I \cdot X_t) + \left[(1-\rho_1) \cdot \tau^{n-1} + \rho_1 V^t\right]\right] \tag{18}$$

Here, X_t represents the new status of the zebra in the first phase, z_p denotes the leading solution, r_1 denotes the random factor in interval (0, 1), I denotes the energy factor of leading solution ranges (1, 2), τ^{n-1} represents the ability to follow the leading solution with pheromone values of ants, n is the construction step, ρ_1 represents the evaporation rate, $\rho_1 = \frac{X^t - X^{t-1}}{t}$, X^t, X^{t-1} denotes the position at current and previous iterations, and v^t denotes the speed of the solution. In Equation (20), the first half shows the leading behavior of the leader solution of zebras, and the second half shows the following behavior of ants.

https://doi.org/10.3390/app16021039

Phase 2: $if\ \phi < 1$ Escape strategy: The second phase shows that the solutions' defensive strategies against attackers are employed to update the position of population members in the search space. It is assumed that one of the following two conditions occurs with the same probability.

(i) Defensive strategy: $if\ F(X_{att}) \geq F(X_t)$: Here, $F(X_t)$ denotes the fitness value of X_t the solution, and $F(X_{att})$ means the fitness value of the attacker solution.

In FZ, when a solution is threatened by an opponent with better fitness, it has to implement defense mechanisms to avoid the attacker to remain safe. Trailing and attractiveness characteristics help the strategy to guide the solution to shift towards the center of the population as a way of enhancing the strategy. In this way, the weaker solutions can stay in the safety phase, where they are given a chance to continuously improve their fitness, while the stronger solutions continue with the actual optimization without compromising the randomness and hence the diversification of the search.

$$X^{t+1} = \frac{\left(X_t + c(2r_2 - 1)\left(1 - \frac{t}{t_{\max}}\right)X_t + \tau_{n+1}\right)}{2} \tag{19}$$

Here, τ_{n+1} denotes the updated following ability of the solution, τ_n denotes the pheromone matrix (the ability to stay in the population), c denotes the constant number with value 0.01, and r_2 represents the probabilistic term in the range (0, 1).

(ii) Offensive strategy: $if\ F(X_{att}) < F(X^t)$: In FZ, if a solution has a fitness value greater than that of the attacker, it goes for an attack in order to safeguard other inferior solutions in the population. In this context, if an attacker targets a weaker solution, all of the stronger solutions converge towards the attacker. This coordinated movement helps to distract and tire down the attacker while creating a shield around the weaker solution. This forces the weaker member to be shielded from all threats that may be from other members of the group, as the stronger members work as a group to shield it. This offensive strategy not only eliminates the impact of the attacker but also increases the strength of the weakest solution in the herding effect. The collective defense mechanism guarantees that the members who have gotten stuck at a certain phase can still survive, along with contributing to the subsequent search process; this makes the entire optimization algorithm stronger.

$$X^{t+1} = X^t + r_2(X_{worst} - IX_t) \tag{20}$$

Here, X_{worst} means the worst solution in the population. Thus, it can be pointed out that in the overall framework of FZ, the model under consideration pays special attention to the efforts aimed at minimizing the loss of weights to improve the accuracy of the prediction. This is done by including the loss function in the objective function during the training phase. As optimization progresses, the solutions move towards the set of parameters that produces the most minimal loss, which enables the model to generate accurate and reliable predictions. This continuous reduction in loss not only enhances the forecast accuracy of the model but also increases the model's overall performance and reliability in terms of demand forecasting in the fashion and apparel retail business.

Termination: Finally, the algorithm terminates the iterative process upon reaching the optimal solution, with the condition $(t < t_{\max})$; otherwise, the solution is re-evaluated to obtain the optimal solution. Hence, the global solution is declared for updating the hyperparameters of the model, resulting in improved demand forecasting performance. Figure 3 shows the flowchart of the proposed Formicary Zebra Optimization, and Algorithm 1 provides the pseudocode of FZ optimization.

https://doi.org/10.3390/app16021039

Algorithm 1. Pseudocode for the Formicary Zebra Optimization model.

Start
Initializing Random Solution.
Evaluate Fitness Function
The Objective Function
If $(\phi \geq 1)$
Foraging Strategy
$X^{t+1} = 0.5\lfloor X_t + r_1(z_p - IX_t)\rfloor + 0.5\lfloor(1-\rho_1)\cdot\tau^{n-1} + \rho_1 v_t\rfloor$
Else If $(F(X_{att}) \geq F(X_t))$
Defensive Strategy
$X^{t+1} = \frac{(X_t + c(2r_2-1)(1-\frac{t}{t_{\max}})X_t + \tau_{n+1})}{2}$
Else
Offensive Strategy
$X^{t+1} = X^t + r_2(X_{worst} - IX_t)$
While $t < t_{\max}$
Declare Global Best Solution
Else
Return To Fitness Evaluation
Terminate The Process
End While
End

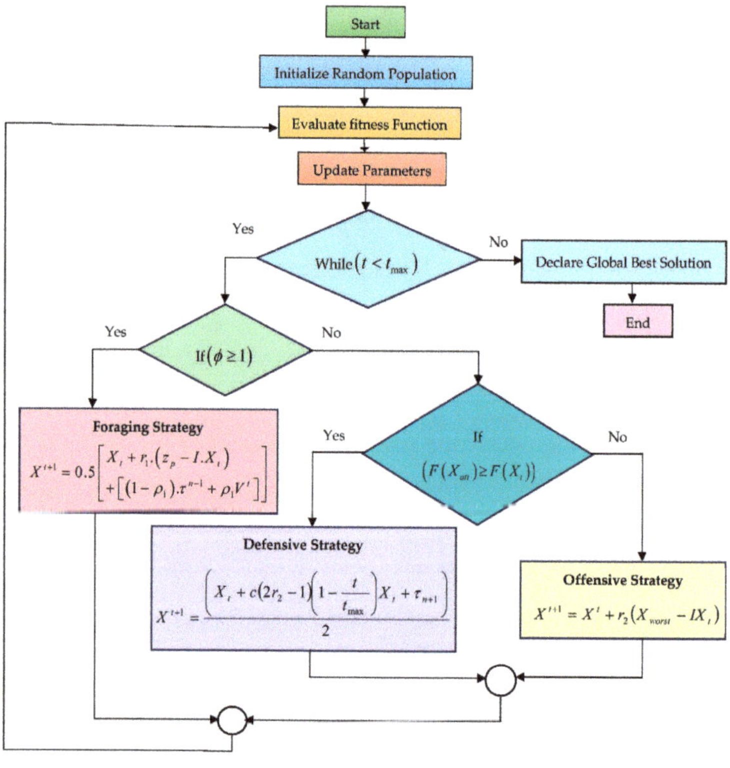

Figure 3. Flowchart for the proposed Formicary Zebra Optimization.

https://doi.org/10.3390/app16021039

4. Result and Discussion

The results obtained with the implementation of the proposed model for demand forecasting are described in this section.

4.1. Experimental Setup

In order to use the FZ-DACR model for demand forecasting, it is executed on Windows 10 OS with 8 GB of RAM in the Python 3.7 programming language. The initial parameter settings of the proposed framework involve a batch size of 32, a dropout rate of 0.5, a linear activation function, loss function "MSE", a learning rate of 0.01, epochs of 500, a population size of 100, and default optimizer Adam. In the proposed research, the training and testing split percentage of 90:10 is used for the validation of the proposed model.

4.2. Dataset Description

Flipkart Grid Software Development Challenge 2.0 dataset [26]: This dataset was collected from the GitHub website accessed in April 2024, which contains more detailed data that are extracted from a range of e-commerce sites, some of which are specialized in selling dresses, skirts, and t-shirts. For the specific subject of dresses, the attributes might encompass different styles, fabrics, measurements, colors, prices, customers' feedback, and ratings. Skirts are also described using other attributes that divide them based on design, length, types of fabrics, prices, and customers' feedback. Specific items in T-shirts are style, size, color, type of material used, price, and customers' reviews on similar products. These clothing types are all well-represented in this extensive database and employed in many machine learning processes for recommendations, trend analysis, profitable pricing, and consumer behavior prediction. In addition, image data provide additional attributes and visual cues that are critical for accurate forecasting. The study utilized samples from the three distinct datasets, including 327 from the dress dataset, 296 from the skirt dataset, and 296 from the T-shirt dataset. These combined samples were used to train and validate the demand forecasting model.

4.3. Experimental Results

Figure 4 displays the experimental outcome produced by the proposed FZ-DACR model. The experimental results display the various processes utilized for demand forecasting, where the input image data are applied to the pre-processing phase. Furthermore, the output received from segmentation and feature extraction is explained as follows.

Figure 4. *Cont.*

https://doi.org/10.3390/app16021039

Figure 4. Experimental result obtained using the proposed FZ-DACR model.

Figure 5 illustrates the graph embedding features obtained using the proposed FZ-DACR model based on the dress dataset, the skirt dataset, and the T-shirt dataset for demand forecasting. The round balls represent the individual items and the similarity between the items are established through the link (straight line connecting the items), while the numbers represents the similarity score between the items.

4.4. Comparative Methods

To demonstrate the advancement of the FZ-DACR model, this evaluation compares a number of existing techniques, such as ML [19], SVM [20], ANN [8], CNN [29], non-linear NNR [3], and MNN [30]. Furthermore, the comparative analysis of the proposed FZ-DACR

https://doi.org/10.3390/app16021039

model with other existing techniques in terms of Training Percentage (TP) Analysis and k-fold analysis is discussed in this section.

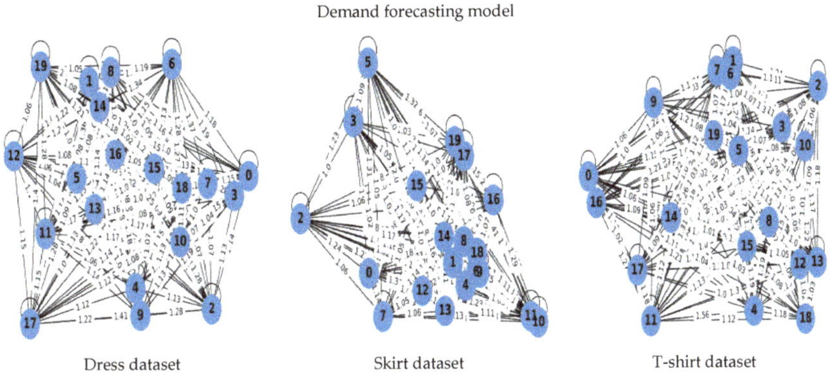

Figure 5. Graph embedding output obtained using the proposed FZ-DACR model.

4.4.1. Comparative Analysis Based on Dress Dataset with TP Analysis

The FZ-DACR model demonstrated the lowest MAE in demand forecasting compared to the MNN model, showing a slight 19.62% increase and achieving the lowest MAE of 1.34, as indicated in Figure 6a.

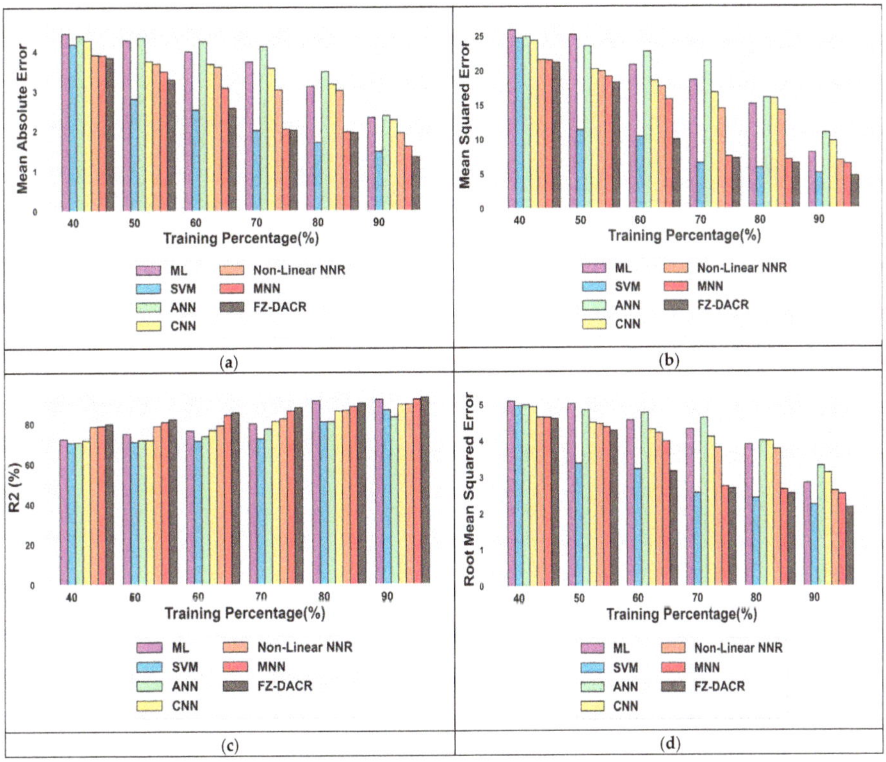

Figure 6. Comparative analysis based on the dress dataset with TP analysis. (**a**) MAE; (**b**) MSE; (**c**) R2; (**d**) RMSE.

https://doi.org/10.3390/app16021039

In Figure 6b, the FZ-DACR model exhibited the lowest MSE in demand forecasting compared to the MNN model, surpassing it by 36.92% and attained the lowest MSE of 4.70, with a TP of 90.

Regarding demand forecasting, the FZ-DACR model illustrated in Figure 6c performed 3.13% better than the MNN model. With a TP of 90, it achieved an R2 of 93.13%, surpassing existing techniques.

The FZ-DACR model displayed the least RMSE in demand forecasting compared to the MNN model in Figure 6d, outperforming it by 16.90% and achieving the lowest RMSE of 2.17, with a TP of 90.

4.4.2. Comparative Analysis Based on Dress Dataset with K-Fold Analysis

The comparative evaluation of the proposed model with existing techniques in terms of k-fold analysis is depicted in Figure 7. The FZ-DACR model demonstrated the lowest MAE in demand forecasting compared to the MNN model, showing a slight decrease of 0.44 and achieving the lowest MAE of 1.38, as indicated in Figure 7a.

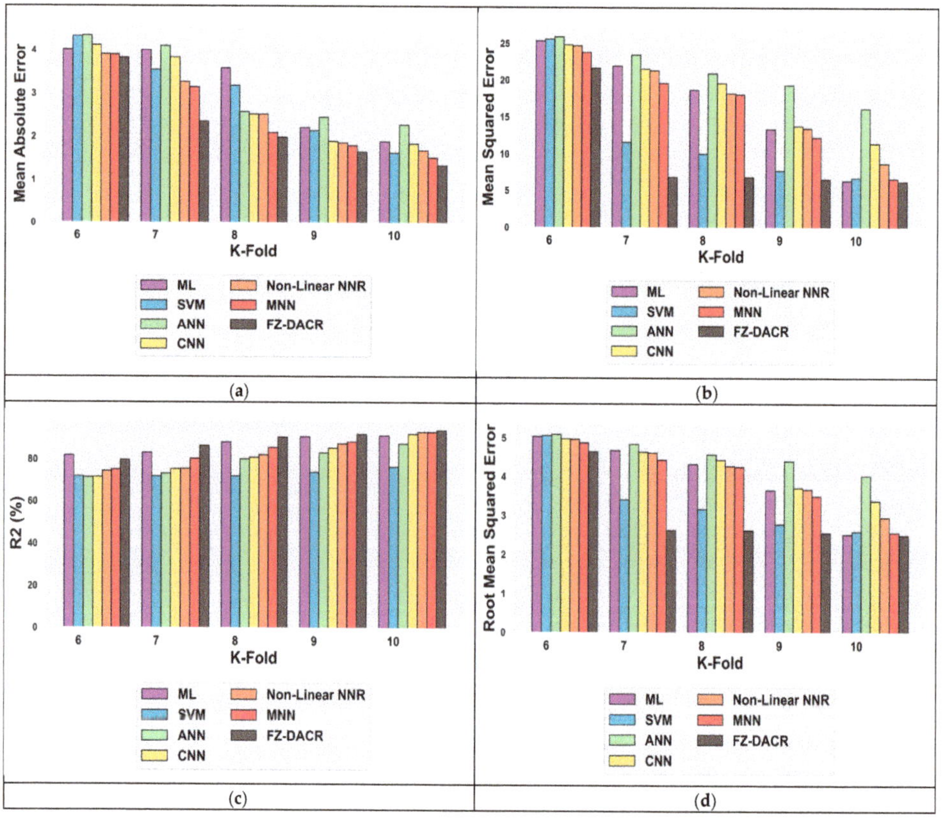

(a) (b) (c) (d)

Figure 7. Comparative analysis based on the dress dataset with k-fold analysis. (**a**) MAE; (**b**) MSE; (**c**) R2; (**d**) RMSE.

In Figure 7b, the FZ-DACR model exhibited the lowest MSE in demand forecasting compared to the MNN model, surpassing it by a difference of 0.185 and attaining the lowest MSE of 4.94, with k fold 10.

Regarding demand forecasting, the FZ-DACR model illustrated in Figure 7c obtained 1.59% better results than the MNN model. With k-fold 10, FZ-DACR achieved an R2 of 93.65%, surpassing existing techniques.

The FZ-DACR model displayed the least RMSE in demand forecasting compared to the MNN model in Figure 7d, outperforming it with an error difference of 0.041 by achieving the lowest RMSE of 2.22, with k-fold 10.

4.4.3. Comparative Analysis Based on SKIRT Dataset with TP Analysis

The FZ-DACR model demonstrated the lowest MAE in demand forecasting compared to the MNN model, showing a slight decrease of 11.53% and achieving the lowest MAE of 1.43, as indicated in Figure 8a.

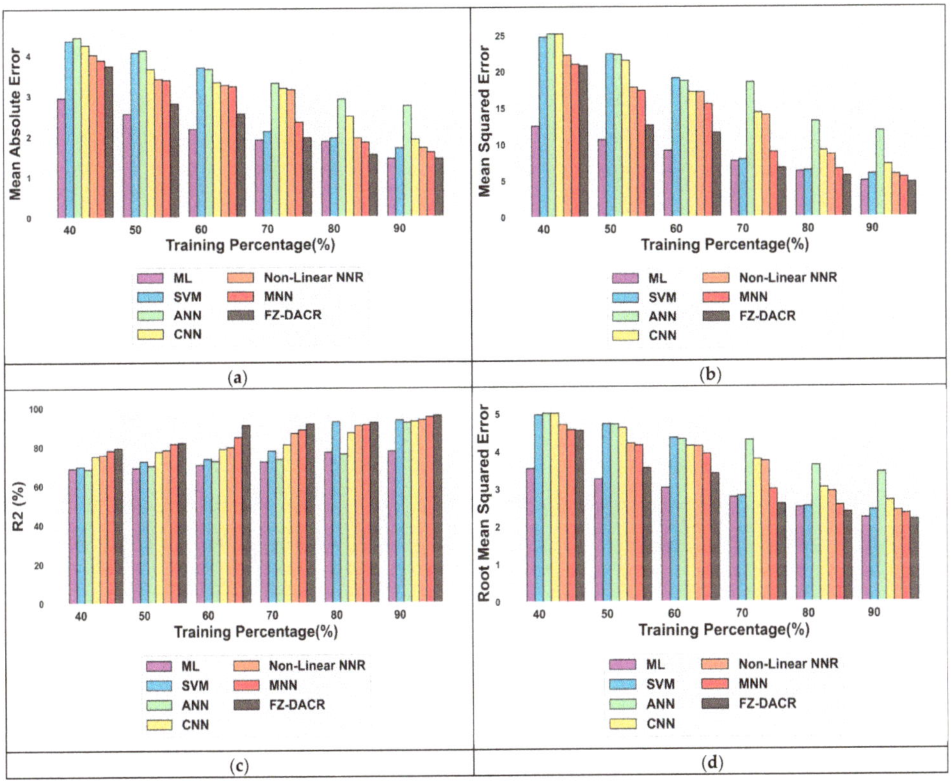

Figure 8. Comparative analysis based on the skirt dataset with TP analysis. (**a**) MAE; (**b**) MSE; (**c**) R2; (**d**) RMSE.

In Figure 8b, the FZ-DACR model exhibited the lowest MSE in demand forecasting compared to the MNN model, surpassing it by 14.97% and attaining the lowest MSE of 4.75, with a TP of 90.

Regarding demand forecasting, the FZ-DACR model illustrated in Figure 8c performed 95.67% better than the MNN model. With a TP of 90, it achieved an improvement of 0.64%, surpassing existing techniques.

The FZ-DACR model displayed the least RMSE in demand forecasting compared to the MNN model in Figure 8d, outperforming it by 7.20% and achieving the lowest RMSE of 2.18, with a TP of 90.

https://doi.org/10.3390/app16021039

4.4.4. Comparative Analysis Based on Skirt Dataset with K-Fold Analysis

The FZ-DACR model demonstrated the lowest MAE in demand forecasting compared to the MNN model, showing a slight decrease of 0.326 and achieving the lowest MAE of 1.66, as indicated in Figure 9a.

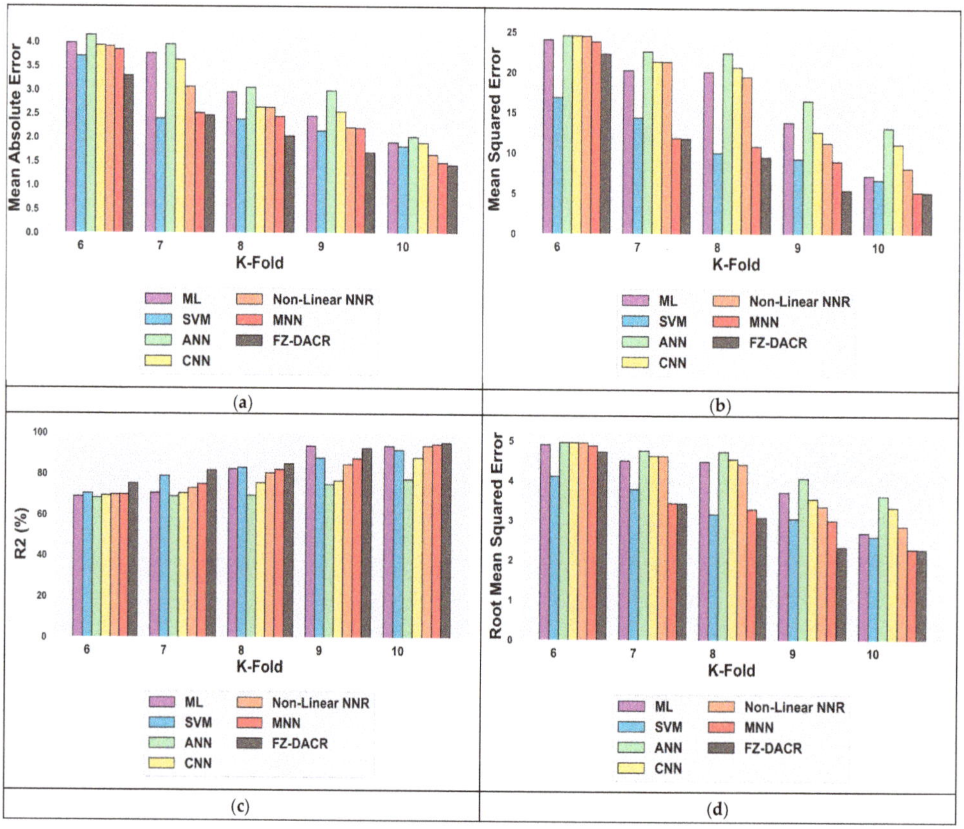

Figure 9. Comparative analysis based on the skirt dataset with k-fold analysis. (**a**) MAE; (**b**) MSE; (**c**) R2; (**d**) RMSE.

In Figure 9b, the FZ-DACR model exhibited the lowest MSE in demand forecasting compared to the MNN model, surpassing it by a difference of 2.23 and attaining the lowest MSE of 5.09, with k-fold 10.

Regarding demand forecasting, the FZ-DACR model illustrated in Figure 9c performed 94.12% better than the MNN model. With k-fold 10, it achieved a relative improvement of 0.83%, surpassing existing techniques.

The FZ-DACR model displayed the least RMSE in demand forecasting compared to the MNN model in Figure 9d, outperforming it with an error difference of 0.44 and achieving the lowest RMSE of 2.25, with k-fold 10.

4.4.5. Comparative Analysis Based on T-Shirt Dataset with TP Analysis

The FZ-DACR model demonstrated the lowest MAE in demand forecasting compared to the MNN model, showing a slight decrease of 11.53% and achieving the lowest MAE of 1.43, as indicated in Figure 10a.

 https://doi.org/10.3390/app16021039

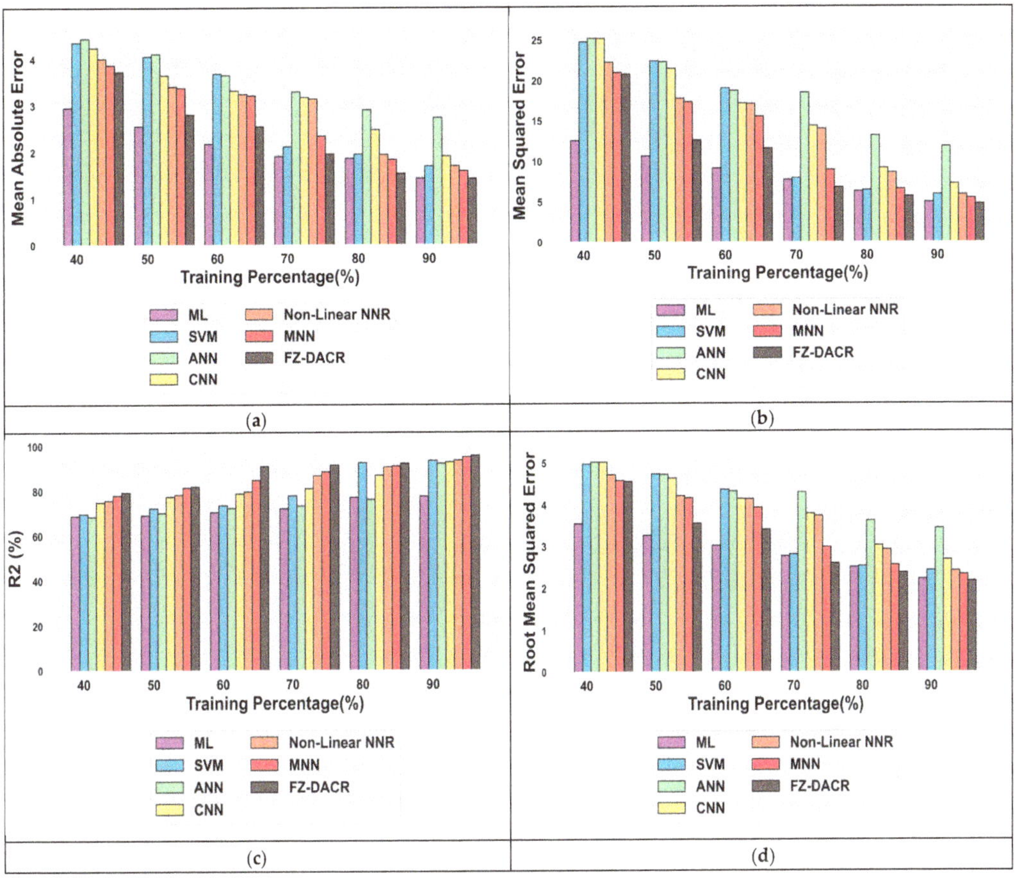

Figure 10. Comparative analysis based on the T-shirt dataset with TP analysis. (**a**) MAE; (**b**) MSE; (**c**) R2; (**d**) RMSE.

In Figure 10b, the FZ-DACR model exhibited the lowest MSE in demand forecasting compared to the MNN model, surpassing it by 14.97% and attaining the lowest MSE of 4.75, with a TP of 90.

Regarding demand forecasting, the FZ-DACR model illustrated in Figure 10c performed 95.67% better than the MNN model. With a TP of 90, it achieved an R2 of 0.64, surpassing existing techniques.

The FZ-DACR model displayed the lowest RMSE in demand forecasting compared to the MNN model in Figure 10d, outperforming it by 7.20% and achieving the lowest RMSE of 2.18, with a TP of 90.

4.4.6. Comparative Analysis Based on T-Shirt Dataset with K-Fold Analysis

The FZ-DACR model demonstrated the lowest MAE in demand forecasting compared to the MNN model, showing a slight decrease of 1.522 and achieving the lowest MAE of 1.46, as indicated in Figure 11a.

https://doi.org/10.3390/app16021039

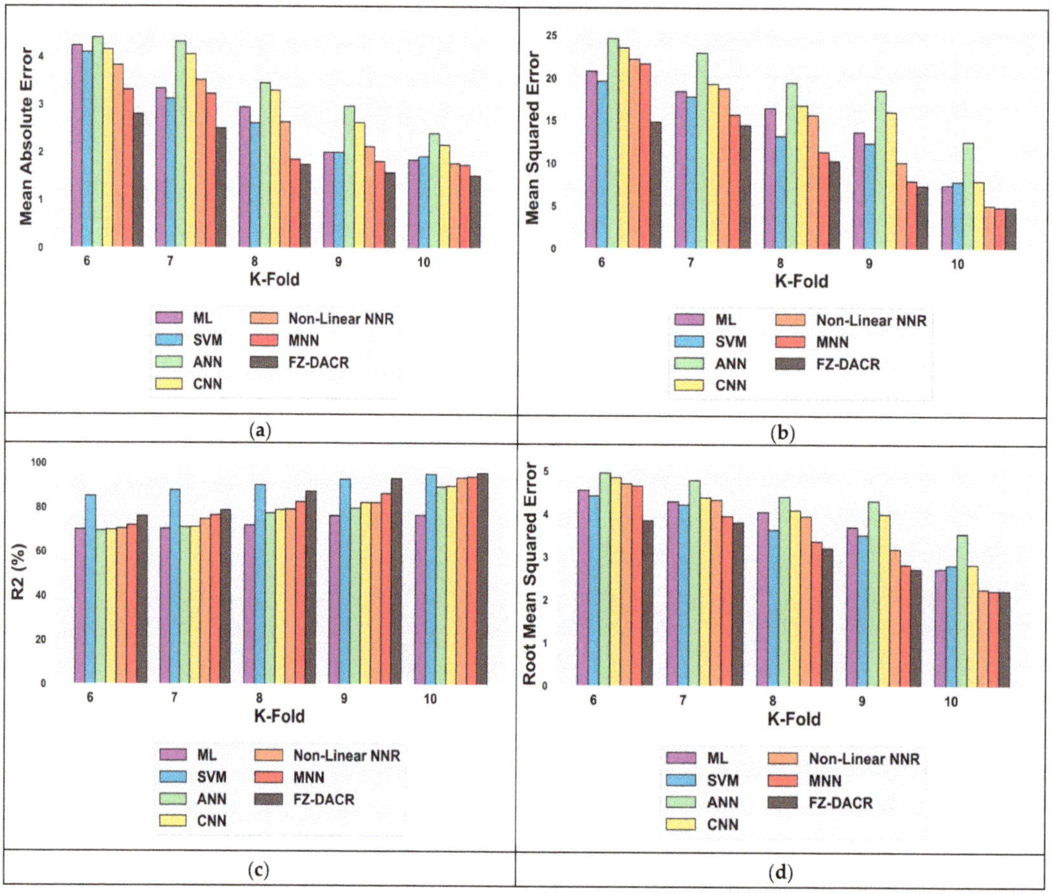

Figure 11. Comparative analysis based on the T-shirt dataset with k-fold analysis. (**a**) MAE; (**b**) MSE; (**c**) R2; (**d**) RMSE.

In Figure 11b, the FZ-DACR model exhibited the lowest MSE in demand forecasting compared to the MNN model, surpassing it by 0.35 and attaining the lowest MSE of 6.29, with k-fold 10.

Regarding demand forecasting, the FZ-DACR model illustrated in Figure 11c obtained 95.24% better results than the MNN model. With k-fold 10, it achieved a relative improvement of 1.02%, surpassing existing techniques.

The FZ-DACR model displayed the least RMSE in demand forecasting compared to the MNN model in Figure 11d, outperforming it with an error difference of 0.069 and achieving the lowest RMSE of 2.50, with k-fold 10.

4.5. Comparative Discussion

The comparative analysis of the proposed FZ-DACR model was evaluated with other existing models to enhance the performance in demand forecasting. The existing models, such as ML, CNN, non-linear NNR, and MNN, in fashion and apparel demand forecasting include drawbacks of inadequate fashion trend fluctuations, multiple data sources, and temporal dependencies in the sales data. Compared with traditional machine learning methods, such as SVM and ANN, which have limitations of slow learning if

https://doi.org/10.3390/app16021039

new trends and outside forces are involved, CNNs are mostly used for image data and may not make good use of sequential characteristics. The non-linear NNR and MNNs also have some difficulty in optimizing across complex data spaces and do not optimally work with high-dimensional and dynamic data in fashion. However, the proposed FZ-DACR model overcomes these issues more effectively and holistically by incorporating effective pre-processing, feature extraction, and a segmentation process. Flexible problem solving and decision making are achieved through the integration of FZ optimization and a distributed attention mechanism to improve the model's adaptability to a highly dynamic and fluctuating fashion environment. The utilization of a distributed attention mechanism with CNN and RNN models enables the extraction of significant properties from fashion images and texts, which is useful for the styles and preferences of products, which is important for selling apparel over different seasons and trends. Thus, the benefits of these advancements in the proposed model generated more accurate and reliable forecasts that encompass the complex nature of the fashion business. However, the imbalanced data may affect the performance of the model over diverse datasets, which will be overcome in future research. Tables 1–3 include the comparative discussion of the dress dataset, the skirt dataset, and the T-shirt dataset.

Table 1. Comparative discussion using the dress dataset.

Models	TP-90				K-Fold 10			
	MAE	MSE	R2	RMSE	MAE	MSE	R2	RMSE
ML	2.34	8.09	92.2	2.84	1.48	6.00	86.07	2.45
SVM	1.48	5.07	86.89	2.25	1.50	17.47	92.01	4.18
ANN	2.38	10.99	83.25	3.32	2.19	6.73	89.21	2.59
CNN	2.28	9.73	89.75	3.12	2.15	5.41	91.75	2.33
Non-Linear NNR	1.94	6.93	89.89	2.63	1.83	5.13	92.17	2.27
MNN	1.6	6.44	92.25	2.54	1.56	4.96	93.45	2.23
FZ-DACR	1.34	4.7	93.13	2.17	1.38	4.95	93.65	2.22

Table 2. Comparative discussion using the skirt dataset.

Models	TP-90				K-Fold 10			
	MAE	MSE	R2	RMSE	MAE	MSE	R2	RMSE
ML	1.44	4.96	77.59	2.23	3.49	10.41	87.65	3.23
SVM	1.70	5.92	93.40	2.43	1.73	7.74	94.01	2.78
ANN	2.73	11.82	92.16	3.44	2.16	8.81	83.86	2.97
CNN	1.91	7.22	92.84	2.69	2.04	7.93	92.27	2.82
Non-Linear NNR	1.70	5.87	93.56	2.42	1.99	7.33	93.34	2.71
MNN	1.59	5.46	95.06	2.34	1.68	7.00	93.46	2.65
FZ-DACR	1.43	4.75	95.67	2.18	1.66	5.09	94.12	2.26

 https://doi.org/10.3390/app16021039

Table 3. Comparative discussion using the T-shirt dataset.

Models	TP-90				K-Fold 10			
	MAE	MSE	R2	RMSE	MAE	MSE	R2	RMSE
ML	1.44	4.96	77.59	2.23	1.99	7.14	93.01	2.67
SVM	1.70	5.92	93.40	2.43	2.52	10.16	92.87	3.19
ANN	2.73	11.82	92.16	3.44	2.62	8.08	87.52	2.84
CNN	1.91	7.22	92.84	2.69	2.17	7.15	92.86	2.67
Non-Linear NNR	1.70	5.87	93.56	2.42	1.70	6.65	94.27	2.58
MNN	1.59	5.46	95.06	2.34	1.52	6.62	94.89	2.57
FZ-DACR	1.43	4.75	95.67	2.18	1.46	6.29	95.24	2.51

4.6. Statistical Analysis

The statistical analysis was carried out in terms of Best, Mean, Variance, and Standard Deviation (STD) for analyzing the robustness of the reported results evaluated using the performance metrics MAE, MSE, R2, and RMSE. Here, the results of the proposed FZ-DACR method are compared with the existing methods ML, SVM, ANN, CNN, non-linear NNR, and MNN. Tables 4–6 depict the statistical analysis of the proposed model using three datasets and compared with other existing methods.

Table 4. Statistical analysis based on the dress dataset.

Methods		ML	SVM	ANN	CNN	Non-Linear NNR	MNN	FZ-DACR
MAE	Best	4.26	4.00	4.16	4.04	3.93	3.63	3.49
	Mean	3.27	2.68	3.12	2.97	2.84	2.65	2.29
	Variance	0.95	0.65	0.52	0.52	0.60	0.54	0.48
	STD	0.97	0.80	0.72	0.72	0.77	0.73	0.69
MSE	Best	23.37	23.97	24.74	24.03	23.03	17.57	16.52
	Mean	15.61	20.98	17.48	16.46	13.89	10.93	10.21
	Variance	45.24	7.07	43.33	46.24	42.56	26.09	20.51
	STD	6.73	2.66	6.58	6.80	6.52	5.11	4.53
R2	Best	86.07	92.01	89.21	91.75	92.17	93.45	93.65
	Mean	79.91	87.11	81.50	83.27	84.98	86.71	89.60
	Variance	32.10	20.91	54.71	58.27	45.83	45.16	12.49
	STD	5.67	4.57	7.40	7.63	6.77	6.72	3.53
RMSE	Best	4.83	4.90	4.97	4.90	4.80	4.19	4.06
	Mean	3.84	4.57	4.09	3.95	3.61	3.21	3.11
	Variance	0.84	0.08	0.74	0.87	0.87	0.62	0.52
	STD	0.92	0.29	0.86	0.93	0.93	0.79	0.72

4.7. Computational Complexity

The computation time required by the proposed method was demonstrated through a comparative analysis across multiple iterations and compared to existing methods. The results highlight that the proposed method consistently requires significantly less time than other

https://doi.org/10.3390/app16021039

approaches, emphasizing its computational effectiveness. The proposed FZ-DACR method exhibits the lowest computational time of 121.66 s at the 100th epoch, which is significantly faster than all other methods. More specifically, the FZ algorithm adaptively tunes the hyperparameter of the proposed model, resulting in faster convergence. Furthermore, the FZ algorithm ensures the optimal configuration of the proposed architecture and minimizes computational complexity during demand forecasting. In addition, the distributed attention mechanism allows the model to focus on the important features while suppressing the irrelevant features, which minimizes the processing time required for analyzing the irrelevant features. Consequently, the proposed approach minimizes the computational overhead compared to the other baseline techniques utilized for demand forecasting. The computation time analysis for the proposed model is depicted in Table 7 and Figure 12.

Table 5. Statistical analysis based on the skirt dataset.

Methods		ML	SVM	ANN	CNN	Non-Linear NNR	MNN	FZ-DACR
MAE	Best	4.00	3.94	4.29	3.99	3.91	3.87	2.91
	Mean	3.76	2.61	3.52	3.10	2.94	2.80	2.25
	Variance	0.05	0.57	0.61	0.59	0.54	0.67	0.18
	STD	0.22	0.76	0.78	0.77	0.73	0.82	0.42
MSE	Best	24.54	13.31	24.96	22.86	21.70	16.71	16.38
	Mean	17.87	10.18	17.81	15.69	13.05	11.66	10.55
	Variance	30.55	3.81	38.30	31.72	23.95	11.35	16.42
	STD	5.53	1.95	6.19	5.63	4.89	3.37	4.05
R2	Best	87.65	94.01	83.86	92.27	93.34	93.46	94.12
	Mean	79.20	82.31	75.01	78.19	80.95	84.65	88.11
	Variance	39.17	69.66	31.65	68.88	78.76	78.93	52.28
	STD	6.26	8.35	5.63	8.30	8.87	8.88	7.23
RMSE	Best	4.95	3.65	5.00	4.78	4.66	4.09	4.05
	Mean	4.17	3.18	4.15	3.89	3.55	3.38	3.18
	Variance	0.44	0.09	0.59	0.56	0.43	0.25	0.41
	STD	0.66	0.30	0.77	0.75	0.66	0.50	0.64

Table 6. Statistical analysis based on the T-shirt dataset.

Methods		ML	SVM	ANN	CNN	Non-Linear NNR	MNN	FZ-DACR
MAE	Best	4.35	4.28	4.35	4.25	4.17	3.68	2.81
	Mean	3.25	3.83	3.67	3.04	2.75	2.55	2.17
	Variance	0.85	0.45	0.44	0.48	0.68	0.5	0.25
	STD	0.92	0.67	0.66	0.69	0.83	0.71	0.5
MSE	Best	20.96	20.19	23.81	22.68	22.52	22.22	16.79
	Mean	13.62	14.12	18.14	16.85	15.19	13.80	11.93
	Variance	20.32	18.78	31.46	28.86	27.82	28.50	13.97
	STD	4.51	4.33	5.61	5.37	5.27	5.34	3.74

https://doi.org/10.3390/app16021039

Table 6. *Cont.*

Methods		ML	SVM	ANN	CNN	Non-Linear NNR	MNN	FZ-DACR
R2	Best	93.01	92.87	87.52	92.86	94.27	94.89	95.24
	Mean	82.36	87.09	76.45	79.27	82.26	85.34	88.05
	Variance	48.55	43.65	53.79	67.12	67.21	59.98	63.66
	STD	6.97	6.61	7.33	8.19	8.20	7.74	7.98
RMSE	Best	4.58	4.49	4.88	4.76	4.75	4.71	4.10
	Mean	3.64	3.72	4.20	4.04	3.83	3.64	3.41
	Variance	0.38	0.32	0.54	0.54	0.53	0.53	0.32
	STD	0.62	0.56	0.73	0.73	0.73	0.73	0.57

Table 7. Comparison of computational time of FZ-DACR.

Methods	Computational Time (s)
ML	124.45
SVM	123.05
ANN	123.12
CNN	123.68
Non-Linear NNR	124.18
MNN	124.21
FZ-DACR	121.66

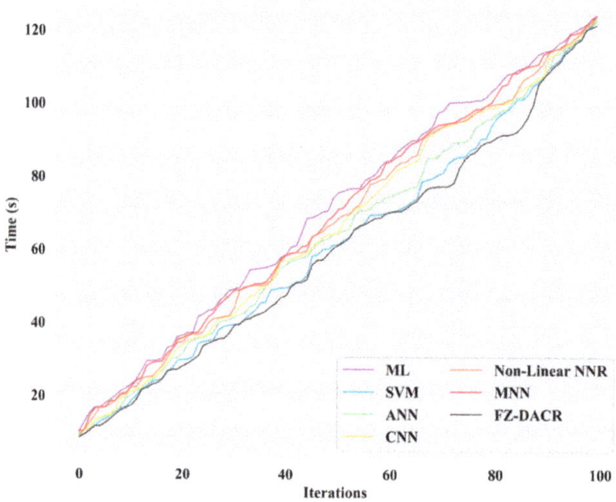

Figure 12. Computational complexity.

4.8. Convergence Analysis

Convergence analysis was carried out to analyze the convergence rate required to find the optimal solution during the iterative process. The convergence analysis of the proposed FZ algorithm is evaluated against several established metaheuristic algorithms, including Particle Swarm Optimization (PSO), Genetic Algorithms (GA), Ant Colony Optimization (ACO), and the Zebra Optimization Algorithm (ZOA), as depicted in Figure 13. From the

https://doi.org/10.3390/app16021039

convergence analysis, the proposed FZ algorithm is observed to have a minimum loss of 0.000001 at the 98th epoch, which is less compared with other optimization techniques. At the same time, the existing PSO algorithm obtained a loss of 0.0027, the GA algorithm obtained a loss value of 0.0026, the ZOA algorithm reached a loss of 0.0002, and the ZOA algorithm gained a loss of 0.0002, respectively. Specifically, the incorporation of the unique traits in the FZ algorithm, offering exploration in the large search space, accelerates the convergence speed while obtaining an optimal solution. As a result, the proposed model minimizes loss and outperforms the other existing algorithms.

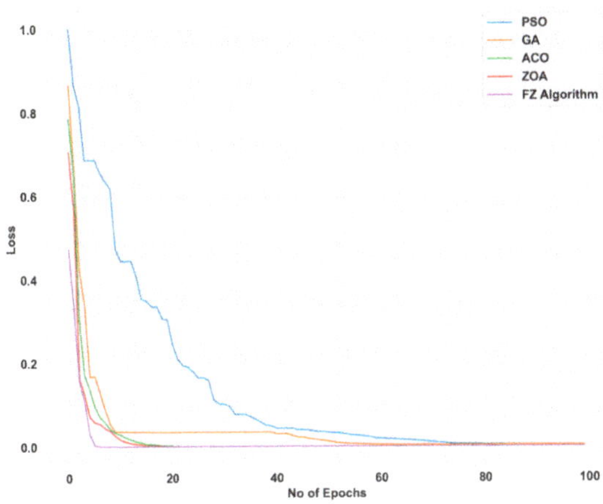

Figure 13. Convergence analysis.

4.9. Sensitivity Analysis

Sensitivity analysis was conducted to analyze how uncertainty in the input variables influences the system's performance in predicting demands, especially in terms of MSE. The sensitivity analysis carried out for the proposed FZ-DACR model over the baseline technique of MNN is shown in Figure 14. Furthermore, the sensitivity analysis shows that the proposed approach attained the lowest MSE of 4.7, while the existing MNN attained an MSE of 6.44. Likewise, the proposed model outperformed all of the existing models. Furthermore, the proposed FZ-DACR approach attained high sensitivity, which assists in maintaining robustness throughout the demand forecasting process.

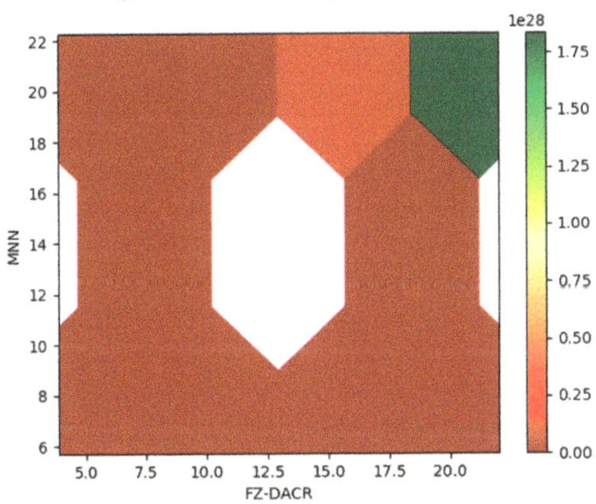

Figure 14. Sensitivity analysis.

https://doi.org/10.3390/app16021039

4.10. Ablation Study

An ablation study was performed to examine the performance of the model after removing or altering different components. Figure 15 shows the results of the ablation study, which assessed the FZ-DACR model's performance by varying the different components. More specifically, the FZ-DACR's performance was compared to that of the CNN-RNN, DACR model. Furthermore, the results of the models are evaluated in terms of the MSE. The FZ-DACR model achieved the lowest MSE of 4.70, which was significantly reduced over the CNN-RNN by 2.95 and DACR by 0.75. Moreover, the FZ-DACR model provides superior performance, as it integrates the strength of CNN-RNN, a distributed attention mechanism, and the FZ algorithm for demand forecasting. Specifically, the combination of the CNN and RNN algorithms assists in extracting the context and hierarchical patterns present in the input product images and historical sales data. Furthermore, the distributed attention mechanism helps the model focus on the important region and suppress the irrelevant features in the sequence data representations, utilizing the attention weights to boost forecasting performance. As a result, the FZ-DACR model attains better performance for demand forecasting. Moreover, the significance of every component utilized in the proposed model for improving demand forecasting is highlighted in this ablation study.

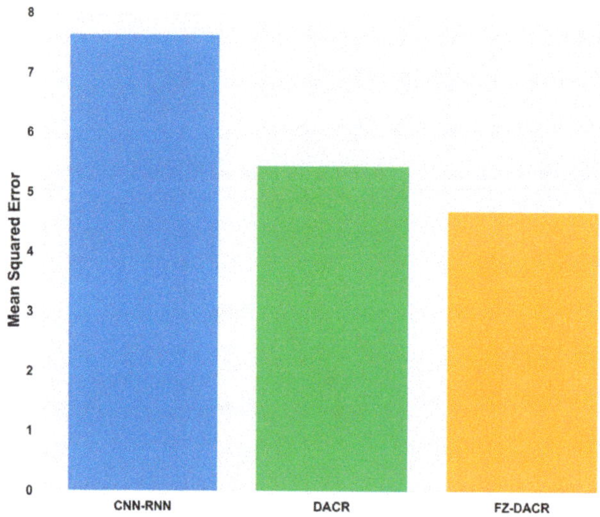

Figure 15. Ablation study.

4.11. Predictive Analysis for Demand Forecasting

The proposed FZ-DACR model is developed for demand forecasting in the fashion industry, and so the dress dataset, the skirt dataset, and the T-shirt dataset were used for training and testing the model. Specifically, the input data were trained with 90% of the data and validated with 10% of the data. The prediction analysis shows that the proposed model predicted the maximum number of fashion items, with only a small difference observed. Furthermore, the graph portraying the historical demand data and forecasting demands using multiple color shades is shown in Figure 16. From the figure, the yellow color wave represents the historical fashion data, the green color represents the actual fashion demands, and the blue wave denotes the predicted fashion demands, while the orange color represents the training data. Thus, the proposed FZ-DACR model

https://doi.org/10.3390/app16021039

effectively predicts demand, offering valuable support for demand forecasting within the fashion industry.

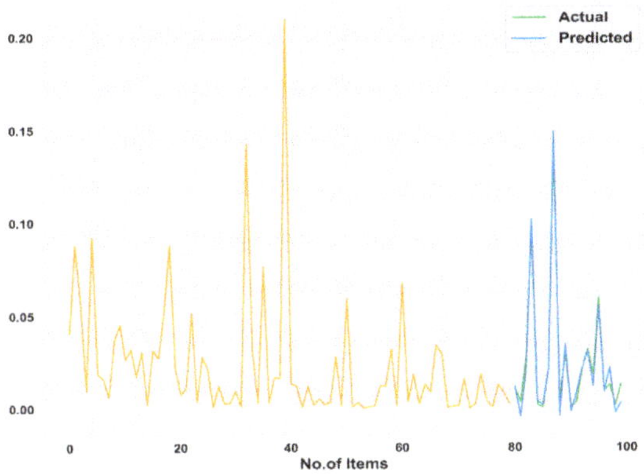

Dress dataset

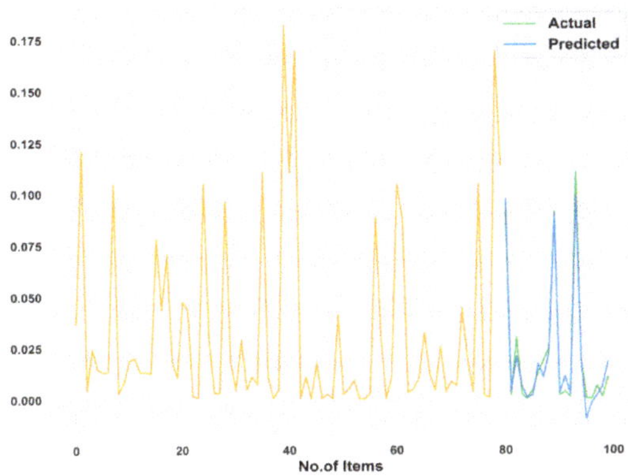

Skirt dataset

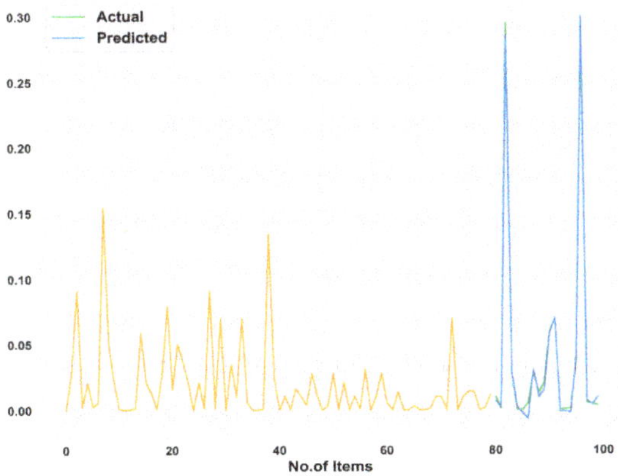

T-shirt dataset

Figure 16. Prediction output vs. actual demands.

https://doi.org/10.3390/app16021039

5. Conclusions

This research offers the prospect of the combined use of FZ optimization and the proposed DACR model to forecast demand in the fashion and apparel retail industry. This approach solves the existing models' issues, including fashion trend fluctuations, data heterogeneity, and consumer preferences with a dynamic nature. By integrating the FZ's adaptive optimization, the weights and bias parameters of the model are tuned to find an optimal solution. Through effective feature extraction and sequential learning, the proposed model improves predictive precision and work productivity. Initially, the proposed approach examines the best way to obtain information from input review text and images and discover more precise demand data. By acquiring this information, the proposed model determines the actual demand and hence resolves the uncertainty. According to the results, the models attain the lowest error values of metrics when utilizing the dress dataset, with MAE of 1.34, MSE of 4.7, R2 of 93.3, and RMSE of 2.17. However, the imbalanced data may affect the model's performance across diverse datasets, and real-time data will be addressed in future research. The implications of the research in the fashion and apparel industry help to price products based on demand. In the future, the proposed model can be extended through real-time implementation to improve the productivity of the model. Although the proposed model with a complex architecture effectively extracts the non-linear representation but adds model complexity. In the future, the effective balance between model complexity and its ability to represent non-linear patterns should be considered to enhance overall performance.

Author Contributions: All authors have made substantial contributions to conception and design, revising the manuscript, and the final approval of the version to be published. Also, all authors agreed to be accountable for all aspects of the work in ensuring that questions related to the accuracy or integrity of any part of the work are appropriately investigated and resolved. All authors have read and agreed to the published version of the manuscript.

Funding: This research did not receive funding from any institution.

Data Availability Statement: The data used in this study can be accessed at the Flipkart Grid Software Development Challenge 2.0 Dataset link, https://github.com/avmand/GriD_Fashion/blob/master/CurrentTrends/final_code/dress-flipkart-final-final.csv, accessed on 20 April 2024.

Conflicts of Interest: No potential conflicts of interest were reported by the authors.

References

1. Seyedan, M.; Mafakheri, F.; Wang, C. Order-up-to-level inventory optimization model using time-series demand forecasting with ensemble deep learning. *Supply Chain. Anal.* **2023**, *3*, 100024. [CrossRef]
2. Giri, C.; Chen, Y. Deep learning for demand forecasting in the fashion and apparel retail industry. *Forecasting* **2022**, *4*, 565–581. [CrossRef]
3. Giri, C.; Thomassey, S.; Balkow, J.; Zeng, X. Forecasting new apparel sales using deep learning and nonlinear neural network regression. In Proceedings of the 2019 International Conference on Engineering, Science, and Industrial Applications (ICESI), Tokyo, Japan, 22–24 August 2019; IEEE: New York, NY, USA, 2019; pp. 1–6.
4. Deng, C.; Liu, Y. A deep learning-based inventory management and demand prediction optimization method for anomaly detection. *Wirel. Commun. Mob. Comput.* **2021**, *2021*, 9969357. [CrossRef]
5. Skenderi, G.; Joppi, C.; Denitto, M.; Cristani, M. On the use of learning-based forecasting methods for ameliorating fashion business processes: A position paper. In Proceedings of the International Conference on Pattern Recognition, Montréal, QC, Canada, 21–25 August 2022; Springer Nature: Cham, Switzerland, 2022; pp. 647–659.
6. Giri, C.; Thomassey, S.; Zeng, X. Customer analytics in the fashion retail industry. In *Functional Textiles and Clothing*; Springer: Singapore, 2024; pp. 349–361.
7. Punia, S.; Shankar, S. Predictive analytics for demand forecasting: A deep learning-based decision support system. *Knowl.-Based Syst.* **2022**, *250*, 109956. [CrossRef]

https://doi.org/10.3390/app16021039

8. Güven, İ.; Şimşir, F. Demand forecasting with color parameter in the retail apparel industry using artificial neural networks (ANN) and support vector machines (SVM) methods. *Comput. Ind. Eng.* **2020**, *147*, 106678. [CrossRef]
9. Rizqi, Z.U.; Chou, S.Y. Neuroevolution reinforcement learning for multi-echelon inventory optimization with delivery options and uncertain discount. *Eng. Appl. Artif. Intell.* **2024**, *134*, 108670. [CrossRef]
10. Kharfan, M.; Chan, V.W.K.; Firdolas Efendigil, T. A data-driven forecasting approach for newly launched seasonal products by leveraging machine-learning approaches. *Ann. Oper. Res.* **2021**, *303*, 159–174. [CrossRef]
11. Vo, T.T.B.C.; Le, P.H.; Nguyen, N.T.; Nguyen, T.L.T.; Do, N.H. Demand Forecasting and Inventory Prediction for Apparel Products using the ARIMA and Fuzzy EPQ Model. *J. Eng. Sci. Technol. Rev.* **2021**, *14*, 80–89. [CrossRef]
12. Loureiro, A.L.; Miguéis, V.L.; Da Silva, L.F. Exploring the use of deep neural networks for sales forecasting in fashion retail. *Decis. Support Syst.* **2018**, *114*, 81–93. [CrossRef]
13. Rizqi, Z.U.; Khairunisa, A.; Maulani, A. Financial assessment on designing inventory policy by considering demand, lead time, and defective product uncertainties: A Monte Carlo simulation. *Indones. Sch. Sci. Summit Taiwan Proc.* **2021**, *3*, 36–42. [CrossRef]
14. Christopher, M.; Lee, H. Mitigating supply chain risk through improved confidence. *Int. J. Phys. Distrib. Logist. Manag.* **2004**, *34*, 388–396. [CrossRef]
15. Punia, S.; Nikolopoulos, K.; Singh, S.P.; Madaan, J.K.; Litsiou, K. Deep learning with long short-term memory networks and random forests for demand forecasting in multi-channel retail. *Int. J. Prod. Res.* **2020**, *58*, 4964–4979. [CrossRef]
16. Christopher, M.; Towill, D. An integrated model for the design of agile supply chains. *Int. J. Phys. Distrib. Logist. Manag.* **2001**, *31*, 235–246. [CrossRef]
17. Battista, C.; Schiraldi, M.M. The Logistic Maturity Model: Application to a fashion firm. *Int. J. Eng. Bus. Manag.* **2013**, *5*, 1–11. [CrossRef]
18. Marvaniya, S.; Singh, J.; Galichet, N.; Otieno, F.O.; De Mel, G.; Weldemariam, K. Encoding Seasonal Climate Predictions for Demand Forecasting with Modular Neural Network. *arXiv* **2023**, arXiv:2309.02248. [CrossRef]
19. Feizabadi, J. Machine learning demand forecasting and supply chain performance. *Int. J. Logist. Res. Appl.* **2022**, *25*, 119–142. [CrossRef]
20. Villegas, M.A.; Pedregal, D.J.; Trapero, J.R. A support vector machine for model selection in demand forecasting applications. *Comput. Ind. Eng.* **2018**, *121*, 1–7. [CrossRef]
21. Zhang, C.; Zhu, F.; Lv, Y.; Ye, P.; Wang, F.Y. MLRNN: Taxi demand prediction based on multi-level deep learning and regional heterogeneity analysis. *IEEE Trans. Intell. Transp. Syst.* **2021**, *23*, 8412–8422. [CrossRef]
22. Xu, Z.; Lv, Z.; Li, J.; Sun, H.; Sheng, Z. A novel perspective on travel demand prediction considering natural environmental and socioeconomic factors. *IEEE Intell. Transp. Syst. Mag.* **2022**, *15*, 136–159. [CrossRef]
23. Rathipriya, R.; Abdul Rahman, A.A.; Dhamodharavadhani, S.; Meero, A.; Yoganandan, G.J.N.C. Demand forecasting model for time-series pharmaceutical data using a shallow and deep neural network model. *Neural Comput. Appl.* **2023**, *35*, 1945–1957. [CrossRef] [PubMed]
24. Chandriah, K.K.; Naraganahalli, R.V. RNN/LSTM with modified Adam optimizer in a deep learning approach for automobile spare parts demand forecasting. *Multimed. Tools Appl.* **2021**, *80*, 26145–26159. [CrossRef]
25. Zohdi, M.; Rafiee, M.; Kayvanfar, V.; Salamiraad, A. Demand forecasting based on machine learning algorithms on customer information: An applied approach. *Int. J. Inf. Technol.* **2022**, *14*, 1937–1947. [CrossRef]
26. Available online: https://github.com/avmand/GriD_Fashion/blob/master/CurrentTrends/final_code/dress-flipkart-final-final.csv (accessed on 20 April 2024).
27. Trojovská, E.; Dehghani, M.; Trojovský, P. Zebra optimization algorithm: A new bio-inspired optimization algorithm for solving optimization problems. *IEEE Access* **2022**, *10*, 49445–49473. [CrossRef]
28. Dorigo, M.; Birattari, M.; Stutzle, T. Ant colony optimization. *IEEE Comput. Intell. Mag.* **2006**, *1*, 28–39. [CrossRef]
29. Joseph, R.V.; Mohanty, A.; Tyagi, S.; Mishra, S.; Satapathy, S.K.; Mohanty, S.N. A hybrid deep learning framework with CNN and Bi-directional LSTM for store item demand forecasting. *Comput. Electr. Eng.* **2022**, *103*, 108358. [CrossRef]
30. Nguyen, H.H.; Chan, C.W. Multiple neural networks for a long-term time series forecast. *Neural Comput. Appl.* **2004**, *13*, 90–98. [CrossRef]

https://doi.org/10.3390/app16021039

MDPI AG
Grosspeteranlage 5
4052 Basel
Switzerland
Tel.: +41 61 683 77 34

Applied Sciences Editorial Office
E-mail: applsci@mdpi.com
www.mdpi.com/journal/applsci

www.ingramcontent.com/pod-product-compliance
Lightning Source LLC
Chambersburg PA
CBHW061826210926
51850CB00006B/817